CLINICAL CHEMISTRY
IN
DIAGNOSIS AND TREATMENT

OTHER PUBLISHED WORK
OF RELATED INTEREST

Multiple-choice Questions on
"Clinical Chemistry in Diagnosis and Treatment"
(Second edition)

P. R. Fleming M.D., F.R.C.P.

Clinical Chemistry
in
Diagnosis and Treatment

JOAN F. ZILVA

M.D., B.Sc., F.R.C.P., F.R.C.Path., D.C.C. (Biochem.)

Professor of Chemical Pathology, Charing Cross and Westminster Medical School, London; Honorary Consultant in Chemical Pathology, Westminster Hospital, London

P. R. PANNALL

M.B., B.Ch. (Witwatersrand), F.F.Path. (S.A.), F.R.C.Path., F.R.C.P.A., F.A.A.C.B.

Director in Clinical Chemistry, The Queen Elizabeth Hospital, Adelaide; Senior Clinical Lecturer, University of Adelaide

Fourth Edition

LLOYD-LUKE (MEDICAL BOOKS) LTD

49 NEWMAN STREET
LONDON
1984

FIRST EDITION 1971
Reprinted 1972
Reprinted 1973
SECOND EDITION 1975
Reprinted 1978
THIRD EDITION 1979
Reprinted 1981
Reprinted 1982
Reprinted 1983
FOURTH EDITION 1984

(This work has also been translated into the Turkish, Spanish, Italian and Serbo-Croat languages)

PRINTED AND BOUND IN ENGLAND BY
HAZELL WATSON AND VINEY LIMITED
MEMBER OF THE BPCC GROUP
AYLESBURY, BUCKS
ISBN 0 85324 201 1

FOREWORD

THIS book aims at giving, within a single cover, all the relevant biochemical and pathological facts and theories necessary to the intelligent interpretation of the analyses usually performed in departments of Clinical Chemistry or Chemical Pathology. The approach is firmly based on general principles and the authors have gone to great trouble to ensure that the development of ideas is logical and easy to follow. A basic knowledge of medicine and of elementary biochemistry is assumed but, given this, the reader should be able to understand even the more involved metabolic inter-relationships without undue difficulty and should thereafter be in a strong position to apply this knowledge in medical practice.

I have appreciated very much the opportunity of reading this book during its preparation and feel that it represents a distinctly novel approach to the interpretation of biochemical data. In my opinion it could be read with advantage by all the categories of reader mentioned in the Preface, and moreover I believe that they will enjoy the experience.

N. F. MACLAGAN

January, 1971

PREFACE TO THE FOURTH EDITION

MUCH of the text has again been extensively revised, either because of advances in knowledge since the last edition, or because some readers found the previous text less clear than we had hoped. We have included a new chapter introducing the important subject of Drug Monitoring.

Readers studying for advanced examinations often ask us to expand the text. These requests must be weighed against the danger of so increasing the scope and size of the book that we lose sight of our original aim of writing for medical students. We hope that we have reached a reasonable compromise.

We are again indebted for helpful criticism to more people than we have space to acknowledge. Drs. Paul Eldridge, Shruti Gandhi and Philip Nicholson have read and commented on every chapter. Amongst others who have been equally helpful in their suggestions about several chapters are, in alphabetical order, Dr. Stephen Bangert, Mr. John Chappell, Drs. John Connelly, Gurdeep Dhatt, Michael Dunne, Julia Heptenstall, Stephen Holmes, John Kerin, William Shatwell, Sudarshini Gnana Sunderam and Maurice Wellby. Drs. Allan Kerr-Grant and Andrew Paull made helpful contributions to Chapter XIII, and Professor J. R. Hobbs, Mr. David Perry, and Drs. Pamela Riches and John Whicher to Chapter XIV. Mrs. Joanne Payne has borne with us through the typing of several drafts and the final preparation of the manuscript, and Miss Anne Jeffries of the Department of Medical Photography and Illustration prepared many of the diagrams. The protein electrophoretic strip in Fig. 29 was prepared by Dr. Pamela Riches and the staff of the Protein Reference Unit at Westminster Hospital and Fig. 30 by Mr. John Chappell; they were photographed by the Department of Medical Photography and Illustration of the Westminster Hospital and Medical School and the Department of Clinical Photography, the Queen Elizabeth Hospital, Adelaide, respectively. Members of the Drug Information section of the Westminster Hospital Pharmacy have checked the details of intravenous fluids and other pharmaceutical preparations.

We thank the publishers for their continuing co-operation.

JFZ
PRP

January, 1984

PREFACE TO THE FOURTH EDITION

PREFACE TO THE FIRST EDITION

THIS book is intended primarily for medical students and junior hospital staff. It is based on many years practical experience of undergraduate and postgraduate teaching, and of the problems of a routine chemical pathology department. It is written for those who learn best if they understand what they are learning. Wherever possible, explanations of the facts are given: if the explanation is a working hypothesis (like, for instance, that for the ectopic production of hormones) this is stressed, and where no explanation is known, this is stated. Our experience of teaching and our discussions with students have led us to believe that many of them are willing to read a slightly longer book if it gives them a better understanding than a shorter one. Electrolyte and acid-base balance have been discussed in some detail because, in our experience, these are the most common problems of chemical pathology met with by junior clinicians and ones in which there are often dangerous misunderstandings. Some subjects, such as the porphyrias and conditions of iron overload, are discussed in greater detail than is necessary for undergraduate examinations: however, the incidence of these is high in some areas of the world, and an elementary source of reference seemed to be needed.

We have tried to stress the clinical importance of an understanding of the subject: by including in the chapter appendices some details of treatment, difficult to find together in other books, we hope that this one may appeal to clinicians as well as to students. Two chapters are included on the best use of a laboratory, including precautions which should be taken in collecting specimens and interpreting results. As we have stressed, pathologists and clinicians should work as a team, and full consultation between the two should be the rule. The diagnostic tests suggested are those which we have found, by experience, to be the most valuable. For instance, in the differential diagnosis of hypercalcaemia we find the steroid suppression test very helpful, phosphate excretion indices fallible and tedious to perform and estimation of urinary calcium useless.

Our own junior staff have found the drafts helpful in preparing for the Part 1 examination for Membership of the Royal College of Physicians, as well as for the primary examination for Membership of the Royal College of Pathologists, and our Senior Technicians are using it to study for the Special Examination for Fellowship in

Chemical Pathology of the Institute of Medical Laboratory Technology. We feel that those studying for the primary examination for the Fellowship of the Royal College of Surgeons might also make use of it. It could provide a groundwork for study for the final examination for the Membership of the Royal College of Pathologists in Chemical Pathology, and for the Mastership in Clinical Biochemistry.

Initially each chapter should be worked through from beginning to end. For revision purposes there are lists and tables, and summaries of the contents of each chapter. The short sub-indices in the Table of Contents should facilitate the use of the book for reference.

Appendix A lists some analogous facts which we hope may help understanding and learning. The list must be far from complete, and the student should seek other examples for himself.

We wish to thank Professor N. F. Maclagan for his unfailing encouragement and his helpful advice and criticisms. We are also indebted to a great many other people, foremost amongst whom we should mention Dr. J. P. Nicholson who has read and commented on the whole book, and Professor M. D. Milne, Professor D. M. Matthews, Mr. K. B. Cooke and Dr. B. W. Gilliver for helpful advice and criticism on individual chapters. Many registrars and senior house officers in the Department, in particular Dr. Krystyna Rowland, Dr. Elizabeth Small, Dr. Nalini Naik and Dr. Noel Walmsley, have been closely involved in the preparation of the book and have provided invaluable suggestions and criticisms. Students, too, have read individual chapters for comprehensibility, and we would particularly like to thank Mr. B. P. Heather, Mr. J. Muir and Miss H. M. Merriman for helpful comments. Mr. C. P. Butler of the Westminster Hospital Pharmacy was most helpful during the preparation of the sections on therapy. Mrs. Valerie Moorsom and Mrs. Marie-Lise Pannall, with the help of Mrs. Brenda Sarasin and Miss Barbara Bridges, have borne with us during the typing of the drafts and the final transcript. The illustrations were prepared by Mr. David Gibbons of the Department of Medical Photography and Illustration of the Westminster Medical School.

Finally, we would like to thank the publishers for their co-operation and understanding during the preparation of this book.

JFZ
PRP

May, 1971

CONTENTS

UNITS IN CHEMICAL PATHOLOGY

RESULTS in chemical pathology have been expressed in a variety of units. For instance, electrolytes were usually quoted in mEq/l, protein in g/100 ml and cholesterol in mg/100 ml. The units used might vary from laboratory to laboratory: calcium might be expressed as mg/100 ml or mEq/l; in Britain, urea results were expressed as mg/100 ml of *urea*, whereas in the United States it is usual to report mg/100 ml of *urea nitrogen*. This situation can be confusing and with patients moving, not only from one hospital to another, but from one country to another, dangerous misunderstandings could arise.

Système International d'Unités (SI Units)

International standardisation is obviously desirable; such standardisation has already been introduced into many branches of science and technology.

The main recommendations for chemical pathology are as follows:

1. Where the molecular weight (MW) of the substance being measured is known, the unit of quantity should be the *mole* or submultiple of a mole.

$$\text{Number of moles (mol)} = \frac{\text{wt. in g}}{\text{MW}}$$

In chemical pathology millimoles (mmol), micromoles (μmol) and nanomoles (nmol) are the most common units.

2. The unit of volume should be the *litre*. Units of concentration are therefore mmol/l, μmol/l or nmol/l.

Examples

1. *Results previously expressed as mEq/l*

$$\text{Number of equivalents (Eq)} = \frac{\text{wt. in g}}{\text{Equivalent wt.}}$$

$$= \frac{\text{wt. in g} \times \text{valency}}{\text{MW}}$$

(a) In the case of univalent ions, such as sodium and potassium, the units will be numerically the same. A sodium of 140 mEq/l becomes 140 mmol/l.

(b) For polyvalent ions, such as calcium or magnesium (both divalent), the old units are numerically divided by the valency. For instance, a magnesium of 2·0 mEq/l becomes 1·0 mmol/l.

2. *Results previously expressed as mg/100 ml*

If results were previously expressed in mg/100 ml the method of conversion to mmol/l is to divide by the molecular weight (to convert from mg to mmol), and to multiply by 10 (to convert from 100 ml to a litre). Thus effectively the previous units are divided by 1/10 of the molecular weight. For instance, the molecular weight of urea is 60, and of glucose 180. A urea value of

60 mg/100 ml and a glucose value of 180 mg/100 ml are both equivalent to 10 mmol/l.

The factor of 10 is, of course, only used for concentrations. The total amount of urea excreted in 24 hours in mg is numerically 60 times that in mmol.

Exceptions

1. *Units of pressure* (e.g. mm Hg) are expressed as pascals (or kilopascals—kPa). 1 kPa = 7·5 mmHg, so that a PO_2 of 75 mmHg is 10 kPa. Pascals are SI units.

2. *Proteins*. Body fluids contain a complex mixture of proteins of varying molecular weights. It is therefore recommended that the gram (g) be retained, but that the unit of volume be the litre. Thus a total protein of 7·0 g/100 ml becomes 70 g/l.

3. The expression 100 ml is to be expressed as *decilitre* (dl).

4. *Enzyme units* are not to be changed yet. Note that *the definition of international units for enzymes does not state the temperature of the reaction* (p. 369).

5. Some constituents, such as IgE and some hormones, are still expressed in "international" or other special units.

At the moment different laboratories are at different stages of implementation. In this book we have adopted the following policy.

1. Where the old and new units are numerically the same, we have given only the new units (e.g. for sodium and potassium).

2. For proteins we give only g/l.

3. Where it is generally accepted that the new units be adopted we have given these, with the equivalent old units in brackets.

A conversion table for some of the commoner results is given opposite.

Note that:

1 mol	= 1 000 mmol
1 mmol (10^{-3} mol)	= 1 000 μmol
1 μmol (10^{-6} mol)	= 1 000 nmol (nanomoles)
1 nmol (10^{-9} mol)	= 1 000 pmol (picomoles).

Some Approximate Conversion Factors for SI Units

	From SI Units	To SI Units
Bilirubin	μmol/l $\times$ 0·058 $=$ mg/dl	mg/dl $\div$ 0·058 $=$ μmol/l
Calcium		
Plasma	mmol/l $\times$ 4 $=$ mg/dl	mg/dl $\div$ 4 $=$ mmol/l
Urine	mmol/24 h $\times$ 40 $=$ mg/24 h	mg/24 h $\div$ 40 $=$ mmol/24 h
Cholesterol	mmol/l $\times$ 39 $=$ mg/dl	mg/dl $\div$ 39 $=$ mmol/l
Cortisol		
Plasma	nmol/l $\times$ 0·036 $=$ μg/dl	μg/dl $\div$ 0·036 $=$ nmol/l
Urine	nmol/24 h $\times$ 0·36 $=$ μg/24 h	μg/24 h $\div$ 0·36 $=$ nmol/24 h
Creatinine		
Plasma	μmol/l $\times$ 0·011 $=$ mg/dl	mg/dl $\div$ 0·011 $=$ μmol/l
Urine	μmol/24 h $\times$ 0·11 $=$ mg/24 h	mg/24 h $\div$ 0·11 $=$ μmol/24 h
Gases		
PO_2 PCO_2	kPa $\times$ 7·5 $=$ mmHg	mmHg $\div$ 7·5 $=$ kPa
Glucose	mmol/l $\times$ 18 $=$ mg/dl	mg/dl $\div$ 18 $=$ mmol/l
Iron TIBC	μmol/l $\times$ 5·6 $=$ μg/dl	μg/dl $\div$ 5·6 $=$ μmol/l
Phosphorus	mmol/l $\times$ 3 $=$ mg/dl	mg/dl $\div$ 3 $=$ mmol/l
Proteins		
Serum		
Total Albumin Immuno-globulins	g/l $\div$ 10 $=$ g/dl	g/dl $\times$ 10 $=$ g/l
Urine		
Concentration	g/l $\times$ 100 $=$ mg/dl	mg/dl $\div$ 100 $=$ g/l
Daily Output	g/24 h	No change
Urate	mmol/l $\times$ 17 $=$ mg/dl	mg/dl $\div$ 17 $=$ mmol/l
Urea	mmol/l $\times$ 6 $=$ mg/dl	mg/dl $\div$ 6 $=$ mmol/l
5-HIAA HMMA	μmol/24 h $\times$ 0·2 $=$ mg/24 h	mg/24 h $\div$ 0·2 $=$ μmol/24 h
Oestriol	μmol/24 h $\times$ 0·3 $=$ mg/24 h	mg/24 h $\div$ 0·3 $=$ μmol/24 h
Faecal "Fat"	mmol/24 h $\times$ 0·3 $=$ g/24 h	g/24 h $\div$ 0·3 $=$ mmol/24 h

ABBREVIATIONS USED IN THE BOOK OR IN COMMON USE

ACP	Acid Phosphatase
ACTH	Adrenocorticotrophic Hormone (corticotrophin)
ADH	Antidiuretic Hormone (Pitressin: arginine vasopressin: AVP)
ALA	5-Aminolaevulinate
ALP	Alkaline Phosphatase
ALS	Aldolase
ALT	Alanine Transaminase (= SGPT)
AMS	α-Amylase
APRT	Adenine Phosphoribosyl Transferase
APUD	Amine-Precursor Uptake and Decarboxylation
AST	Aspartate Transaminase (= SGOT)
AVP	Arginine Vasopressin (antidiuretic hormone: ADH)
BJP	Bence Jones Protein
BUN	Blood Urea Nitrogen $\left(\text{in mg/dl} = \dfrac{28}{60} \times \text{blood urea in mg/dl}\right.$ $\left. \text{or } 2{\cdot}8 \times \text{blood urea in mmol/l}\right)$
CBG	Cortisol-Binding Globulin (Transcortin)
CC	Cholecalciferol
CK	Creatine Kinase (= CPK)
CoA	Coenzyme A
CPK	Creatine Phosphokinase (= CK)
CRF	Corticotrophin Releasing Factor
CSF	Cerebrospinal Fluid
DDAVP	1-Deamino-8-D-arginine-vasopressin (desmopressin acetate)
1,25-DHCC	1,25-Dihydroxycholecalciferol
DIT	Di-iodotyrosine
DNA	Deoxyribonucleic Acid
DOC	Deoxycorticosterone
DOPA	Dihydroxyphenylalanine
DOPamine	Dihydroxyphenylethylamine
ECF	Extracellular Fluid
EDTA	Ethylene Diamine Tetra-acetate (Sequestrene)
EM Pathway	Embden-Meyerhof Pathway (Glycolytic Pathway)
ESR	Erythrocyte Sedimentation Rate
FAD	Flavin Adenine Dinucleotide
FFA	Free Fatty Acids (= NEFA)
FMN	Flavin Mononucleotide
FSH	Follicle-Stimulating Hormone (follitropin)

FSH-LHRH	Follicle-Stimulating and Luteinising Hormone-Releasing Hormone
FTI	Free Thyroxine Index
GFR	Glomerular Filtration Rate
GGT	γ-Glutamyltransferase (γ-Glutamyltranspeptidase: γGT)
GH	Growth Hormone (somatotropin)
GMD	Glutamate Dehydrogenase
Gn-RH	Gonadotrophin-Releasing Hormone
GOT	Glutamate Oxaloacetate Transaminase ($=$ AST)
G-6-P	Glucose-6-Phosphate
G6PD	Glucose-6-Phosphate Dehydrogenase
GPT	Glutamate Pyruvate Transaminase ($=$ ALT)
GTT	Glucose Tolerance Test
HBD	Hydroxybutyrate Dehydrogenase
HBsAg	Hepatitis B Surface Antigen
25-HCC	25-Hydroxycholecalciferol
HCG	Human Chorionic Gonadotrophin
HDL	High-Density Lipoprotein
hGH	Human Growth Hormone
HGPRT	Hypoxanthine Guanine Phosphoribosyl Transferase
5HIAA	5-Hydroxyindole Acetic Acid
HMMA	4-Hydroxy-3-Methoxymandelic Acid ($=$ VMA)
HPL	Human Placental Lactogen
5HT	5-Hydroxytryptamine ($=$ Serotonin)
5HTP	5-Hydroxytryptophan
ICD	Isocitrate Dehydrogenase
ICF	Intracellular Fluid
ICSH	Interstitial Cell-Stimulating Hormone ($=$ LH)
IDDM	Insulin-Dependent Diabetes Mellitus
Ig	Immunoglobulin
LATS	Long-Acting Thyroid Stimulator
LCAT	Lecithin Cholesterol Acyl Transferase
LD	Lactate Dehydrogenase ($=$ LDH)
LDH	Lactate Dehydrogenase ($=$ LD)
LDL	Low-Density Lipoprotein
LH	Luteinising Hormone ($=$ lutropin; ICSH)
LH-RH	LH-Releasing Hormone
MEA	Multiple Endocrine Adenopathy ($=$ pluriglandular syndrome)
MIT	Mono-iodotyrosine
α-MSH	Melanocyte-Stimulating Hormone (melanotropin)
NAD	Nicotinamide Adenine Dinucleotide
NADP	Nicotinamide Adenine Dinucleotide Phosphate
NEFA	Non-Esterified Fatty Acids ($=$ FFA)
NIDDM	Non-Insulin-Dependent Diabetes Mellitus
5'NT	5'-Nucleotidase ($=$ NTP)
NTP	5'-Nucleotidase ($=$ 5'NT)
17-OGS	17-Oxogenic Steroids

11-OHCS	11-Hydroxycorticosteroids ("Cortisol")
17-OHCS	17-Hydroxycorticosteroids (17-oxogenic steroids)
OP	Osmotic Pressure
PBG	Porphobilinogen
PIF	Prolactin-Release Inhibiting Factor (prolactostatin)
PP factor	Pellagra Preventive Factor (nicotinamide: niacin)
PRPP	Phosphoribosyl Pyrophosphate
PTH	Parathyroid Hormone ("Parathormone")
RF	Releasing Factor (= RH; liberin)
RH	Releasing Hormone (= RF; liberin)
RNA	Ribonucleic Acid
RU	Resin Uptake (of T_3 or T_4)
SG	Specific Gravity
SGOT	Serum Glutamate Oxaloacetate Transaminase (= AST)
SGPT	Serum Glutamate Pyruvate Transaminase (= ALT)
SHBD	Serum Hydroxybutyrate Dehydrogenase (= HBD)
T_3	Tri-iodothyronine
T_4	Thyroxine (Tetra-iodothyronine)
TBG	Thyroxine Binding Globulin
TBPA	Thyroxine-Binding Prealbumin
TBW	Total Body Water
TCA cycle	Tricarboxylic Acid Cycle (= Krebs' Cycle or Citric Acid Cycle)
TIBC	Total Iron-Binding Capacity (usually measure of transferrin)
TP	Total Protein
TPP	Thiamine Pyrophosphate
TRF	Thyrotrophin-Releasing Factor
TRH	Thyrotrophin-Releasing Hormone
TSH	Thyroid-Stimulating Hormone (Thyrotrophin)
UDP	Uridine Diphosphate
UTP	Uridine Triphosphate
VLDL	Very Low-Density Lipoprotein
VMA	Vanillyl Mandelic Acid (= HMMA)
WDHA	Watery Diarrhoea, Hypokalaemia and Achlorhydria (= Verner-Morrison syndrome)
Z-E syndrome	Zollinger-Ellison Syndrome

Chapter I

THE KIDNEYS: RENAL CALCULI

THE KIDNEYS

THE kidneys excrete waste products of metabolism and play an essential homeostatic role by adjusting the body balance of water and solute. Normal function depends on:

the integrity of glomeruli and tubular cells
a normal blood supply
normal secretion and feed-back control of hormones acting on the kidney.

Other functions of the kidney, which will not be dealt with further in this chapter, are:

the production of erythropoietin—a hormone stimulating erythropoiesis. The student is referred to textbooks of haematology for further details
the production of renin (p. 38)
the conversion of 25-hydroxycholecalciferol to the active 1,25-dihydroxycholecalciferol (p. 254).

PASSIVE FILTRATION

About 200 litres of plasma ultrafiltrate enter the tubular lumen each day, mainly by glomerular filtration but also through the spaces between tubular cells. This flow of fluid from the blood depends on the fact that the hydrostatic pressure in the renal capillaries is higher than that in the lumen: any factor reducing this pressure gradient will lower the filtration rate.

Because this is a passive process, the filtrate contains diffusible constituents at almost the same concentrations as plasma. For example, at normal plasma concentrations about 30 000 mmol of sodium, 800 mmol of potassium, 300 mmol of ionised calcium, 1000 mmol (180 g) of glucose and 800 mmol (48 g) of urea would be filtered daily in 200 litres. The very large volume of filtrate allows adequate elimination of waste products such as urea, but unless the bulk of water and essential solute were reclaimed, death from water and electrolyte depletion alone would occur in a few hours.

Protein and protein-bound substances are filtered in only small amounts by the normal kidney; most of that filtered is reabsorbed.

The colloid osmotic pressure (p. 36) of plasma is therefore slightly higher than that of tubular fluid and tends to oppose the filtration due to the hydrostatic pressure gradient. This osmotic effect is so weak that it can usually be ignored, but it should be remembered that overenthusiastic intravenous infusion may dilute plasma proteins enough to cause an abnormally high filtration rate, and inappropriate loss of some of the infused fluid (p. 47).

TUBULAR FUNCTION

Changes in filtration rate alter the total amount of water and solute filtered, but not the composition of the filtrate.

Although about 200 litres of plasma are filtered, only about 2 litres of urine are formed each day. The composition of urine differs markedly from that of plasma (and therefore filtrate); the concentrations of individual constituents not only vary independently of each other, but also vary widely as body requirements alter. This reabsorption of about 99 per cent of filtered volume, and adjustment of individual solute, indicates that tubular cells have carried out selective *active* transport of solute against physicochemical gradients. Active transport needs an energy supply, usually from adenosine triphosphate (ATP), and is affected by cell death, enzyme poisons and hypoxia, which impair the production of ATP by oxidative phosphorylation.

Transport of charged ions will tend to produce an electrochemical gradient which would inhibit further transport. This is minimised by two processes:

Isosmotic transport occurs mainly in the proximal tubule and reclaims the bulk of filtered constituents essential for the body. Active transport of one ion leads to passive movement of an ion of the opposite charge in the same direction along the electrochemical gradient. For instance, isosmotic reabsorption of sodium (Na^+) depends on the availability of diffusible negatively charged ions (such as Cl^-). The process is "isosmotic" because the active transport of solute causes equivalent movement of water in the same direction. Isosmotic transport also occurs in the distal part of the nephron, but is of less importance at that site.

Ion exchange occurs mainly in the more distal parts of the nephron, and at this site is important for fine adjustment after bulk reabsorption has taken place. Ions of the same charge, usually cations, are exchanged and neither electrochemical nor osmotic gradients are created. There is therefore insignificant net movement of anion or water. For example, Na^+ may be reabsorbed in exchange for potassium (K$^+$) or hydrogen ion (H$^+$) secretion. Na$^+$ and H$^+$ exchange also occurs proximally, but that site is more important for bicarbonate reclamation than for fine adjustment (see Chapter IV).

Some other substances, such as phosphate and urate, can be secreted into, as well as reabsorbed from, the lumen.

All body cells carry out both types of ion transport, but in most cells the pumps are uniformly distributed on the membrane surrounding the cell and solute passes into or out of the cell. In cells of the renal tubule, the intestine and many secretory organs, the pumps are located on the membrane on one side of the cell and pass solute between lumen and blood.

Waste products such as urea are not significantly handled by tubular cells. Almost all filtered urea is passed in the urine, although a small amount diffuses back passively with water. The urinary *concentration* of urea depends on the amount of water reabsorbed in excess of urea.

Reclamation in the Proximal Tubule

Over 70 per cent of the filtered *sodium* and ionised *calcium* and almost all the *potassium* is actively reabsorbed in the proximal tubule. Many inorganic anions follow the electrochemical gradient, and reabsorption of sodium is limited by the availability of *chloride*—the most abundant diffusible anion in the filtrate (p. 114). *Bicarbonate* is almost completely recovered (though not strictly reabsorbed) following exchange of sodium and hydrogen ions (see Chapter IV). Specific active transport mechanisms result in almost complete reabsorption of *glucose*, *urate* and *amino acids*. There is incomplete *phosphate* reabsorption, and its presence in tubular fluid is important for buffering: inhibition of its reabsorption by parathyroid hormone (PTH) may occur here, or at a more distal site, and accounts for the hypophosphataemia of PTH excess (p. 253).

Isosmotic reabsorption of 70 to 80 per cent of filtered *water* in the proximal tubule depends on the solute transport described above.

Thus almost all the reutilisable nutrients and the bulk of the electrolytes and water are reclaimed in the proximal tubule. Almost all the filtered metabolic waste products which cannot be reused by the body, such as urea and creatinine remain in the lumen.

Fine homeostatic adjustment of water and solute takes place distal to the proximal tubule. We will discuss renal handling of water, and then outline the control of solute secretion by the distal tubule and collecting duct.

Water Reabsorption: Urinary Concentration and Dilution

Water is always reabsorbed *passively* along an osmotic gradient. However, *active* solute transport is necessary to produce this gradient. There are two main processes involved in water reabsorption;

isosmotic reabsorption of water in the proximal tubule;
differential reabsorption of water and solute in the loop of Henle, distal tubule and collecting duct.

Isosmotic reabsorption of water in the proximal tubule.—The nephron as a whole reabsorbs 99 per cent of the filtered water, about 70 to 80 per cent (that is 140 to 160 litres a day) being returned to the body by the proximal tubules.

The proximal tubules pass through the renal cortex and their walls are freely permeable to water. We have seen that active solute reabsorption from the filtrate is accompanied by passive reabsorption of an osmotically equivalent amount of water. Blood flow is brisk in this area and solute and water are removed rapidly. As water and solute reabsorption are almost concurrent, fluid entering the loop of Henle, though much reduced in volume, is still almost isosmotic and this process cannot adjust extracellular osmolality; it merely reclaims the bulk of filtered water and solute.

Differential reabsorption of water and solute in the loop of Henle, distal tubule and collecting duct.—Normally between 40 and 60 litres of water a day enter the loops of Henle. Not only is this volume further reduced to about 2 litres, but, if changes in extracellular osmolality are to be corrected, the proportion of water reabsorbed must be varied according to the body's needs. At extremes of water intake urinary osmolality can vary from about 40 to about 1400 mmol/kg. (These figures should be compared with the normal for plasma, and therefore for glomerular filtrate, of about 290 mmol/kg.) As the proximal tubule cannot dissociate water and solute reabsorption, the adjustment must occur between the end of the proximal tubule and the end of the collecting duct.

It is generally agreed that two mechanisms are involved:

Countercurrent multiplication is an *active* process occurring in the *loop of Henle*, whereby high medullary osmolality is created, and urinary osmolality is reduced. This acts in the absence of ADH, and a dilute (hypo-osmolal) urine is produced.

Countercurrent exchange is a *passive* process, only occurring in the *presence of ADH*, whereby water without solute is reabsorbed from the *distal tubules and collecting ducts* into the *ascending vasa recta* along the osmotic gradient created by

multiplication; by this means the urine is concentrated and the plasma diluted.

Countercurrent multiplication.—The most generally held theory considers that this occurs in the loops of Henle, solute being actively pumped from the ascending to the descending limb while fluid is flowing through the loop. There is experimental evidence for a chloride pump at this site, but a sodium pump would have the same effect.

Fluid entering the descending limb from the proximal tubule is almost isosmolal – that is, is of the same osmolality as that in the general circulation. This is normally a little under 300 mmol/kg, and for ease of discussion we will use the figure 300 mmol/kg.

Suppose that the loop has been filled, no pumping has taken place, and the fluid in the loop is stationary. Osmolality throughout the loop and the adjacent medullary tissue will be about 300 mmol/kg.

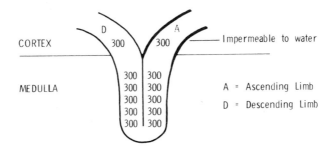

Suppose 1 mmol of solute per kg is pumped from the ascending limb (A) into the descending limb (D), the fluid column remaining stationary.

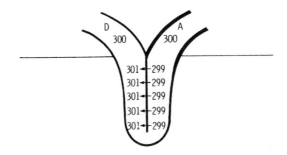

If this pumping were continued and there were no flow, limb D would become very hyperosmolal and limb A equally hypo-osmolal.

Let us now suppose that the fluid flows so that each figure "moves two places".

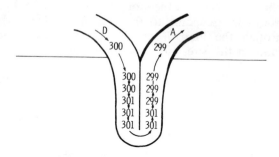

As this happens more solute is pumped from limb A to limb D.

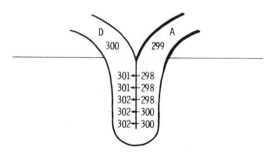

If the fluid again flows "two places", then the situation will be:

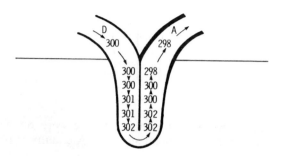

If these steps occur simultaneously and continuously, the wall of the *distal part* of limb A being *impermeable to water*, the consequences would be:

> increasing osmolality in the tips of the loops. As the *walls of the loops are permeable* to water and solute, osmotic equilibrium would be reached with all the surrounding tissues and the deeper layers of the medulla including the blood in the vasa recta, which will also be of increasing osmolality;
> hypo-osmolal fluid leaving the ascending limb.

The final result might be:

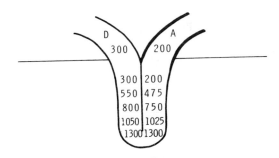

In the absence of ADH the walls of the distal tubules and collecting ducts are impermeable to water, no further change in osmolality occurs, and hypo-osmolal urine would be passed.

Countercurrent exchange is essential, *together with multiplication*, for *concentration of urine*. It can only occur in the presence of ADH, and depends on the "random" apposition of collecting ducts and ascending vasa recta, a result of the close anatomical relations of *all* medullary constituents (Fig. 1) (apposition to descending vasa recta will also occur, but this will have little effect on urinary osmolality). The action of ADH makes the walls of the distal part of the tubule and the collecting ducts permeable to water, which then passes along the osmotic gradient created by multiplication; urine is thus concentrated as the collecting ducts pass into the increasingly hyperosmolal medulla. The increasing concentration of the fluid as it passes down the ducts would reduce the osmotic gradient if it did not meet even more concentrated blood flowing in the opposite (countercurrent) direction. The gradient is thus maintained, and water can continue to be reabsorbed until urine reaches the osmolality of the deepest layers (four or five times that of plasma). The diluted blood is carried towards the cortex and soon enters the general circulation, thus tending to reduce plasma osmolality.

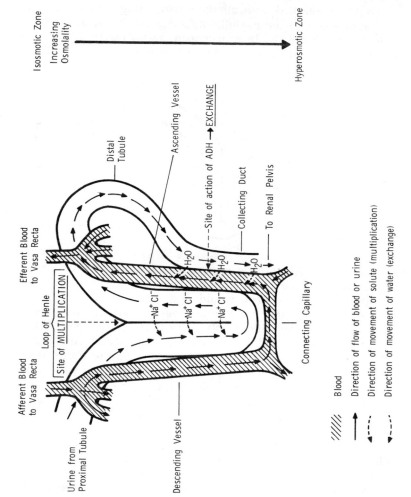

FIG. 1.—The countercurrent mechanism.

Whether this is the exact mechanism or not, both concentration and dilution of urine depend on active processes, which may be deranged if tubules are damaged.

Let us now look at the process in more detail as it works at extremes of water intake, bearing in mind that the tendency to a change in *plasma* osmolality is normally rapidly corrected.

Water load.—A high water intake dilutes the extracellular fluid and the fall in osmolality *cuts off ADH* production (p. 39). As the walls of the collecting ducts are then impermeable to water, *countercurrent multiplication* is acting *alone*, and as we have seen, a dilute urine will be produced. However, plasma osmolality would not be corrected unless the hyperosmolality created by multiplication could be carried into the general circulation.

It has been found that during a maximal water diuresis the osmolality at the tips of the papillae may only reach about 600 mmol/kg, rather than the maximum of about 1400 mmol/kg. Since increasing the circulating volume increases renal blood flow, the *more rapid flow in the vasa recta* may tend to "wash out" medullary hyperosmolality, returning some of the solute, without extra water, to the circulation. Thus, not only is more water than normal lost in the urine, but more solute is also "reclaimed" into the general circulation. Medullary hyperosmolality, and therefore the ability to concentrate the urine maximally, may only be fully restored several days after a prolonged water load is stopped. Results of a urine concentration test in a case of suspected polydipsia must be interpreted with this fact in mind (p. 88).

Water restriction.—Water restriction, by increasing plasma osmolality, *leads to ADH production* and allows countercurrent exchange. Reduced circulating volume results in *sluggish flow in the vasa recta*, allowing build-up of the medullary hyperosmolality produced by multiplication and therefore increasing this exchange.

Osmotic diuresis.—Under physiological circumstances by far the largest contribution to the osmolality of the glomerular filtrate comes from sodium and its associated anions, and active removal of sodium in the proximal tubule is followed by passive reabsorption of water. Suppose that another osmotically active substance is circulating in significant amounts, that this substance is freely filterable at the glomerulus, but that it cannot be reabsorbed either actively or passively to any significant extent in the proximal tubule. *Mannitol*, which cannot diffuse significantly through cell walls, is such a substance. As water reabsorption takes place with that of sodium the mannitol will become more concentrated, and its osmotic effect will inhibit further water diffusion. Since less water is reabsorbed a larger volume than usual will enter the loop of Henle and flow through the rest of the tubular system. The rate of water reabsorption in the distal tubule and collecting duct is limited, and water diuresis will result. Note that urine leaving the proximal tubule is still isosmotic with blood, but that the contribution to the osmotic pressure of the tubular fluid from sodium is less than that in blood; the difference is made up by mannitol. A little *urea* diffuses back in the proximal tubule and most *glucose* is actively reabsorbed. However, if these substances are filtered at high concentrations, less is absorbed and they then act as osmotic diuretics (see p. 40).

Homeostatic Solute Adjustment in the Distal Tubule and Collecting Duct

Sodium reabsorption in exchange for *hydrogen ion* occurs throughout the nephron. In the proximal tubule the main effect of this exchange is reclamation of filtered bicarbonate. In the distal tubule and collecting duct filtered bicarbonate has usually been reclaimed, and the process is more likely to be associated with net generation of bicarbonate to replace that lost in extracellular buffering, and so with fine adjustment of hydrogen ion homeostasis. Aldosterone stimulates the exchange. The possible mechanisms are discussed in Chapter IV.

Sodium reabsorption in exchange for *potassium* in the distal nephron is stimulated by aldosterone; the most important stimulus to aldosterone secretion is mediated via renal blood flow, and this method of sodium reabsorption is part of the homeostatic mechanism of sodium and water balance (Chapter II).

Potassium and hydrogen ions in tubular cells compete for secretion in exchange for sodium ions.

About 30 per cent of filtered ionised *calcium* is not reclaimed from the proximal tubule. Much of this is reabsorbed at distal sites, possibly in the loop of Henle, and this reabsorption may be stimulated by parathyroid hormone and inhibited by loop diuretics such as frusemide (furosemide). Only about 2 per cent of filtered calcium appears in the urine.

The site of *urate* secretion has not been clarified, but may be in the distal tubule.

In Summary

The very *large daily volume of filtrate* allows *waste products to be excreted* at a rate equal to their production.

Most of the filtered water, electrolytes and reusable metabolites are *reclaimed* from the *proximal tubule*.

Fine homeostatic adjustments are made in the more *distal nephron,* often under hormonal control.

CHEMICAL PATHOLOGY OF KIDNEY DISEASE

Different parts of the nephron are closely associated anatomically, and are dependent on a common blood supply. Renal dysfunction of any kind affects all parts of the nephron to some extent, although sometimes either glomerular or tubular dysfunction is predominant (see "Renal Circulatory Insufficiency", p. 16 for an example of predominant glomerular dysfunction, and p. 19 for examples of predominant tubular dysfunction). The net effect of renal disease on plasma and urine depends on the proportion of glomeruli and tubules affected, and on the number of nephrons involved. However, it is probably easier to understand renal disease if we start by considering hypothetical individual nephrons, first with a low glomerular filtration rate (GFR) and normal tubular function, and then with tubular damage but a normal GFR.

Reduced GFR with Normal Tubular Function

If the proximal tubular cells are capable of reabsorbing a normal total amount of solute, and therefore water, a large proportion of the reduced amount of filtrate will be reclaimed by isosmotic processes. This, by itself, will reduce urine volume.

In the subject with a low GFR there is often a stimulus to secretion of antidiuretic hormone this, acting on the distal nephron, would cause water to be reabsorbed in excess of solute. Urinary volume would be further reduced, and urinary osmolality increased well above that of plasma. The high osmolality is mainly due to substances not actively handled by the tubules. For example, the urinary urea concentration will be high. This distal response will *only occur in the presence of ADH*; in its absence even normal nephrons will form a dilute urine.

The amounts of urea and creatinine excreted depend almost entirely on the GFR. If excretion fails to balance production, plasma levels rise.

Phosphate and urate are released by cell breakdown. Plasma levels rise because less is filtered. Most of the reduced amount reaching the proximal tubule can be reabsorbed and the capacity for distal secretion is impaired if the filtered volume is low: these factors further contribute to high plasma concentrations.

A large proportion of the reduced amount of filtered sodium is reabsorbed by isosmotic mechanisms: less than usual is available for exchange with hydrogen and potassium ions (potassium can be reabsorbed proximally).

This has two important results:

reduction of hydrogen ion secretion throughout the nephron. Bicarbonate can only be reclaimed by hydrogen ion secretion (p. 97), and there will be a low plasma bicarbonate concentration;

reduction of potassium secretion in the distal tubule with potassium retention.

Only if the low GFR is due to a low renal blood flow will systemic aldosterone secretion be maximal (p. 38); any sodium reaching the distal tubule will then be almost completely reabsorbed in exchange for H^+ and K^+, and urinary sodium concentration will be low.

Thus the findings in venous plasma and urine from the affected nephron will be:

Plasma

high urea and creatinine concentrations;
low bicarbonate concentration, with low pH;
hyperkalaemia;
hyperuricaemia and hyperphosphataemia.

Urine

reduced volume:
(a) *only if renal blood flow is low* (stimulating aldosterone secretion), a low (appropriate) sodium concentration
(b) *only if ADH secretion is stimulated*, a high (appropriate) urea concentration, and therefore a high osmolality.

Reduced Tubular with Normal Glomerular Function

Damage to the tubular cells impairs the adjustment of the composition and volume of the urine.

Failure of solute reabsorption in the proximal tubule impairs isosmotic water reabsorption at this site. The countercurrent mechanism may also be impaired and the ability to respond to ADH is reduced. Large volumes of dilute urine are passed.

The tubules cannot secrete hydrogen ion and therefore cannot reabsorb bicarbonate normally and cannot acidify the urine.

The response to aldosterone of exchange mechanisms involving sodium is impaired and the urine contains an inappropriately high concentration of sodium relative to the renal blood flow.

Potassium reabsorption in the proximal tubule is impaired and plasma potassium levels may be low.

Reabsorption of glucose, phosphate, urate and amino acids is impaired. The plasma phosphate and urate levels may be low.

Thus the finding in venous plasma and urine from the nephron would be:

Plasma

normal urea and creatinine levels;
low bicarbonate with low pH;
hypokalaemia;
hypophosphataemia and hypouricaemia.

Urine

increased volume:
(a) *even if renal blood flow is low* there is a relatively high (inappropriate) sodium concentration;
(b) *even if ADH secretion is stimulated* there is a relatively low (inappropriate) urea concentration, and therefore osmolality.

CLINICAL SYNDROMES OF RENAL DISEASE

There is a spectrum of conditions in which the proportions of tubular and glomerular dysfunction vary. The findings will depend on the relative contributions from these two factors, and an attempt has been made to indicate this in Table I. The dotted line indicates that the proportions are variable.

When the GFR falls below about 30 per cent of normal, substances almost unaffected by tubular action such as *urea and creatinine* will be retained with a consequent rise in their plasma concentrations.

The degree of retention of *potassium, urate and phosphate* will depend on the balance between the degree of glomerular retention and the degree of loss due to failure of proximal tubular reabsorption: at the glomerular end of the spectrum so little is filtered that, despite failure of reabsorption, plasma levels rise; at the tubular end of the spectrum glomerular retention is more than balanced by impaired reabsorption of filtered potassium, urate and phosphate. Similarly, the urine volume depends on the balance between the volume filtered, and the proportion reabsorbed by the tubules (remember that normally 99 per cent of filtered water is reabsorbed). At the glomerular end of the spectrum nothing is filtered and the patient is anuric: at the tubular end, although filtration is reduced, tubular reabsorption is so impaired that the patient suffers from polyuria.

While *plasma levels* of urea and creatinine depend largely on glomerular function, *urinary concentrations* depend almost entirely on

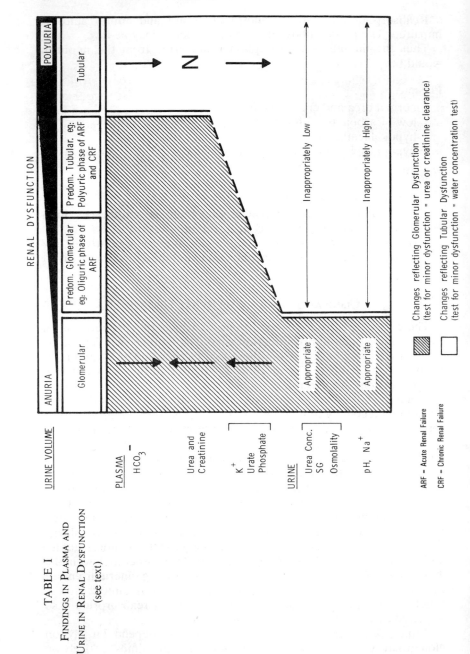

TABLE I

FINDINGS IN PLASMA AND
URINE IN RENAL DYSFUNCTION
(see text)

tubular function. However little is filtered at the glomerulus, the composition of the filtrate is that of a plasma ultrafiltrate. Any variation from this in urine passed is due to tubular activity. The fewer tubules working, the nearer urine concentrations will be to those of plasma. Urine concentrations *inappropriate to the state of hydration* suggest tubular damage, whatever the degree of glomerular damage.

A low plasma *bicarbonate* level is a constant finding. Consequent metabolic acidosis may aggravate the hyperkalaemia of glomerular dysfunction.

Plasma sodium *concentrations* are not primarily affected by renal disease.

Acute Oliguria

In the adult oliguria is, by definition, a urine output of less than 400 ml a day and usually reflects a low glomerular filtration rate.

Acute oliguria without significant renal damage, but with a low GFR, is due to mechanical factors reducing the hydrostatic pressure gradient between the renal capillaries and tubular lumen.

A low intracapillary pressure is the commonest cause of this syndrome. It is known as *renal circulatory insufficiency* ("prerenal uraemia") and may be due to:

intravascular depletion
    (a) of whole blood volume (haemorrhage);
    (b) of plasma volume ("dehydration");
reduced pressure due to "shock".

A rise in tubular luminal pressure as a primary cause of oliguria is rarer. The cause is usually but not always clinically obvious. It may be due to intra- or extrarenal obstruction to outflow ("postrenal urae-mia").

    Intrarenal obstruction may be due to blockage of the tubular lumen by haemoglobin, myoglobin, and very rarely urate or calcium. Other causes of intrarenal obstruction such as casts and oedema of tubular cells are usually the result of true renal damage.

    Extrarenal obstruction may be due to calculi, neoplasms, strictures or prostatic hypertrophy, any of which may cause sudden obstruction. The finding of a palpable bladder in the male is most likely to be due to prostatic hypertrophy, although there are other causes of urethral obstruction.

Early correction of the cause of an acute reduction of GFR due to mechanical factors rapidly increases the urine output. The longer it remains untreated the greater the danger of ischaemic or pressure damage to the kidney.

Acute oliguria due to renal damage often follows one of the above conditions, or may be due, for example, to:

septicaemic damage;
ingestion of a variety of poisons;
acute glomerulonephritis.

Acute glomerulonephritis usually occurs in children. The history of sore throat and the finding of red cells in the urine usually make the diagnosis obvious.

Septicaemia should be considered when the cause of oliguria is obscure.

The commonest problem in the differential diagnosis of acute oliguria is to distinguish between renal circulatory insufficiency and the renal damage which may have followed it.

Renal circulatory insufficiency is probably the commonest cause of low GFR with relatively normal tubular function, and is most commonly due to reduction of the circulating volume by haemorrhage or dehydration. The many other causes of the "shock" syndrome include acute myocardial infarction, acute intra-abdominal lesions (for example, rupture of an ectopic pregnancy, acute pancreatitis or perforation of a peptic ulcer) and intravascular haemolysis (including that due to mismatched blood transfusion). If blood pressure (or more accurately renal blood flow) is restored within a few hours the condition is reversible, but the longer it persists the greater the danger of ischaemic renal damage. Since most glomeruli are involved but tubular function is relatively normal, the systemic findings in plasma and urine are those described on p. 12 as due to a low GFR in a single nephron.

The patient will usually be hypotensive and possibly clinically dehydrated and, in addition to the laboratory findings listed above, there may be haemoconcentration (p. 41). Uraemia due to renal dysfunction is aggravated if there is increased protein breakdown, due either to tissue damage, or to the presence of blood in the gastro-intestinal tract or in large haematomas: the liberated amino groups of amino acids are converted to urea in the liver, and intravenous amino-acid infusion may have the same effect. Increased tissue breakdown also aggravates hyperkalaemia, hyperuricaemia and hyperphospha-taemia.

Acute oliguric renal failure often follows a period of reduced GFR due to renal circulatory insufficiency. The oliguria is probably due to glomerular damage and reduced cortical blood flow, aggravated by back-pressure on the glomeruli due to obstruction of tubular flow by oedema. At this stage, as indicated in Table I, the findings are at the glomerular end of the spectrum; the fact that the tubules are damaged is evident if the urinary concentrations are inappropriate – a fact which may sometimes be used to differentiate acute oliguria due to true renal damage from that due to renal circulatory insufficiency. However, laboratory tests are rarely necessary. When the blood pressure and hydration of a patient with renal circulatory insufficiency are restored the urine output will increase. If this fails to occur, or if the oliguric patient is already normally hydrated and normotensive, renal damage is likely. If laboratory tests are used, their limitations must be understood. *If the renal blood flow is still low* (the patient is still dehydrated or hypotensive), aldosterone secretion should be maximal, and functioning tubules will respond appropriately by selective reabsorption of sodium by distal exchange mechanisms. A urinary sodium concentration of less than 30 mmol/l is usually taken to indicate adequate renal tubular, and therefore overall renal, function. Measurement of urinary osmolality or other indicators of selective water reabsorption are less valuable, since ADH secretion is not invariably stimulated. *Measurement of urinary sodium concentration cannot be used to test tubular function once renal blood flow has been restored*: in the absence of aldosterone the appropriate response of *normal* tubules is a reduction in differential sodium reabsorption. If renal damage is suspected fluid should be given with caution after dehydration has been corrected: there is a danger of overhydration if the glomerular membranes are unable to filter normally despite an adequate hydrostatic gradient.

If the patient recovers oliguria is replaced by *polyuria*. As cortical blood flow increases and as the tubular oedema resolves, glomerular function recovers before tubular function. The findings gradually progress to the tubular end of the spectrum until they approximate to those of "pure" tubular lesions, and as urinary output increases and is further increased by the high osmotic load of urea, the polyuria may cause water depletion. Electrolyte depletion may occur and the initial hyperkalaemia may be replaced by hypokalaemia. Mild acidosis (common to both glomerular and tubular lesions) persists until late. Finally, recovery of the tubules restores renal function to normal.

Chronic Renal Failure

Chronic renal failure may rarely follow an episode of acute oliguric renal failure. It is usually the end result of a variety of conditions which include chronic glomerulonephritis, chronic obstructive uropathy, polycystic disease, renal artery stenosis, and any of the conditions listed below as causes of tubular dysfunction. Sometimes there is no obvious precipitating factor.

In acute oliguric renal disease there is diffuse damage to the kidney involving virtually all nephrons. It is obvious that such a situation would lead to early death and that an untreated patient who survives long enough to develop chronic renal failure must, by contrast, have some functioning nephrons. Histologically it can be shown that not all nephrons are equally affected. Some may be completely destroyed, others may be quite normal, and yet others may have some functions disturbed to a greater extent than others. Some of the effects of chronic renal failure can be explained by this patchy distribution of disease (compare the effects of patchy pulmonary disease, p. 117).

Polyuric phase.—At first glomerular function may be adequate to maintain plasma urea and creatinine levels. As more glomeruli are involved the rate of urea excretion falls and cannot balance the rate of production by protein catabolism: as a consequence the plasma urea level rises and the urea concentration in the filtrate of the functioning nephrons rises. This may cause an osmotic diuresis in these nephrons: in other nephrons tubules may be damaged out of proportion to the glomeruli. Both the tubular dysfunction in nephrons with functioning glomeruli and the osmotic diuresis through intact nephrons contribute to polyuria. If fluid loss is replaced by a high intake, urea excretion can continue through functioning glomeruli at a high rate: the increased excretion through normal glomeruli may eventually balance the reduced permeability of others, and a new steady state is reached at a higher level of plasma urea. At this stage the findings are near the tubular end of the spectrum in Table I, the glomerular dysfunction being indicated by the high plasma urea level. If these subjects are kept well hydrated, they may remain in a stable condition with a moderately raised plasma urea for years. Potassium levels are variable, but tend to be raised.

Oliguric phase.—If nephron destruction continues the condition approximates more and more to the glomerular end of the spectrum, until oliguria precipitates a steep rise in plasma urea and potassium concentrations. This stage is often terminal. Before assuming that the latter is the case, care should be taken to ensure that the sudden rise is not due to electrolyte and water depletion.

The diagnosis of chronic renal failure is usually obvious, and chemical estimations on urine are not, in any case, helpful.

Incidental Abnormal Findings in Renal Failure

In assessing the severity and progress of renal failure it is usual to estimate plasma urea or creatinine and electrolytes (especially potassium). Other abnormalities, although not useful in diagnosis or assessment of renal dysfunction, may be misinterpreted if the cause is not recognised. Plasma *urate* levels rise in parallel with plasma urea, and a high level does not necessarily indicate primary hyperuricaemia (p. 410). Plasma *phosphate* levels also rise, and those of *calcium* fall

(p. 262). In chronic renal failure which has lasted for several years secondary hyperparathyroidism may lead to bone disease with a high level of *alkaline phosphatase*. Hypocalcaemia without evidence of bone disease should only be treated after correction of hyperphosphataemia (p. 273). *Anaemia* is commonly present, and is normocytic and normochromic in type: it does not respond to iron therapy.

Mild Uraemia with Normal Urinary Volume

A slight elevation of plasma urea is a common incidental finding especially in elderly subjects. It almost certainly indicates some degree of renal damage but, unless progressive, is unlikely to need treatment.

In congestive cardiac failure circulation to the kidney may be sufficiently impaired to cause mild uraemia: the clinical findings are those of the primary condition, and haemodilution is often present.

Syndromes Reflecting Predominant Tubular Damage

A group of conditions initially affects tubules more than glomeruli, although scarring often leads to eventual chronic renal failure. The patient may first present complaining of polyuria and be found to have hypokalaemia, hypophosphataemia and hypouricaemia (the tubular end of the spectrum in Table I). The effects, as described on p. 13, are mostly due to the failure of the proximal tubule to reclaim the bulk of filtrate. Glycosuria, phosphaturia and generalised amino-aciduria may be detectable and make up the so-called acquired Fanconi syndrome.

Water and electrolyte depletion may cause renal circulatory insufficiency. If this is prevented by adequate intake plasma urea and creatinine levels are often normal. Potassium supplementation may be necessary.

Tubular cells may be damaged in several ways.

Precipitation of poorly soluble substances in or around them. Predisposing factors include:

hypercalcaemia ⎱
hyperuricaemia ⎰ the commonest causes in adults;

Bence Jones proteinuria.

Amongst many inborn errors (discussed in the relevant chapters and summarised in Chapter XVI), the following may be associated with renal tubular damage because of the intracellular accumulation of a metabolite or other toxic substance:

galactosaemia (galactose-l-phosphate);
hereditary fructose intolerance (fructose-l-phosphate);
Wilson's disease (copper);

cystinosis (cystine);
Fabry's disease (glycosphingolipids).

Infective damage in early pyelonephritis.
Vacuolation of tubular cells due to prolonged hypokalaemia.
Damage by exogenous toxins, which include:

nephrotoxic drugs (for example, phenacetin);
other toxins, including heavy metals.

The **urine concentration test** may be used if tubular damage is suspected. The ability to form a concentrated urine in response to water deprivation depends on normal tubular function (countercurrent multiplication) and on the presence of ADH. Failure of this ability is usually due to renal disease, but if there is any doubt the test can be repeated after administration of ADH (Pitressin) or its synthetic analogue, DDAVP (p. 88).

Nephrotic Syndrome

Increased glomerular permeability occurs in the *nephrotic syndrome*. All but the highest molecular weight plasma proteins can then pass the glomerulus and proteinuria of several grams a day occurs. The main effects are on *plasma proteins* and the subject is discussed more fully in Chapter XIV. Uraemia occurs only in the late stages of the disease when many glomeruli cease to function.

Low Plasma Urea Concentrations

Occasionally the plasma urea concentration is found to be below 3 mmol/l or even 1 mmol/l. The causes of this finding are:

Due to an Increased GFR (common)

Pregnancy (the commonest cause in young women);
Overenthusiastic intravenous therapy (p. 49) (the commonest cause in hospital);
"Inappropriate" secretion of ADH (p. 473).

Due to Decreased Synthesis (rare)

Extensive liver disease;
Low protein intake;
Inborn errors of the urea cycle (infants only).

GLOMERULAR FUNCTION TESTS

Plasma Levels

Plasma urea and creatinine levels depend on the balance between their production and excretion.

Urea is a product of amino acid breakdown in the liver and is therefore derived from protein, either in the diet or in body tissues: the rate of production is accelerated by a high protein diet, by increased endogenous catabolism due to starvation or tissue damage, or by absorption of amino acids and peptides after gastro-intestinal haemorrhage. The capacity of the normal kidney to excrete urea is high, and in the presence of normal renal function only extremely high protein diets can cause rises of plasma urea to levels above the "normal" range. In patients with a gross increase in catabolism due to severe tissue damage or to acute starvation urea levels may rise above "normal", especially as renal function is often mildly impaired in such cases due to renal circulatory factors. (In chronic starvation plasma levels tend to be low due to protein depletion.) In spite of these reservations, a significantly high plasma urea concentration almost always indicates impaired renal function (this is almost certainly so at levels above 15 mmol/l [urea above 90 mg/dl: blood urea nitrogen above 42 mg/dl]). The probability is increased if a cause for renal dysfunction is present, or if there are protein, casts or cells in the urine. In the few cases where doubt remains, measurement of plasma creatinine may resolve it.

Plasma *creatinine* is less affected by diet, because most is produced endogenously by tissue creatine breakdown: although plasma levels would be expected to increase with tissue damage, they do so less than those of urea. In theory, therefore, creatinine would seem preferable to urea estimation as an index of renal function, and for this reason is used by many laboratories for this purpose. However, plasma creatinine levels in normal subjects vary as much as those of urea. In spite of an improvement in the precision and speed of creatinine estimation with the advent of automated methods, that of urea methods is often better, especially at near normal levels. Whether both parameters are routinely measured, or creatinine is reserved for problem cases is largely a matter of local choice.

If urea or creatinine levels are significantly raised, and especially if they are rising, if there is oliguria or a history suggestive of renal disease, or if there is proteinuria, and the urine contains protein, casts, cells or bacteria in significant amounts, the chemical findings can usually be safely assumed to be due to renal impairment. Changes reflect changes in clearance, and the progress of renal disease can be monitored using plasma levels alone.

Clearances

More than 60 per cent of the kidney must be destroyed before either plasma urea or creatinine levels are significantly raised (this is only valid for urea if the patient is taking a normal or low-protein diet, and if there is no excessive protein catabolism). Clearance tests, which should be more sensitive, measure the amount of blood which could theoretically be completely cleared of a substance per minute. Therefore, for example:

Creatinine clearance (ml/minute) =

$$\frac{\text{Urinary [creatinine]} \times \text{Urine volume (ml)}}{\text{Plasma [creatinine]} \times \text{Time of collection (minutes)}}$$

Care should be taken that plasma and urinary creatinine concentrations are expressed in the same units (for example, mmol/l).

If a true estimate of GFR is to be made a substance should be chosen which is excreted solely by glomerular filtration and is not reabsorbed or secreted by the tubules. *Inulin* is thought to be such a substance: inulin is not produced in the mammalian body and measurement of its clearance requires administration either by constant infusion to maintain its level in the blood steady during the period of the test, or by a single injection followed by serial blood sampling so that its concentration at the mid-point of urine collection can be calculated. Such exogenous clearances are not very practicable for routine use.

Substances produced in the body are usually present in a fairly steady circulating concentration for the period of the test, and blood need only be taken at the mid-point of the urine collection. Creatinine or, more rarely, urea clearances are used for clinical purposes. Neither of these substances fulfils the criterion that it is not reabsorbed from or added to the glomerular filtrate by the tubular cells. Small amounts of *urea* diffuse back into the blood stream from the proximal tubule, and urea clearance values are lower than those of inulin: small amounts of *creatinine* are secreted by the tubule and give clearance values higher than those of inulin. The rate of protein breakdown, while it may affect levels of plasma urea, does *not* affect its rate of clearance by the kidney. By custom, creatinine clearances are usually performed but in practice, there is little to choose between them.

If urea clearance is chosen, care should be taken to inhibit the urea-splitting activity of organisms such as *Proteus vulgaris*, which may be present in the urine, by the use of mercury salts as a preservative. If such organisms are present, short periods of collection (for example, three collections of 1 hour) are preferable to a long one (for example,

24 hours). Creatinine can also be destroyed by bacterial action, and estimations should always be done soon after the collection is complete.

It should be noted that clearance tests will give equally low values with renal circulatory insufficiency, or with "post-renal" causes, as with true renal damage, and cannot distinguish between them.

Precision and Validity of Clearances

We have seen that progress of established renal failure can be monitored by plasma levels. Clearances are often used to detect more minor degrees of renal dysfunction. Those under discussion do not measure tubular function.

Several factors render clearances more imprecise and inaccurate than measurement of plasma levels.

> Every laboratory assay has an inherent imprecision. The combined imprecision of two assays is greater than that of one. The clearance calculation uses results of assays on urine as well as plasma.
>
> *The biggest error of any method depending on timed urine collection is in the figure for the urine volume.* Even highly skilled staff and intelligent and highly motivated patients find accurate collection difficult. Collection by uninformed personnel or patients yields misleading results. This source of error is minimised by lengthening the collection period.
>
> Both creatinine and urea may be partially destroyed in stale or infected urine. This error is *increased* by lengthening the period of collection.

It is most unlikely that treatment will be based on low clearances with normal plasma levels: if it is, the above limitations must be understood. We wonder if clearances are ever indicated.

Proposed schemes for investigating patients with oliguria or polyuria are outlined on pp. 82 and 83.

BIOCHEMICAL PRINCIPLES OF TREATMENT OF
RENAL DYSFUNCTION

Oliguric renal failure.—The oliguria of dehydration or haemorrhage, which is due to a reduction of glomerular filtration rate only, should be treated with the appropriate fluid (p. 65).

In oliguric renal failure due to parenchymal damage the aims are:
> to restrict fluid and sodium, giving only enough fluid to replace losses (p. 42);

to provide an adequate non-protein energy source to prevent aggravation of uraemia and hyperkalaemia by increased endogenous catabolism;

to prevent dangerous hyperkalaemia (p. 78).

Diuretics tend to increase renal blood flow. In some cases of acute oliguric renal failure recovery may be hastened by administration of an osmotic diuretic, such as mannitol, or other diuretics such as frusemide (furosemide) or ethacrynic acid in large doses.

In chronic renal failure with polyuria the aim is to replace fluid and electrolytes lost. Sodium and water depletion may aggravate the uraemia.

Haemodialysis or peritoneal dialysis removes urea and toxic substances from the blood stream, and corrects electrolyte balance, by dialysing the patient's blood against fluid containing no urea, and normal plasma concentrations of electrolytes, *ionised* calcium and other plasma constituents. The blood is either passed over a dialysing membrane before being returned to the body, or the folds of the peritoneum are used as a dialysing membrane with their capillaries on one side, and suitable fluid injected into the peritoneal cavity on the other. A relatively slow reduction of urea concentration is preferable to a rapid one because of the danger of cellular overhydration when extracellular osmolality falls abruptly (p. 35). Dialysis is used in cases of potentially recoverable acute oliguric renal failure to tide the patient over a crisis, or as a regularly repeated procedure in suitable cases of chronic renal failure. It may also be used to prepare patients for renal transplantation, and to maintain them until the transplant functions adequately.

A possible scheme for the investigation of renal dysfunction will be found on p. 82.

RENAL CALCULI

Renal stones are usually composed of normal products of metabolism which are present in the normal glomerular filtrate, often at concentrations near their maximum solubility: quite minor changes in urinary composition may cause precipitation of such constituents, whether in the substance of the kidney (see section on tubular damage, p. 19), as crystals, or as calculi. Although this discussion concerns stone formation it should be remembered that crystalluria and parenchymal damage can occur under the same circumstances, and that the treatment of all such conditions is the same.

Conditions Favouring Calculus Formation

A high urinary concentration of one or more constituents of the glomerular filtrate.

A *low urinary volume*, with normal renal function, due to restricted fluid intake or excessive fluid loss over long periods of time (this is particularly common in the tropics). This condition favours formation of most types of stone, especially if one of the other conditions listed below is also present.

An abnormally *high rate of excretion* of the metabolic product forming the stone, due either to an increased level in the glomerular filtrate (secondary to high plasma concentrations) or to a failure of normal tubular reabsorption from the filtrate.

Changes in pH of the urine, sometimes due to bacterial infection, which favour precipitation of different salts at different hydrogen ion concentrations.

Urinary stagnation due to obstruction to urine outflow.

Lack of normal inhibitors.—It has been suggested that normal urine contains an inhibitor, or inhibitors, of calcium oxalate crystal growth that are absent in the urine of some patients with a liability to recurrent calcium stone formation.

Composition of Urinary Calculi

Calcium-containing stones
    Calcium oxalate   ⎱ with or without magnesium
    Calcium phosphate ⎰ ammonium phosphate.
Uric-acid-containing stones.
Cystine-containing stones.
Xanthine-containing stones.

Calculi Composed of Calcium Salts

These account for between 70 and 90 per cent of all renal stones. Precipitation of calcium is favoured by hypercalciuria, and the type of salt depends on urinary pH and on the availability of oxalate or phosphate. All patients presenting with renal calculi should have a plasma calcium estimation performed, and if this is normal it should be repeated at regular intervals.

Hypercalcaemia causes hypercalciuria if renal function is normal, and estimation of urinary calcium in such cases does not help in the diagnosis. The causes and differential diagnosis of hypercalcaemia are discussed on p. 278.

In many subjects with calcium-containing renal calculi the plasma calcium level is normal. It is in such cases that the estimation of the daily excretion of urinary calcium may be useful. The commonest cause of *hypercalciuria with normocalcaemia* is the so-called *idiopathic hypercalciuria*, a name which reflects our ignorance of the aetiology of the condition: because some of these cases may represent an early stage of primary hyperparathyroidism, plasma calcium estimations should be carried out at regular intervals, especially if the plasma phosphate level is low. Any *increased release of calcium from bone*, as in actively progressing osteoporosis (in which loss of matrix causes secondary decalcification) or in prolonged acidosis (in which ionisation of calcium salts is increased), causes hypercalciuria, but rarely hypercalcaemia. One type of renal tubular acidosis (p.111) not only increases the renal load of calcium but, because of the relative alkalinity of the urine, favours its precipitation in the kidney and renal tract: this is a *rare* cause.

An increased excretion of *oxalate* favours the formation of the very insoluble calcium oxalate, even if calcium excretion is normal. The source of the increased oxalate may be the diet: the very rare inborn error of oxalate metabolism, primary hyperoxaluria, should be considered if renal calculi occur in childhood.

It has already been mentioned that *alkaline conditions* favour calcium precipitation, and whereas calcium oxalate stones form at any urinary pH, a high pH favours formation of calcium phosphate: this type of stone is particularly common in chronic renal infection with urease-containing (urea-splitting) organisms (for example, *Proteus vulgaris*), which convert urea to ammonia.

A significant proportion of cases remains in which there is no apparent cause for the calcium precipitation.

Calcium-containing calculi are usually *hard and white*. They are *radio-opaque*. Calcium phosphate stones are particularly prone to form "staghorn" calculi in the renal pelvis, easily visualised by straight X-ray.

Treatment of calcium-containing calculi.—This depends on the cause. Urinary calcium concentration should be reduced:

> by treating the primary condition (especially hypercalcaemia);
> if this is not possible, by reducing intake of calcium in the diet, and possibly by decreasing calcium absorption by administration of oral phosphate (p. 282);
> by reducing the concentration of urinary calcium by maintaining a high fluid intake (unless renal failure is present). It is the concentration rather than the total 24-hour output which determines the tendency to precipitation.

Uric Acid Stones

These account for about 10 per cent of all renal calculi, and are sometimes associated with *hyperuricaemia* (with or without clinical gout). Precipitation is favoured in an *acid urine*. In a large proportion of cases no predisposing cause can be found.

Uric acid stones are usually *small*, *friable* and *yellowish-brown* in colour, but can be large enough to form "staghorn" calculi. They are *radiotranslucent*, but may be visualised by an intravenous pyelogram.

Treatment of hyperuricaemia is discussed on p. 409. If the plasma urate concentration is normal, fluid intake should be kept high and the urine alkalinised. A low purine diet may help to reduce urate production and therefore excretion.

Cystine Stones

These are rare. In normal subjects the concentration of urinary cystine is well within its solubility. In severe cases of the inborn error cystinuria (p. 391) the solubility may be exceeded and the patient may present with *radio-opaque* renal calculi. Like urate, cystine is more soluble in alkaline than acid urine and the principles of treatment are the same as for uric acid stones. Penicillamine can also be used in therapy (p. 391).

Xanthine Stones

These are very uncommon and may be the result of the rare inborn error, xanthinuria (p. 411). Xanthine stones following the use of xanthine oxidase inhibitors, such as allopurinol, have not been reported.

A proposed scheme for the investigation of renal calculi is given on p. 84.

SUMMARY

THE KIDNEYS

1. Normal renal function depends on a normal filtration rate and normal tubular function.
2. A low glomerular filtration rate (GFR) leads to:

oliguria;

uraemia and retention of other nitrogenous end-products including urate, and of phosphate;

a low plasma bicarbonate, with metabolic acidosis;

hyperkalaemia.

3. Tubular damage leads to:

polyuria. The urine is inappropriately dilute and contains an inappropriately high sodium concentration in relation to the patient's state of hydration;
a low plasma bicarbonate, with metabolic acidosis;
hypokalaemia;
hypophosphataemia and hypouricaemia.

4. In most cases of renal disease impairment of glomerular and tubular function coexist. The clinical findings depend on the proportions of each, and the total number of nephrons involved.

5. A low GFR without significant renal damage may be due to a reduced hydrostatic pressure gradient between the capillary plasma and the tubular lumen. This is most commonly due to renal circulatory insufficiency, but may accompany post-glomerular obstruction.

6. In acute oliguric renal damage plasma findings cannot be distinguished from those of renal circulatory insufficiency.

7. The differentiation between the oliguria of renal circulatory insufficiency with relatively normal tubular function and of acute oliguric renal failure is best made on clinical grounds and, if necessary, by estimating the urinary sodium concentration, only when a low renal blood flow is likely.

8. In most cases plasma urea or creatinine levels reflect changes in renal clearance and are adequate for diagnosing and following up cases of renal disease. Tubular function may be tested by assessing the concentrating ability of the kidney.

9. Clearance tests are relatively imprecise and inaccurate.

RENAL CALCULI

1. The formation of renal calculi is favoured by:

a high urinary concentration of the constituents of the calculi. This may be due to oliguria, or a high rate of excretion of the relevant substances;
a urinary pH which favours precipitation of the constituents of the calculi;
urinary stagnation.

2. Calcium-containing calculi account for 70 to 90 per cent of all renal stones. They are most commonly idiopathic in origin but hypercalcaemia, especially that of primary hyperparathyroidism, should be excluded as a cause.

3. Uric acid stones account for about a further 10 per cent of renal calculi.

4. Rare causes are cystinuria, xanthinuria and hyperoxaluria.

FURTHER READING

DE WARDENER, H. E. (1973). *The Kidney*, 4th edit. Edinburgh and London: Churchill Livingstone.

ROSE, G. A. (1982). *Urinary Stones: Clinical and Laboratory Aspects.* Lancaster: M.T.P. Press.

Chapter II

SODIUM AND WATER METABOLISM

THE control of sodium and of water balance are so closely linked that an understanding of one is impossible without an understanding of the other.

Sodium is normally the most abundant extracellular cation and it accounts, together with its associated anions, for most of the osmotic activity of the extracellular fluid (ECF). Osmotic activity depends on *concentration*, and is determined by the relative amounts of water and sodium, rather than the absolute amounts of either. Hypo- or hyper-natraemia is caused by imbalance between the two, and the clinical picture is due to the consequent osmotic changes.

If sodium and water are lost in equivalent amounts the osmolal concentration is unaffected: symptoms are then those of an inadequate circulatory volume. Of course, osmotic and volume changes may occur together.

Transport of sodium, of potassium and of hydrogen ions across cell membranes is often interdependent, and disturbances of potassium and hydrogen ion homeostasis usually accompany those of sodium: changes in associated anions such as chloride and bicarbonate may occur at the same time. The separation of the contents of this chapter from those on potassium and hydrogen ion homeostasis is arbitrary, and for ease of discussion only. A clinical situation should be assessed with all these factors in mind.

WATER AND SODIUM BALANCE

A 70 kg man contains approximately 45 litres of water and 3000 mmol of osmotically active sodium. Maintenance of the total amount depends on the balance between intake and loss. Water and electrolytes are taken in food and drink, and are lost in urine, faeces and sweat: in addition, about 500 ml of water is lost daily in expired air.

LOSS THROUGH THE KIDNEYS AND INTESTINAL TRACT

The kidney and intestine handle water and electrolytes in a very similar way. Net loss through both organs depends on the balance between the volume filtered proximally and that reabsorbed more distally. Anything affecting either passive filtration or epithelial cell function may disturb this balance.

In addition to the approximately 200 litres of water and 30 000 mmol of sodium filtered by the kidney a further 10 litres of water and 1500 mmol of sodium enter the intestinal tract each day. The whole of the extracellular water and sodium could be lost by passive filtration in about 3 hours. Normally 99 per cent of this initial loss is reabsorbed, and net daily losses amount to about 1·5 to 2 litres of water and 100 mmol of sodium in the urine, and 100 ml and 15 mmol in faeces. It is not surprising that failure of absorptive mechanisms causes such extreme disturbances of water and sodium balance.

Passive filtration.—Most of this large amount of luminal water and electrolyte is derived from the plasma by ultrafiltration—in the kidney through the glomerulus, and in both organs through the so-called "tight junctions" between the epithelial cells lining the lumen. Far from being "tight", they are freely permeable to water and only slightly less so to small molecules and ions.

The hydrostatic gradient from plasma to lumen is the most important factor maintaining filtration in the kidney and in the resting small intestine: in the latter the postprandial flow of fluid into the lumen is greatly increased by the temporary increase in luminal osmolality due to partially digested food. Active secretion of digestive juices contributes only a small proportion of the total volume entering the lumen. Overenthusiastic fluid therapy may, by increasing blood flow to the kidney and intestine and by reducing capillary colloid osmotic pressure, cause inappropriate passive fluid loss (p. 47). A high luminal solute concentration may cause an osmotic diuresis (p. 9) or osmotic diarrhoea.

Bulk reabsorption.—Solute transport is accompanied by isosmotic water flow (p. 3). The solute absorbed in the proximal renal tubule and the resting small intestine is mostly sodium and its associated anions. After meals nutrient is the main solute absorbed from the gut lumen. Epithelial cell dysfunction of either organ impairs this isosmotic reclamation of the bulk of filtrate, and may cause inappropriate diuresis or diarrhoea.

Fine adjustment.—Fine adjustment of water and solute, often under hormonal control, occurs in the distal nephron and lower intestine. The effects of antidiuretic hormone (ADH) and aldosterone on the kidney seem the most important clinically.

LOSS IN SWEAT AND EXPIRED AIR

Normal daily water loss in sweat and expired air amounts to about 900 ml: less than 30 mmol of sodium is lost a day in sweat. Although

ADH and aldosterone have some effect on the composition of sweat, loss is primarily controlled by body temperature. Respiratory water loss depends on respiratory rate, and control of this bears no relation to the body requirements for water. Normally loss by sweat and respiration is rapidly corrected by changes in renal and intestinal loss. However, as neither of the former can be controlled to meet sodium and water requirements, they may contribute considerably to abnormal balance when homeostatic mechanisms fail, or in the presence of gross depletion, whether due to poor intake or to excessive loss by other routes.

DISTRIBUTION OF WATER AND SODIUM IN THE BODY

In mild disturbances of water and electrolyte metabolism the total amount of these in the body may be of less importance than their distribution within it.

DISTRIBUTION OF ELECTROLYTES

The body has two main fluid compartments of very different electrolyte compositions. The compartments are:

　the intracellular compartment, in which *potassium* is the predominant cation;

　the extracellular compartment, in which *sodium* is the predominant cation.

The extracellular fluid can be subdivided into:

　interstitial fluid which is of very low protein concentration;
　intravascular fluid (plasma) which contains protein in high concentration.

Distribution of Electrolytes Between Cells and Extracellular Fluid

The intracellular concentration of sodium is less than a tenth of that in the extracellular fluid (ECF), while that of potassium is about thirty times as much. In absolute amounts about 95 per cent of the metabolically active sodium in the body is outside cells, and about the same proportion of potassium is intracellular: energy is needed to maintain these differential concentrations.

Other ions tend to move across cell membranes with sodium and potassium. Hydrogen ion has already been mentioned. For example, magnesium and phosphate are predominantly intracellular, and chloride extracellular, ions. The distribution of all these, and of bicarbonate, will be affected by the same conditions as those for sodium and potassium.

Distribution of Electrolytes Between Plasma and Interstitial Fluid

The vascular endothelium, unlike the cell membrane, is permeable to small ions. Significant concentrations of protein are present in plasma but not in interstitial fluid. Electrolyte concentrations are therefore very slightly higher in the latter to balance the osmotic effect of the protein concentration inside vessels. The difference is small and clinically insignificant, and for practical purposes one can assume that plasma electrolytes are representative of those in the extracellular fluid as a whole.

DISTRIBUTION OF WATER IN THE BODY

A little over half the body water is inside cells. Of the extracellular water about 15–20 per cent is in the plasma. The remainder makes up the extravascular, extracellular interstitial fluid.

The distribution of water across cell walls depends on the *in vivo* osmotic difference between intra- and extracellular fluids: that across blood vessel walls is determined by the balance between the *in vivo* effective osmotic (or oncotic, p. 36) pressure of plasma and the net outward hydrostatic pressure. Correct interpretation of plasma electrolyte results depends on a clear understanding of these factors.

Osmotic Pressure

Net movement of water across a membrane permeable only to water depends on the concentration *difference* of dissolved particles between the two sides. For any given weight/volume concentration, the larger the particle (the higher the molecular weight) the fewer there are in unit volume, and the less osmotic effect they will exert. However, if the membrane is freely permeable to smaller particles as well as to water, these exert no osmotic effect, and the larger molecules become more important in affecting movement of water. To explain water distribution in the body it is important to appreciate these three factors:

      number of particles per unit volume;
      concentration gradient across the membrane;
      particle size relative to membrane permeability.

Plasma Osmotic Pressure: Distribution of Water across Cell Walls

For the definitions of osmolarity and osmolality see p. 36.

Measured plasma osmolality.—Plasma osmotic pressure is usually determined by measuring the depression of its freezing point below that of pure water. The result is a measure of its *total* osmotic pressure—the osmotic effect which would be exerted by the sum of all the dissolved molecules and ions across a membrane permeable only to water. Table II shows that by far the largest contribution to this (90 per cent or more) comes from *sodium and its associated anions* (mainly chloride), the effect of protein being negligible. The only major difference between extravascular fluid and plasma is in their protein content: thus *total plasma osmolality is almost identical with the osmolality of the fluid bathing cells.*

TABLE II

APPROXIMATE CONTRIBUTIONS OF PLASMA CONSTITUENTS TO PLASMA OSMOLALITY

	Osmolality (mmol/kg)	Per cent total
Sodium and anions	270	92
Potassium and anions	7	
Calcium (ionised) and anions	3+	
Magnesium and anions	1+	
Urea	5	8
Glucose	5	
Protein	Approximately 1	
Total	Approximately 292	

Hydrostatic pressure differences across cell membranes are negligible, and cell hydration depends on the osmotic difference between intra- and extracellular fluids. The cell membrane is freely permeable to water, but although different substances diffuse, or are actively transported, across it at different rates, solute always diffuses into cells more slowly than does water. Normally the total intracellular osmolality, due predominantly to potassium and associated anions, equals that of the ECF, mostly due to sodium and associated anions, and there is no *net* movement of water into or out of cells. In pathological circumstances rapid changes of extracellular solute concentration affect cell hydration: slower changes, by allowing time for redistribution of solute, have less effect.

Because *sodium* and its associated anions account for at least 90 per cent of plasma osmolality in the normal subject (Table II), rapid changes of sodium concentration affect cell hydration, a rise causing cellular dehydration and a fall cellular overhydration. Normal levels of *urea* and *glucose* contribute very little to measured plasma osmolality. However, concentrations 15-fold or more above normal can occur in severe uraemia and hyperglycaemia, and these solutes then contribute significantly. Urea diffuses into cells slowly (about 100 000 times more slowly than water), and during prolonged uraemia its osmotic *effect* is reduced by this fact: acute changes can alter cell hydration. Glucose is actively transported into cells, but once there it is rapidly metabolised: intracellular levels remain low, and severe hyperglycaemia has a marked influence on cell hydration. Although uraemia and hyperglycaemia can cause cellular dehydration, the contribution of normal concentrations of urea and glucose to plasma osmolality is so small that reduced levels of these solutes, unlike those of sodium, do not cause cellular overhydration.

Because the *speed* of change is even more important than the absolute level, over-zealous treatment of established hypernatraemia (with hypotonic fluids), uraemia (by haemodialysis), or hyperglycaemia (with large doses of insulin) may produce dangerous cerebral cellular overhydration.

Concentrations of other solutes of low osmolal concentration, such as those of calcium, potassium or magnesium, rarely vary by a factor of more than 3, even in pathological states; they do not cause significant osmolality changes.

Substances not transported into cells, such as mannitol, can be infused to reduce cerebral oedema. They can also be used as osmotic diuretics. (Hypertonic glucose or urea can also be used in this way.) They are included in the results of total plasma osmolality measurements.

A plasma alcohol level of 100 mg/dl contributes about 20 mmol/kg to osmolality.

Calculated plasma osmolarity.—Ideally we need to know the osmotic *effect* due to the concentration *difference* between intra- and extracellular fluid. Even if it were possible to measure plasma osmolality with 100 per cent accuracy we could still only roughly gauge cell osmolality from a knowledge of the length of history, and of the permeability of the cell membrane to the plasma solute contributing most to the change. In almost all circumstances plasma osmolarity, calculated from sodium, urea and glucose concentrations, and interpreted with intelligence, is at least as clinically valuable as measurement of true plasma osmolality: this method has the advantage of identifying the solute responsible. Of the many formulae proposed, we find the one given below satisfactory and simple.

If all measurements are in mmol/l, the *approximate* total osmolarity will be:

$$2[Na^+] + [urea] + [glucose]$$

The factor of 2 applied to the sodium concentration allows for associated anions, and assumes complete ionisation. The omission of potassium partly "corrects" for incomplete ionisation.

This calculation is *not* valid:

if an unmeasured osmotically active substance, such as mannitol or alcohol is circulating;

if there is gross hyperproteinaemia or lipaemia (see section on "Osmolarity and Osmolality").

Plasma Colloid Osmotic Pressure: Distribution of Water across Capillary Walls

The distribution of water across capillary walls is unaffected by electrolyte concentration, but affected by the osmotic effect of *plasma proteins*.

The maintenance of blood pressure depends on the retention of intravascular water at a hydrostatic pressure higher than that of interstitial fluid. Hydrostatic pressure therefore tends to force fluid into the extravascular space. In the absence of any opposing effective osmotic pressure across vascular walls water would be lost rapidly from the intravascular compartment.

Unlike the cell membrane, the capillary wall is permeable to small molecules; sodium therefore exerts almost no osmotic effect at this site. The smallest molecule present intravascularly at significant concentration, and which is in very low concentration extravascularly, is albumin (MW 65 000). The normal capillary wall is poorly permeable to it. Albumin concentration is therefore the most important factor opposing the net outward hydrostatic pressure: the higher molecular weight proteins although together present in much the same weight/volume concentration as albumin, contribute much less to this effect because of their larger size. This effective osmotic pressure across blood vessel walls is the *colloid osmotic* or *oncotic pressure*.

Because proteins contribute negligibly to measured plasma osmolality (Table II), *measurement of osmolality cannot be used to assess osmotic effects across capillary walls.*

Units of Measurement of Osmotic Pressure: Osmolarity and Osmolality

Concentrations of molecules can be expressed in two ways:

in molarity, the number of moles (or mmol) *per litre of solution*;

in molality, the number of moles (or mmol) *per kg of solvent*.

If the molecules are dissolved in pure water at concentrations such as are found in biological fluids these two figures will differ very little. Plasma, however, is a complex solution. It also contains large molecules such as proteins, and the total volume of solution (water + protein) is greater than that of solvent (water only). The small molecules are dissolved only in the water, and at a protein concentration of 70 g/l the volume of water is about 6 per cent less than that of total solution (that is, the molality will be about 6 per cent greater than the molarity). Most methods of measuring individual ions such as sodium, measure them in molarity (mmol/l).

Osmotic pressure can also be expressed in two ways:

in *osmolarity* expressed as mmol *per litre of solution*;

in *osmolality* expressed as mmol *per kg of solvent.*

(The term milliosmol (mosmol) has been used to express osmolarity and osmolality: it has been recommended that this terminology be abandoned.)

The usual method of measuring plasma osmotic pressure by freezing point depression depends on osmo*lal*ity, whereas calculated osmotic pressure from dissolved ions and molecules is expressed as osmo*lar*ity. The osmometer therefore measures osmotic concentration in *extracellular water*. It is this osmo*lal*, rather than osmo*lar*, concentration which exerts its effect across cell walls and which is concerned in homeostatic mechanisms.

Under normal circumstances, although measured osmolality should be higher than calculated osmolarity (because of the protein content of plasma), little difference is found between the two figures. This is because incomplete dissociation of, for example, NaCl to Na^+ and Cl^- reduces the osmotic effect by almost the same amount as the volume occupied by protein raises it. Calculated osmolarity is then an adequate approximation to true osmolality. *This may not be true in gross hyperlipaemia or hyperproteinaemia* when protein or lipid contribute much more than 6 per cent to measured plasma volume. The mo*lar* concentration of sodium measured by flame photometry may then give results significantly lower than the true concentration in plasma water and measurement of osmolality is advantageous: this may not apply if the mo*lal* concentration of sodium is measured using some ion-selective electrodes. As urine does not normally contain either protein or lipids, *direct comparison between urine and blood can only be made by measurement of osmolality of both.* Calculation of urinary osmolarity is not feasible because of the considerable variation in concentration of different, usually unmeasured, solutes, and osmolality should always be measured.

CONTROL OF SODIUM AND WATER METABOLISM

The following is a simplified account of a complex situation and the student should bear this in mind.

CONTROL OF SODIUM

If there is a mechanism which controls sodium intake in man it is of little importance when compared with the thirst mechanism which controls water intake. The most important factor controlling sodium loss is the mineralocorticoid hormone, aldosterone.

Aldosterone

Aldosterone is secreted by the zona glomerulosa of the adrenal cortex (p. 141). It affects sodium-potassium and sodium-hydrogen ion exchange across *all* cell membranes. We shall concentrate on its effect on renal tubular cells, but we should bear in mind that it also affects faecal sodium loss, and the distribution of electrolytes in the body.

In the distal part of the nephron aldosterone increases sodium reabsorption from the luminal fluid in exchange for potassium or hydrogen ion. The net result is retention of sodium in excess of water, and loss of potassium and hydrogen ions. In the presence of high levels of circulating aldosterone *urinary sodium concentrations are low*.

Many factors have been implicated in the feed-back control of aldosterone secretion. Such factors as local electrolyte concentration in the adrenal and the kidney are probably of little clinical or physiological importance compared with the effect on the adrenal of the renin-angiotensin system.

The Renin-Angiotensin System

Renin is a proteolytic enzyme secreted by a cellular complex situated near the renal glomeruli (and therefore called the juxtaglomerular apparatus). In the blood stream it acts on a renin substrate (an α_2-globulin) to form *angiotensin I*. This decapeptide is further split by a peptidase, located predominantly in the lungs, to *angiotensin II*. This peptide hormone has two main systemic actions:

it acts directly on blood vessel walls, causing vasoconstriction and so probably helps to maintain blood pressure;

it stimulates the cells of the zona glomerulosa to secrete aldosterone.

The most important stimulus to renin production seems to be reduced renal blood flow (possibly changes in the mean blood pressure in the renal afferent arterioles are the actual stimuli). Poor renal blood flow is often associated with an inadequate systemic blood pressure and the two effects of angiotensin II ensure that this is corrected:

vasoconstriction *may* raise the blood pressure before the circulating volume can be restored;

sodium retention occurs due to the action of aldosterone: as we shall see later, this will usually be followed by water retention with restoration of the circulating volume.

We may make an oversimplified statement based on the effect and control of aldosterone secretion. Aldosterone causes sodium retention. The stimulus to its secretion is renin, which is controlled, effectively, by circulating blood volume. Therefore, *blood volume controls net sodium retention.*

Natriuretic hormone.—There is some evidence that a third factor, probably hormonal, may be concerned in the control of sodium excretion in certain circumstances. It is said to be secreted in response to an expansion of plasma volume, and may inhibit sodium reabsorption in the proximal tubule. It has been suggested that it fails to be secreted in oedematous states.

In later discussion we will assume the renin-aldosterone mechanism to be of overriding importance in the control of sodium excretion.

CONTROL OF WATER

Both intake and loss of water are controlled by the osmotic gradient across membranes of cells in hypothalamic centres. These centres, which are closely related anatomically, control *thirst* and secretion of the peptide *antidiuretic hormone* (ADH:arginine vasopressin); both are stimulated by an osmotic gradient which causes a flow of water out of the cells (a relatively high extracellular osmolality). Both increased water intake due to thirst, and retention of more water than solute when ADH acts on the renal collecting ducts, tend to dilute extracellular osmolality towards normal. An acute increase of extracellular osmolality of only 2 per cent quadruples ADH output, and a similar fall cuts it off completely; this represents a change in sodium concentration of only about 3 mmol/l. More chronic changes may have little or no effect if the osmotic gradient has been minimised by solute redistribution.

Severe hypovolaemia may sometimes stimulate ADH secretion and thirst despite plasma hypo-osmolality. Such processes are inappropriate to the osmolar state: much of the retained water enters cells along the osmotic gradient, and is relatively ineffective in correcting hypovolaemia (p. 46).

Plasma osmolality normally depends largely on its sodium concentration. We have already pointed out that vascular volume controls sodium retention. We can now make the further *simplified* statement that *sodium concentration controls the amount of water in the body*. (It must be stressed that in *uraemia or hyperglycaemia*, or *after infusion of substances such as mannitol*, this is not true.)

Sometimes the effectiveness of ADH is opposed by other factors; for example, during an osmotic diuresis due to glucose, urea, amino acids or mannitol, the urine, although not hypo-osmolal, will contain more water than sodium (p. 10). Patients being fed intravenously, or who, because of tissue damage, are breaking down larger than usual quantities of protein and producing excessive amounts of urea from the released amino acids, may become water-depleted even in the presence of adequate amounts of ADH (p. 52): urinary osmolality will be high.

As in the case of sodium and aldosterone, such non-ADH effects are usually relatively unimportant.

INTERRELATIONSHIP BETWEEN SODIUM AND WATER HOMEOSTASIS

Let us now look more closely at the simplified statements made above concerning the control of ADH and aldosterone secretion. We said that sodium, by its osmotic effect, controls ADH secretion, and water, by its effect on renal blood flow, controls aldosterone secretion. But ADH controls water loss and aldosterone controls sodium loss. The homeostasis of sodium and water is interdependent and this simplified scheme is shown in Fig. 2. The thirst mechanism is not shown here but it should be remembered that an increase of osmolality not only reduces water loss but increases thirst and therefore intake.

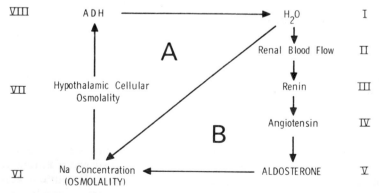

FIG. 2.—Simplified cycle of sodium and water homeostasis. (Note that in uraemia, hyperglycaemia and after infusion of hyperosmolal solutions such as mannitol or amino acids, factors other than sodium cause significant changes in osmolality.)

The diagonal line which divides the rectangle into two triangles, A and B, indicates that changes in hydration can also directly alter sodium concentration.

We shall refer to this scheme repeatedly while explaining changes occurring in pathological states. The reader should bear in mind that it is a simplified scheme and does not allow for the effect of fluid shifts across cell walls on extracellular concentrations and volume, for changes in osmotically active solutes other than sodium, nor for redistribution of solute in chronic disturbances of extracellular osmolality.

ASSESSMENT OF SODIUM AND WATER BALANCE

Although it is easy to measure the intake of water and sodium of patients receiving liquid feeds (either oral or intravenous), it is less easy to do so when a solid diet is being taken. Fortunately, accurate measurement is rarely necessary in such cases.

Renal loss is also easy to measure but that in formed faeces, sweat and expired air ("insensible loss") is more difficult to assess and may be important when homeostatic mechanisms have failed, in the presence of abnormal losses by extrarenal routes, in unconscious patients and in infants (see p. 52). The aim should be to ensure that such subjects are normally hydrated, and then to keep them "in balance". The medical attendant has the difficult task of imitating normal homeostasis without the aid of the sensitive and interlinked mechanisms of the body.

ASSESSMENT OF STATE OF HYDRATION

Assessment of the state of hydration of a patient depends on observation of his clinical state, and on laboratory evidence of haemoconcentration or haemodilution. It must be stressed that both these methods are crude, and that quite severe disturbances of water balance can occur before they are obvious clinically or by laboratory methods.

In extracellular fluid loss (other than that due to haemorrhage), water and often electrolytes are lost from the vascular compartment, and the concentrations of large molecules and of cells rise: there is therefore a rise in all plasma protein fractions, in haemoglobin level, and in the haematocrit reading (*haemoconcentration*). Conversely, in overhydration these concentrations fall (*haemodilution*). These findings, of course, may also be affected by pre-existing abnormalities of protein or red cell concentrations, and while changes are almost always more informative than single readings, it must be remembered that both protein and haemoglobin levels may change following, for example, blood transfusion, or because of the primary disease.

The problem of assessing hydration can be a difficult one, but can usually be resolved if the history and clinical and laboratory findings are all taken into account. Occasionally, measurement of urinary sodium concentration may help (p. 47).

ASSESSMENT OF FLUID BALANCE

By far the most important measurement in assessing changes in day-to-day fluid balance is that of intake and output. "Insensible loss" is usually assumed to be about 900 ml per day, but this must be balanced against "insensible" production of about 500 ml water daily by metabolism. The *net* "insensible loss" is therefore the difference between these two—about 400 ml a day. A normally hydrated patient, unable to control his own balance, should be given this basic volume of fluid daily with, in addition, the volume of losses measured (urine, vomitus, etc.) during the preceding 24 hours. If he is thought to be abnormally hydrated, fluid intake should be adjusted until hydration is restored. Inevitably the intake for any day must be calculated from the output in the preceding 24 hours. This is adequate when the patient starts in a normal state of hydration.

It should be remembered that a pyrexial patient may lose a litre or more of fluid in sweat, and that if he is also overbreathing, or on a respirator, respiratory water loss can be considerable. In such cases the allowance of 400 ml daily for insensible loss may not be enough.

Many very ill patients are incontinent of urine, and measurement of even this volume may be impossible. Changes in body fluid may be assessed by daily weighing (1 litre of water weighs 1 kg). Over short periods of time changes in solid body weight will be small, and alteration in weight can be assumed to be due to changes in fluid balance. Unfortunately, very ill patients may be unable to sit in weighing chairs. Some hospitals have weighing beds, in which the patient is weighed with the bed and bedclothes. In this situation, care must be taken to keep the weight of the bedclothes constant.

Assessment of fluid balance in severely ill patients may present grave problems. Every attempt must be made to make *accurate* measurements of fluid intake and loss. In most circumstances carefully kept fluid charts are of more importance than frequent plasma electrolyte estimations. *Inaccurate charting* is useless, and *may be dangerous*.

ASSESSMENT OF SODIUM STATUS

Sodium is important because of the osmotic effect of its *concentration*. Plasma sodium levels should be monitored while volume is being corrected to ensure that the distribution of the fluid between cells and the extracellular compartment is correct. The presence of other osmotically active solute should be taken into account. It is not necessary to measure the total amount of body sodium.

CLINICAL FEATURES OF WATER AND SODIUM DISTURBANCES

We are now in a position to explain the immediate clinical consequences of water and sodium disturbances. These depend on changes in extracellular osmolality and hence in cellular hydration (sodium) and changes in circulating volume (water).

CLINICAL FEATURES OF DISTURBANCES OF SODIUM CONCENTRATION

Measuring sodium concentration is a poor man's substitute for measuring osmolality. Sodium levels *per se* are not important to the body, but the osmotic gradient is. It is important to understand that the one does not always reflect the other.

If sodium concentration changes without change of other extracellular solute most of the immediate clinical features are due to an osmotic difference across cell walls. Gradual changes, by allowing time for redistribution of diffusible solute such as urea, and therefore for equalisation of osmolality without major shifts of water, may produce little clinical effect.

Hyponatraemia *may* reflect extracellular *hypo-osmolality*, and may therefore cause cellular overhydration. However, it may be appropriate to the osmolal state. For example, acute uraemia, hyperglycaemia, infusion of amino acids or mannitol or high plasma alcohol levels increase extracellular osmolality: the consequent homeostatic dilution of *total* extracellular solute towards a normal osmolality results in hyponatraemia. This therefore reflects partial or complete compensation of hyperosmolality, not hypo-osmolality. In gross hyperlipaemia (especially that due to intravenous lipid feeding) or hyperproteinaemia, "hyponatraemia" found in whole plasma may coexist with a normal sodium concentration in plasma water, and therefore with normal osmolality at cell walls (p. 37). *Infusion of sodium salts into patients with appropriate hyponatraemia is dangerous*. It is also important to exclude artefactual "hyponatraemia" caused by taking blood from the limb into which fluid of low sodium concentration is being infused (p. 500).

In the cases in which hyponatraemia reflects true hypo-osmolality overhydration of cerebral cells may cause *headache, confusion* and, later, *fits*.

Hypernatraemia *always* reflects extracellular *hyperosmolality* with the danger of cellular dehydration. Because solute other than sodium is present in very low concentration in normal plasma (Table II, p. 34) severe hypoglycaemia, for example, could only reduce plasma osmolality by about 5 in 280 mmol/kg. (Hyperglycaemia, by contrast, can increase it by as much as 50 mmol/l.) Significant hypernatraemia cannot, therefore, be appropriate to the osmolal state.

The clinical effects of dehydration of cerebral cells are *thirst, mental confusion* and, later, *coma*.

Hyponatraemia is a much more common finding than hypernatraemia and is often appropriate. The number of cases in which it indicates true hypo-osmolality is probably similar to the incidence of hypernatraemia, which always reflects hyperosmolality.

DISTURBANCES OF SODIUM AND WATER METABOLISM

Disturbances of sodium and water balance are most commonly due to excessive losses from the body, usually of gastro-intestinal fluid. These losses may be inappropriately increased if the volume or composition of infused fluid is incorrect: repletion will then be relatively ineffective, and the increase of gastro-intestinal excretion may aggravate the clinical abnormalities.

More rarely the primary defect is due to excessive or deficient secretion of aldosterone or ADH.

WATER AND SODIUM DEFICIENCY

Water and sodium are always lost from the body together, but an imbalance between the loss of each is relatively common. This imbalance may be due to the composition of the fluid lost or to the composition of the fluid given to replace it.

The initial effects depend on the composition of the fluid lost *compared with that of plasma.*

If the *sodium concentration is similar, volume depletion* is more likely than changes in plasma sodium concentration (isosmolar depletion).

If the *sodium concentration is much lower,* relatively more water than sodium is lost from plasma and *hypernatraemia* is likely, even when there is little volume depletion.

No body secretion has a significantly higher sodium concentration than plasma. *Hyponatraemia due to predominant sodium depletion is usually the result of inappropriate treatment.* Hyponatraemia is usually due to factors other than sodium depletion.

Subsequent effects depend on the efficiency of homeostatic mechanisms, and on the availability and composition of fluid replacement.

ISOSMOLAR VOLUME DEPLETION

CAUSES OF ISOSMOLAR FLUID LOSS

All small intestinal secretions between the duodenum and ileocaecal junction, the biliary and pancreatic secretions, and the urine passed when tubular function is minimal have sodium concentrations of between 120 and 140 mmol/l. The clinical conditions causing approximately isosmolar fluid loss are therefore:

small intestinal fistulae (including new ileostomies)

small intestinal obstruction ⎫ Fluid accumulating in the
paralytic ileus ⎬ gut has been lost from the ECF,
⎭ like urine in the bladder

severe renal tubular damage with minimal glomerular dysfunction (for example, the recovery phase of acute oliguric renal failure or polyuric chronic renal failure).

Intestinal losses are more likely to produce severe volume depletion than is renal tubular disease.

Results of Isosmolar Fluid Loss

Hypovolaemia, by reducing renal blood flow, causes renal circulatory insufficiency with *oliguria, uraemia* and the other changes described on p. 12. Sodium and water are lost in almost equivalent amounts, and plasma sodium levels are usually normal: the patient may not complain of thirst despite severe volume depletion.

Haemoconcentration, with a raised plasma protein and haemoglobin concentration and haematocrit, confirms that there has been considerable fluid loss. However, its absence does not exclude such loss: there may be pre-existing anaemia or hypoproteinaemia, and in the shocked patient increased capillary permeability may cause hypoalbuminaemia by allowing albumin to diffuse more freely than usual into the interstitial fluid.

Severe hypovolaemia causes *hypotension*.

Isosmolar hypovolaemia is relatively common, and the above findings are those usually recognised as "dehydration". It is important to realise that a much less obvious volume depletion due to loss of water in excess of sodium is equally dangerous (p. 49).

Changes Produced by Homeostatic Mechanisms

Homeostatic responses can only occur if renal tubular function is adequate, and are not found if the isosmolar depletion is due to tubular damage.

The reader should refer to Fig. 2.

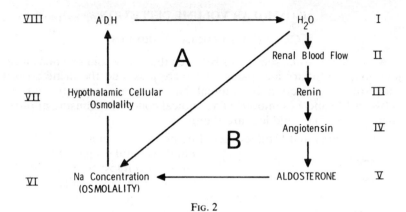

FIG. 2

Triangle B.—The low renal blood flow, by stimulating renin and therefore aldosterone secretion (I–V), causes selective sodium reabsorption (V–VI) and a *low urinary sodium concentration*.

Triangle A.—The tendency of the retained sodium to increase plasma osmolality stimulates ADH secretion (VI–VIII) and water is reabsorbed in amounts equivalent to sodium: this tends to correct volume and keep the plasma sodium concentration normal. Occasionally severe volume depletion stimulates enough ADH secretion, and therefore water retention, to cause mild hyponatraemia: much of this water will move from the depleted and now hyposmolar extracellular compartment into the relatively well-hydrated cells.

Even complete sodium and water retention could not correct extrarenal losses which exceed those in a normal urine output. Sodium and water must be replaced in adequate amounts to provide "substrate" for the kidneys. However, it is important to understand that excessive replacement, especially with fluid of an inappropriate concentration, has undesirable effects. Replacement of isosmolar loss with fluid of low sodium concentration is discussed below as the commonest cause of predominant sodium depletion. Here we will describe the consequences of volume replacement with protein-free fluid.

Effects of Intravenous Volume Replacement

Subjects unable to absorb adequate amounts of oral fluid because of gastro-intestinal loss usually require intravenous replacement. The following discussion applies to such cases, and assumes normal renal tubular function.

Correction of the presenting hypovolaemia can be monitored by clinical observation, measurement of urine output and estimation of serial plasma urea levels. Once the initial loss has been replaced, the plasma urea will fall to normal. At this stage there is the danger that overcorrection will increase intestinal loss of fluid, which may accumulate in the already distended bowel of intestinal obstruction or paralytic ileus.

To explain this fluid loss due to overinfusion we should remember that passive filtration depends on the capillary hydrostatic pressure, and is weakly opposed by the capillary colloid osmotic pressure. Infusion of protein-free fluid increases the hydrostatic gradient, and reduces the opposing osmotic gradient by diluting plasma proteins. The necessary increase in renal filtration during correction of hypovolaemia is inevitably accompanied by a less desirable intestinal loss, and this will be aggravated by overcorrection: "wastage" in the urine occurs for the same reasons but, although uncomfortable for the patient, is of less clinical importance.

Clinical assessment is inefficient in detecting when volume repletion is complete, but not excessive. Measurement of urinary sodium concentration may help. If, during the period of known hypovolaemia, it was less than about 30 mmol/l, tubular function is adequate and further measurements can be used to assess the presence of circulating aldosterone. A very low urinary sodium concentration suggests that renal blood flow is still low enough to stimulate maximal renin secretion, and infusion should be increased: a urinary sodium concentration much higher than 30 mmol/l in a patient with adequate tubular function suggests overcorrection and the need to slow the infusion. Urinary sodium may sometimes need to be monitored in this way until intestinal obstruction is relieved. In cases where all losses can be measured further maintenance of normal balance can be based on *accurate* fluid balance charts.

PREDOMINANT SODIUM DEPLETION

Effects of the Composition of Infused Fluid

As has been mentioned, no body secretion has a sodium concentration significantly higher than that of plasma. Predominant sodium depletion is almost always due to intravenous infusion. The composition of the fluid is even more important than the volume.

It is common practice to infuse patients with excessive isosmolar loss, or postoperatively, with fluid such as "dextrose saline" which contains about 30 mmol/l of sodium. Although the glucose in the infused fluid renders it isosmolar despite the low sodium concentration, the glucose is metabolised after infusion, and both plasma sodium and osmolality are diluted by the remaining hypo-osmolar fluid. Homeostatic mechanisms are brought into play which tend to correct this hypo-osmolality.

The reader should again refer to Fig. 2.

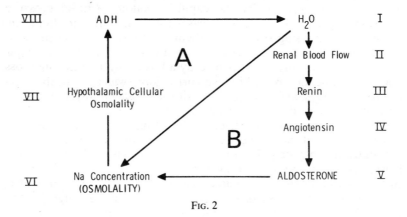

Fig. 2

Triangle A. These mechanisms tend to correct hypo-osmolality (and therefore to safeguard normal cell hydration).

ADH secretion is cut off in response to hypo-osmolality (VI–VIII).
Water is lost in the urine (VIII–I).

This would tend to restore osmolality at the expense of volume, as shown in the diagonal line I–VI. However, if the volume is maintained by replacing this urinary loss with effectively hypotonic fluid, hypo-osmolality (as shown by hyponatraemia) persists.

Sodium depletion may even be aggravated by this regime. If the urinary loss of fluid were uncorrected loss of fluid would stimulate aldosterone secretion (I–V): sodium would be retained (V–VI), and the sequence of events depicted in Triangle A would be reversed, so that the balance of both sodium and water would be corrected. However, expanding plasma volume by infusion cuts off aldosterone secretion, and sodium is lost in the urine despite sodium depletion and hypo-osmolality. The net effect is restoration of circulating volume by infusion, but cellular overhydration due to hypo-osmolality.

The clinical signs are those of hypo-osmolality (p. 44).

The laboratory findings are:

hyponatraemia;

passage of a large volume of dilute urine (due to inhibition of ADH secretion);

and if fluid intake is excessive:

haemodilution;

a low plasma urea concentration due to the high GFR (excessive intravenous infusion is one of the commonest causes of a low plasma urea);

a high urinary sodium concentration (due to inhibition of aldosterone secretion).

It is not, therefore, surprising that postoperative hyponatraemia is so often found. No serious harm is done in subjects with normal renal function, because the mechanisms described above rapidly correct both osmolality and sodium and water balance when the infusion is stopped. However, by alternating "dextrose saline" and isotonic saline, it is our experience that most patients can maintain relatively normal plasma sodium and urea levels, whilst undergoing a reassuring diuresis: the danger of overloading the circulation is minimal if renal function is normal.

In seriously ill patients, with impairment of homeostatic mechanisms, infusion of "dextrose saline" is even more likely to cause hyponatraemia.

PREDOMINANT WATER DEPLETION

This syndrome is due to loss of water in excess of loss of sodium. It is usually the result of the loss of fluid with a lower concentration of sodium than that of plasma, or of deficient water intake. In *sweat*, *gastric juice* and *diarrhoea stools* the concentrations of sodium are about half those in plasma; in *diabetes insipidus*, during an *osmotic diuresis*, or in the rare inborn error associated with failure of the renal tubules to respond to ADH, *urine* of low sodium concentration is passed. *As hyperosmolality due to predominant water loss causes thirst, effects are only seen if water is not available or cannot be taken in adequate quantities.*

The clinical situations associated with predominant water loss are:

water deficiency in the presence of normal homeostatic mechanisms.

    Excessive fluid loss:

        loss of excessive amounts of sweat;

        loss of gastric juice;

        loss of fluid stools of low sodium content
            (usually in infantile gastroenteritis);

        excessive respiratory loss, especially during artificial
            respiration;

        loss of fluid from extensive burns.

    Deficiency of fluid intake:

        inadequate water supply, or mechanical
            obstruction to its intake;

failure of homeostatic mechanisms for water retention.

    inadequate response to thirst mechanism (for example in comatose patients and infants). It is doubtful if significant water depletion can develop if thirst mechanisms are normal and water is available;

    deficiency of ADH (diabetes insipidus);

    overriding of ADH action by osmotic diuresis;

    failure of renal tubular cells to respond to high levels of ADH (nephrogenic diabetes insipidus).

In most clinical states associated with predominant water depletion more than one of these factors is responsible.

Predominant water depletion with normal homeostatic mechanisms.—The reader should refer to Fig. 2 during the following explanation.

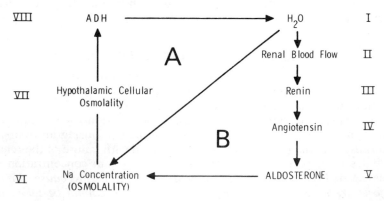

FIG. 2

Triangle B—Immediate effects:

loss of water in excess of sodium increases the plasma sodium
concentration and osmolality (diagonal line I–VI);

reduction of circulating volume reduces renal blood flow and
stimulates aldosterone production (I–V). Sodium is retained
and the rise in plasma sodium is aggravated (V–VI).

Triangle A—Compensatory effects:

Increased plasma osmolality stimulates:

thirst, increasing water intake if water is available and if the
patient can respond to it (not shown on diagram);

ADH secretion (VI–VIII). Urinary volume falls (VIII–I) and
water loss through the kidney is minimised.

If adequate amounts of water are available depletion is rapidly
corrected. In the absence of intake, or in the presence of continued
loss, the mechanisms in triangle A cannot function effectively and
homeostasis breaks down. Hypernatraemia is therefore an early
finding in water depletion: it is important to realise that it may be
present before the more common clinical signs of volume depletion
are obvious.

The clinical signs are those of:

hyperosmolality (p. 44);

oliguria due to ADH secretion;

later, signs of volume depletion become apparent (p. 45).

The laboratory findings are:

hypernatraemia;

haemoconcentration (due to fluid depletion);

mild uraemia (due to volume depletion and hence low GFR);

a urine of low volume, and high osmolality and urea concentration
(due to the action of ADH);

low urinary sodium concentration (in response to high aldosterone levels).

Failure of homeostatic mechanisms for water retention.—These
syndromes are relatively rare.

Diabetes insipidus may be due to pituitary or hypothalamic damage
caused by head injury or during hypophysectomy, or to invasion of
the region by tumour. It may be idiopathic in origin.

Hereditary nephrogenic diabetes insipidus is a rare inborn error of
renal tubular function in which ADH levels are high but the tubules
cannot respond to it: the newborn infant passes large volumes of urine
and rapidly becomes dehydrated. Urinary loss is difficult to assess at
this age, and the cries of thirst may be misinterpreted.

The reader should refer again to Fig. 2 and to the immediately
preceding section.

The first stage is identical with that described in the previous section. Because the mechanisms shown in Triangle A are not functioning, compensation cannot occur, and the point is soon reached at which intake cannot adequately replace loss.

The clinical signs and findings are those described in the previous section, with the exception that there is *polyuria*, not oliguria, and that the urine is of *low osmolality and urea concentration*. These findings are the result of ADH deficiency. If any doubt remains as to the diagnosis, water may be deliberately withheld (see p. 87). In the absence of ADH, or when tubular cells fail to respond to it, concentration of the urine fails to occur and plasma osmolality rises. In the case of true ADH deficiency a concentrated urine will be passed if exogenous ADH (as DDAVP or Pitressin) is given. In nephrogenic diabetes insipidus maximal amounts of ADH are already circulating, and administration of it will not influence water reabsorption.

The unconscious patient.—The syndrome of predominant water depletion with hypernatraemia is seen most commonly in the unconscious or confused patient or in the infant with gastro-enteritis or pneumonia. In such subjects there is usually more than one cause of water depletion.

1. They are often pyrexial. Loss of hypotonic *sweat* is increased.

2. They may be *overbreathing* because of pneumonia, acidosis or brain stem damage. Water loss in expired air is increased.

3. Humidifiers on *respirators* may be inadequate and artificially respired patients, who tend to be hyperventilated, may become water depleted.

4. They may be given hypertonic intravenous infusions, for example to provide nutrient (dextrose or amino acids). There may be tissue damage and hence breakdown of protein to urea. The coma may be due to diabetes mellitus with glycosuria. All these factors will not only contribute further to plasma hyperosmolality but will also cause an *osmotic diuresis* which, by overriding the effect of ADH, causes further water loss.

5. There may be true diabetes insipidus due to head injury.

6. *The subject cannot respond to hyperosmolality by drinking.*

In the presence of factors 4 and 5 a high urine volume contributes to water depletion (and is *not* an indication of "good" hydration) and the extent of loss may not be realised if the subject is incontinent. Homeostatic mechanisms may not be capable of responding to hyperosmolality (5 and 6) and hypernatraemia is an early finding. It may occur before the more usual clinical signs of volume depletion are evident and is dangerous because of the resulting cellular dehydration. Monitoring the cumulative fluid balance (p. 82) of patients at risk of developing hypernatraemia helps detection of water depletion early enough to allow preventive measures to be taken.

Such a syndrome is also seen in unconscious or confused patients with extensive *burns*. The water in the exuded ECF evaporates, while some of the electrolyte is reabsorbed into the circulation. Tissue breakdown and intravenous feeding contribute to an osmotic diuresis.

Hyperosmolar saline should never be used as an emetic in cases of poisoning. Movement of water into the gut along the osmotic gradient, and absorption of some of the sodium, can cause marked hypernatraemia: the vomiting or unconscious patient cannot respond to the hyperosmolality by drinking. Death is a common consequence.

Failure of Homeostatic Mechanisms for Sodium

The homeostatic mechanisms associated with diabetes insipidus produce osmolality changes early (p. 52). By contrast, the sodium loss due to aldosterone deficiency initiates homeostatic reactions which tend to maintain osmolality at the expense of volume. The conditions described in this section are rare.

Addison's disease, the least rare of these conditions, causes hypoaldosteronism, and is discussed more fully on p. 148. Primary renin deficiency has been reported. "Pseudo" Addison's disease, due to failure of otherwise normal tubular cells to respond to aldosterone, is exceedingly rare.

Loss of sodium in excess of water reduces plasma osmolality and cuts off ADH secretion (Triangle A, Fig. 2). Water is lost until osmolality is corrected: the patient becomes hypovolaemic.

The homeostatic mechanisms of Triangle B are not functioning. Osmolality is therefore maintained until late by water loss; in advanced

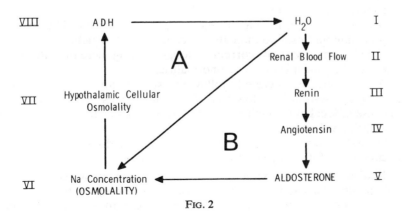

FIG. 2

cases the stimulatory effect of hypovolaemia on ADH secretion may override this cycle, and hypo-osmolality may occur.

The clinical signs are:

  those of fluid depletion (p. 45);

  later those of hypo-osmolality (p. 44).

The laboratory findings are:

  haemoconcentration (due to fluid depletion);

  renal circulatory insufficiency with mild uraemia (due to volume depletion);

  later hyponatraemia;

  inappropriately high urinary sodium concentration in the face of volume depletion (due to aldosterone deficiency).

WATER AND SODIUM EXCESS

In the presence of normal homeostatic mechanisms an excess of water and sodium is of relatively little importance, because it is rapidly corrected. These syndromes are commonly associated with failure of homeostatic mechanisms.

Fluid excess with no change in osmotic difference across cell walls, causes *hypertension* and overloading of the heart, with *cardiac failure.* *Oedema* occurs when albumin levels are low, because the reduced oncotic pressure together with increased intravascular hydrostatic pressure causes passage of water into interstitial fluid. Laboratory findings are characteristic of *haemodilution* and, unless there is renal glomerular failure, the plasma urea level tends to be low.

Secondary aldosteronism: oedema.—Any of the conditions already described in the sections on water and sodium depletion, in which aldosterone secretion is stimulated following reduction in renal blood flow could, strictly, be called secondary aldosteronism. The term is more commonly used to indicate the conditions in which, because the initial abnormality is not corrected, long-standing increased aldosterone secretion itself produces abnormalities.

Aldosterone is secreted following stimulation of the renin-angiotensin system by a low renal blood flow. This may occur either because of local abnormalities in renal vessels or because of a reduced circulating volume.

Secondary aldosteronism, in the usually accepted sense of the word, occurs in the following conditions:

Redistribution of extracellular fluid, leading to a reduction of plasma volume in the presence of normal or high total extracellular fluid volume. These conditions are due to a reduced plasma oncotic pressure, and are therefore associated with low plasma albumin levels. *Oedema is present.* Such conditions are:

    liver disease (impaired aldosterone catabolism aggravates the hyperaldosteronism);

    nephrotic syndrome;

    protein malnutrition.

Damage to the renal vessels, reducing renal blood flow. These conditions are usually *not associated with oedema:*

    essential hypertension;

    malignant hypertension;

    renal hypertension (for example, renal artery stenosis).

Cardiac failure. In this case two factors may cause low renal blood flow. Firstly, the cardiac output may be low, with poor renal perfusion pressure. Secondly, high intravascular hydrostatic pressure on the venous side of the circulation may cause redistribution of fluid and *oedema.* Aldosterone catabolism is also impaired.

The mechanisms in Fig. 2 are brought into play:

    reduced renal blood flow stimulates aldosterone secretion (I–V);

    sodium retention stimulates ADH secretion (V–VIII) and therefore *water retention* (VIII—I).

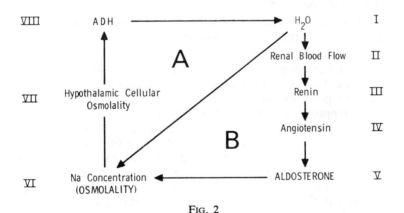

FIG. 2

Intravascular volume tends to be restored, but in conditions with hypo-albuminaemia or in cardiac failure more fluid passes into the interstitial fluid and the cycle restarts. A vicious circle is set up in which circulating volume can only be maintained by water retention and oedema results. In non-oedematous states hypertension occurs.

This cycle should only stimulate water retention in parallel with sodium retention. However, many patients with oedematous secondary aldosteronism have hyponatraemia. It is possible that loss of fluid from the vascular compartment into the intersitial fluid reduces the intravascular volume enough to stimulate ADH secretion which is inappropriate to the osmolality. Whether or not this is the true explanation, it must be remembered that in the presence of oedema the total amount of sodium in the body is high, even if there is hyponatraemia (see Table III).

Hypokalaemia is less common in secondary than in primary aldosteronism. The reason for this discrepancy is not clear, but it may be due to a reduction in the amount of sodium reaching the distal tubule (p. 11). Hypokalaemia is more readily precipitated by therapy with "loop" diuretics in secondary hyperaldosteronism than in normal subjects.

The clinical features of these cases are those of the primary condition.

The findings are:

a normal or low plasma sodium concentration;

a low urinary sodium excretion;

findings due to the primary abnormality, for example hypoalbuminaemia, uraemia, etc.

Predominant Excess of Water

Water overloading may occur in circumstances in which normal homeostasis has failed, or is over-ridden:

in *renal glomerular failure*, when fluid of low sodium concentration has been replaced in excess of that lost. Fluid balance should be carefully controlled in such patients;

in the presence of "*inappropriate*" *ADH secretion* (the term "inappropriate" is used in this book to describe continued secretion of a hormone under conditions in which it should normally be cut off). ADH, or a peptide with ADH-like activity, can be manufactured by malignant tissue of non-endocrine origin (see Chapter XXII for a further discussion). "Inappropriate" secretion (possibly from the pituitary or hypothalamus itself) occurs in a variety of other conditions, including infections. Such ADH production is not under normal osmotic feedback control and therefore continues in the presence of low extracellular osmolality. This fact is evidenced by the production of a relatively concentrated urine in the presence of dilute plasma;

during intravenous administration of the posterior pituitary hormone *oxytocin* (Syntocinon, Pitocin) to induce labour. Oxytocin has an antidiuretic effect similar to that due to ADH. If glucose (5 per cent) is used as a carrier, the glucose is metabolised and the net effect is water retention. If infusion is prolonged dangerous hyponatraemia and hypo-osmolality may develop: a careful watch should be kept on the patient's fluid balance and plasma sodium concentration, and some of the oxytocin should be given in isotonic saline.

If we refer again to Fig. 2 we can see how excessive intake is normally corrected.

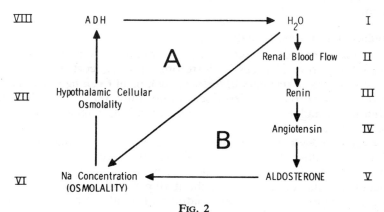

FIG. 2

Triangle B. Immediate effects:

> *excess water* tends to *lower plasma sodium* (diagonal arrow), (I–VI);
> *increased renal blood flow* cuts off aldosterone production, increasing urinary sodium loss and therefore further decreasing plasma sodium (V–VI).

Triangle A. Compensatory effects:

> *ADH is cut off* (VI–VIII) and large volumes of *dilute urine* are passed (VIII–I).

When renal glomerular function is poor, in the presence of "inappropriate" ADH secretion (which is not subject to osmotic feedback control), or if the fluid contains oxytocin, the mechanisms in triangle A are impaired. The second stage cannot take place.

The clinical consequences are:

> those of water excess;
> if overhydration is rapid, those of hypo-osmolality.

In patients with "inappropriate" ADH secretion the hypo-osmolality is usually of gradual onset, allowing time for redistribution of solute across cell walls, and clinical symptoms may be absent despite a very low plasma sodium concentration. Administration of oxytocin in 5 per cent glucose, by contrast, reduces osmolality within a few hours, with the serious danger of cerebral overhydration (p. 35).

The findings are:
  haemodilution;
  hyponatraemia.

If the cause is glomerular failure there will be uraemia. If it is due to "inappropriate" ADH secretion the plasma urea level will tend to be low.

Predominant Excess of Sodium

Predominant sodium excess is most commonly due to an "inappropriate" secretion of excess aldosterone or other corticosteroids in *Conn's syndrome (primary aldosteronism)*, or in *Cushing's syndrome.* Cushing's syndrome is described more fully on p. 145. Sodium retention stimulates water retention, minimising changes in plasma sodium concentration.

Primary aldosteronism (Conn's syndrome).—About half the cases of primary aldosteronism (excess aldosterone secretion not subject to normal feedback control) are due to a benign aldosterone-secreting adenoma of the adrenal cortex; about 10 per cent are multiple. Most of the remaining cases are associated with bilateral nodular hyperplasia of the adrenal glands.

The reader should refer again to Fig. 2:

excess aldosterone causes *urinary sodium retention* (V–VI);

the *increased sodium concentration* stimulates ADH secretion. (VI–VIII) and water retention (VIII–I);

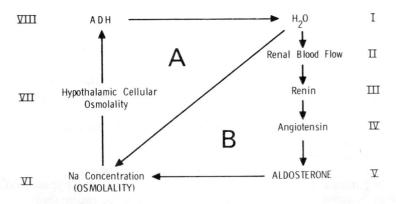

FIG. 2

water retention tends to return plasma sodium concentrations to normal (diagonal arrow, I–VI);

triangle B is out of action and aldosterone secretion cannot be cut off. This tends to maintain *plasma sodium levels at or near the upper end of the normal range*;

continued action of aldosterone causes sodium retention at the expense of potassium loss, and *potassium depletion occurs*.

The typical clinical features are:

those of volume excess. Patients are hypertensive but rarely oedematous;

those of hypokalaemia (p. 77).

Findings are:

hypokalaemia (due to excess aldosterone);

a high plasma bicarbonate (for explanation see p. 71);

a plasma sodium in the high normal range or just above it;

a low urinary sodium concentration in the early stages. However, sodium excretion may rise later, possibly because of hypokalaemic proximal tubular damage (p. 20).

Hypokalaemic alkalosis occurs in potassium depletion from almost any cause. However, the association of these findings in a patient without an obvious cause for potassium loss (such as administration of purgatives or diuretics) and with hypertension should suggest the diagnosis of primary aldosteronism. Further suggestive evidence would be the finding of high aldosterone with low renin levels (in all cases of *secondary* aldosteronism *both* are high). These estimations are not available in all routine laboratories, but in the rare cases in which the diagnosis of primary aldosteronism remains a possibility after careful clinical assessment they can be performed in special centres.

Urinary Sodium Estimation

Estimation of the daily loss of sodium in urine and other fluids with a view to quantitative replacement is not only unnecessary, but may be dangerous. Urinary sodium excretion is not related to body sodium content but to renal blood flow. Replacement of the small amount of sodium lost in the hypovolaemic patient will be inadequate, and of the large amount in the volume-expanded patient excessive. Cell hydration must be protected by giving fluid containing the correct proportion of sodium to water to maintain *plasma concentrations* normal; the total *volume* of this fluid must be determined by the state of hydration.

TABLE III

HYPOTHETICAL NUMERICAL EXAMPLES TO ILLUSTRATE CONDITIONS ASSOCIATED WITH NORMAL AND ABNORMAL PLASMA SODIUM CONCENTRATIONS

		Example	ECF (litres)	ECNa (mmol)	Plasma [Na] (mmol/l)	Haemoconc. or dilution	Clinical Features	Principle of Treatment
Normonatraemia (Equivalent Na and H_2O changes)	Normal	—	20	2800	140	None	—	—
	$Na + H_2O$ depletion	Early Addison's disease	15↓	2100↓	140	Conc. (Urea ↑)	Volume depletion (p. 45)	Treat cause Isosmolal saline
	$Na + H_2O$ excess	Non-oedematous secondary aldosteronism	30↑	4200↑	140	Dil.	Volume excess (p. 54)	Treat cause Restrict $Na + H_2O$
Hyponatraemia (Relative H_2O excess)	H_2O excess	Inappropriate ADH secretion	25↑	2800	112↓	Dil. (Urea N or ↓)	Volume excess (p. 54) Hypo-osmolality (p. 44)	Restrict H_2O

Na depletion	Diarrhoea with H$_2$O replacement	*20*	2240↓	112↓	None	Hypo-osmolality (p. 44)	Give isomolal or hyper-osmolal saline
Na+H$_2$O excess	Oedematous secondary aldosteronism (commonest form)	30↑	3360↑	112↓	Dil.	Volume excess (p. 54) Hypo-osmolality (p. 44)	Restrict Na and H$_2$O. Diuretics.
Na+H$_2$O depletion	Late Addison's disease	15↓	1680↓	112↓	Conc. (Urea ↑)	Volume depletion (p. 45) Hypo-osmolality (p. 44)	Give steroids + isomolal or hyperosmolal saline
Hypernatraemia (Relative Na excess) — H$_2$O depletion	Unconscious patient	18↓	*2800*	155↑	Mild conc.	Hyper-osmolality (p. 44)	Give hypo-osmolal fluid *slowly*
Sodium excess	Excessive intake usually in infants (*Very rare*)	*20*	3100↑	155↑	None	Hyper-osmolality (p. 44)	Remove Na by dialysis

Values in first line taken as "normals" to which other are related. "Normal" values in italics throughout.
No account has been taken of shifts of fluid across cell walls which would slightly reduce changes in plasma sodium levels.
Note especially relatively slight reduction in volume associated with hypernatraemia.

Estimation of urinary sodium *concentration* on a random specimen may be of value to monitor volume requirements (p. 47). More rarely it may help to differentiate renal circulatory insufficiency from renal damage (p. 17).

THE CLINICAL SIGNIFICANCE OF PLASMA SODIUM CONCENTRATIONS

Table III summarises the situations in which hypernatraemia and hyponatraemia may be found, and indicates some aspects of the clinical picture which may help to differentiate the causes. The two important points to remember are:

hypernatraemia almost invariably indicates water depletion;

hyponatraemia is more commonly due to water excess than to sodium depletion: if this is so, the body content of sodium may be normal, but is more often high. If it is due to the primary sodium depletion of untreated Addison's disease the patient will be volume depleted, both on clinical and laboratory findings, with mild to moderate uraemia.

Plasma sodium levels down to about 120 mmol/l occur in many ill patients and, unless there is evidence of volume depletion, require no treatment. In states in which plasma osmolality is contributed to significantly by other substances, as in severe uraemia and hyperglycaemia, mild hyponatraemia may be "appropriate".

(*The student is urged to read the indications for electrolyte estimation on p. 84.*)

BIOCHEMICAL BASIS OF TREATMENT OF SODIUM AND WATER DISTURBANCES

It cannot be stressed too strongly that treatment should not be based on plasma sodium concentrations alone. Hyponatraemia *per se* only rarely requires treatment. Assessment of the clinical history, findings indicating haemoconcentration or haemodilution, and plasma urea level help to unravel the cause of the low sodium concentration by indicating the state of hydration and renal function; this knowledge will determine whether correction of the sodium level is indicated. Rarely it may be useful to know the urinary sodium concentration, or osmolality. Hypernatraemia, on the other hand, should always be treated by *slow* infusion of hypotonic fluid.

Table III uses hypothetical values to illustrate the various combinations of disturbances of sodium and water metabolism and their treatment. Solutions available for intravenous use are given on p. 65.

These rules can only be a guide to treatment in often complicated clinical situations. They may, however, help to avoid some of the more dangerous errors of electrolyte therapy.

When homeostatic mechanisms have failed, especially in renal failure, normal hydration should be maintained according to the principles outlined on p. 42.

SUMMARY

1. Homeostatic mechanisms for sodium and water are interlinked. Potassium and hydrogen ions often take part in exchange mechanisms with sodium.

2. Distribution of fluid between cells and extracellular fluid depends on the osmotic difference between the intra- and extracellular fluid. Changes in this are usually due to changes in sodium concentrations.

3. Distribution of fluid between the intravascular and interstitial compartments depends on the balance between the hydrostatic pressure and the effective plasma osmotic pressure: the latter depends largely on albumin concentration.

4. Aldosterone secretion is the most important factor affecting body sodium content.

5. Aldosterone secretion is controlled by the renin-angiotensin mechanism, which responds to changes in renal blood flow.

6. ADH secretion is the most important factor affecting body water excretion.

7. ADH secretion is controlled by plasma osmolality. Plasma osmolality depends mainly on sodium concentration.

8. Clinical effects of disturbances of water and sodium metabolism are due to:

(a) changes in extracellular osmolality, dependent largely on sodium concentration. In pathological conditions urea and glucose concentrations can be important;

(b) changes in circulating volume.

FURTHER READING

De Wardener, H. E. (1982). The natriuretic hormone. *Ann. clin. Biochem.*, **19**, 137–140.

Tan, S. Y. and Mulrow, P. J. (1980). Aldosterone in hypertension and edema. In: *Metabolic Control and Disease*, 8th edit., pp. 1501–1533. P. K. Bondy and L. E. Rosenberg, eds. Philadelphia: W. B. Saunders Co.

WEITZMAN, R., VORHERR, H. and KLEEMAN, C. R. (1980). Water metabolism and the neurohypophyseal hormones. In: *Metabolic Control and Disease*, 8th edit., pp. 1241–1323. P. K. Bondy and L. E. Rosenberg, eds. Philadelphia: W. B. Saunders Co.

ZILVA, J. F. (1982). Fluid and electrolyte disturbances and their management. In: *Intestinal Obstruction*, pp. 23–37. H. Ellis, ed. New York: Appleton-Century-Crofts.

TABLE IV

SOME ELECTROLYTE-CONTAINING FLUIDS FOR INTRAVENOUS ADMINISTRATION

	Electrolytes (mmol/l)				Glucose mmol/l (g/dl)	Ca mmol/l	Approximate osmolarity relative to plasma
	Na	K	Cl	HCO$_3$			
Saline							
"Normal" (physiological) saline	154	—	154	—	—	—	×1
Twice "Normal"	308	—	308	—	—	—	×2
Half "Normal"	77	—	77	—	—	—	×$\frac{1}{2}$
Fifth "Normal"	31	—	31	—	—	—	×$\frac{1}{5}$
"Dextrose" Saline							
	77	—	77	—	140 (2·5)	—	×1
	31	—	31	—	222 (4·0)	—	×1
	77	—	77	—	278 (5·0)	—	×1$\frac{1}{2}$
	154	—	154	—	278 (5·0)	—	×2
	154	—	154	—	556 (10·0)	—	×3
Sodium Bicarbonate							
1·4%	167	—	—	167	—	—	×1
2·8%	334	—	—	334	—	—	×2
*8·4%	1000	—	—	1000	—	—	×6

* Most commonly used bicarbonate solution. Note marked hyperosmolarity. It should only be used if very strongly indicated.

TABLE IV (continued)

	Electrolytes (mmol/l)				Glucose mmol/l (g/dl)	Ca mmol/l	Approximate osmolarity relative to plasma
	Na	K	Cl	HCO$_3$			
Complex Solutions							
Ringer's Solution	147	4·2	156	—	—	2·2	×1
Darrow's Solution	121	34·8	103	53 (as lactate)	—	—	×1
Hartmann's Solution	131	5·4	112	29 (as lactate)	—	1·8	×1
Sodium lactate 1/6 molar	167	—	—	167 (as lactate)	—	—	×1

TABLE V

Some Electrolyte-Containing Amino Acid Solutions (All Hyperosmolar)

	Electrolytes Na K Cl mmol/l			Glucose (G) or Sorbitol (S) mmol/l (g/dl)	Ca mmol/l	Mg mmol/l	Nitrogen g/l	Other constituents mmol/l	kJ/l (Calories/l)
"Aminofusin L600"	40	30	14	550 (10 as S)	—	5	7·6	Acetate 10	2500 (595)
"Aminoplex 5"	35	28	43	690 (12·5 as S)	—	—	5·0	Ethanol 1086, Malate 103	4200 (1000)
"Aminoplex 12"	35	30	67	—	—	2·5	12·4	Acetate 5	1300 (315)
"Synthamin 14"	73	60	70	—	—	—	14·3	Acetate 14, H$_2$PO$_4$ 30	1600 (380)
"Vamin 9 Glucose"	50	20	55	556 (10 as G)	2·5	1·5	9·4	—	2700 (643)

TABLE VI
ELECTROLYTE-FREE DEXTROSE SOLUTIONS FOR INTRAVENOUS ADMINISTRATION (USUALLY USED AS AN ENERGY SOURCE)

Dextrose	Glucose (mmol/l)	Osmolarity	kJ/l	(Calories/l)
5%	277	Isosmolar	860	(205)
10%	555	Hyperosmolar	1720	(410)
20%	1110	Hyperosmolar	3440	(820)
40%	2220	Hyperosmolar	6880	(1640)

TABLE VII
OTHER HYPEROSMOLAR SOLUTIONS (USUALLY USED AS OSMOTIC DIURETICS, OR TO REDUCE CEREBRAL OEDEMA)

		Concentration (mmol/l)	kJ/l
Mannitol	10%	550	0
	20%	1100	0
Urea	4%	667	0 (But may also contain fructose
	30%	5000	0 10% giving 1720 kJ/l)

Chapter III

POTASSIUM METABOLISM:
DIURETIC THERAPY

THE total amount of potassium in the body is about 3000 mmol.

Potassium is predominantly an *intracellular* ion, and only about 2 per cent of the body content is in the extracellular fluid. The plasma potassium level is therefore a poor indicator of the total amount in the body. This apparent disadvantage is more theoretical than real, because it is *plasma potassium concentrations* that are of immediate importance in therapy; either hyperkalaemia or hypokalaemia, if severe, is dangerous, and must be treated whatever the state of the intracellular potassium. However, some assessment of the overall situation should be made to anticipate therapeutic needs.

The low extracellular concentration of potassium ions is important for normal neuromuscular activity and cardiac action; in this respect it resembles calcium and magnesium.

FACTORS AFFECTING PLASMA POTASSIUM CONCENTRATIONS

The predominantly intracellular location of potassium provides a reservoir of the ion, and the plasma potassium concentration, unlike that of sodium, is relatively little affected by changes in water balance. The hyperkalaemia often associated with dehydration is due more to renal retention (p. 12) than directly to haemoconcentration.

Potassium enters and leaves the extracellular compartment by three main routes:

the intestine
the kidney
   the glomerulus
   the tubular cells
the membranes of *all* other body cells.

The Intestine

Potassium is absorbed throughout the small intestine. Dietary intake replaces urinary and faecal loss, and amounts to 100 mmol/day or less.

Potassium leaves the extracellular compartment in all intestinal secretions, usually at concentrations near or a little above those in plasma. The total daily loss into the gut lumen amounts to about 60 mmol. Most of this potassium is reabsorbed, together with the dietary intake, and less than 10 mmol is present in formed faeces. As in the case of sodium, excessive intestinal loss of potassium in diarrhoea stools, in ileostomy fluid or via fistulae is mainly derived from fluid entering the intestinal lumen from the body rather than from dietary intake. However, prolonged starvation can cause potassium deficiency and hypokalaemia.

The Kidney

Filtrate.—The concentration of potassium filtered is almost the same as in plasma. Because of the very large volume, daily loss by this route would be about a third of the body content (about 800 mmol) if there were no tubular regulation. The net loss, although very variable, is about 10 per cent of this.

The tubules.—Potassium is almost completely reabsorbed in the *proximal tubule* and renal tubular damage can cause potassium depletion.

Potassium is secreted in the *distal tubule* and *collecting duct* in exchange for *sodium*. Hydrogen ions compete with potassium for this exchange, which is stimulated by *aldosterone*. If the proximal tubule is functioning, potassium loss in the urine depends on three factors:

the amount of *sodium available for exchange*. This depends on the *filtration rate, filtered sodium load*, and *sodium reabsorption in the proximal tubule and loop of Henle*. The latter process is inhibited by many diuretics (p. 76).

the relative *amounts of hydrogen and potassium ions* present in the cells of the distal tubule and collecting ducts, and the *ability to secrete H^+ in exchange for Na^+* (inhibited during therapy with carbonate dehydratase inhibitors and in some types of renal tubular acidosis).

the circulating *aldosterone* level. This is increased following water loss (which usually accompanies intestinal loss of potassium) (p. 38) and in almost all conditions requiring diuretic therapy.

The Cell Membrane

Potassium is the predominant intracellular cation and is continuously lost from the cell down a concentration gradient. This loss is opposed by the "sodium pump" situated at the cell surface. This pumps sodium out of the cell in exchange for potassium and hydrogen ions.

In most circumstances the shift of potassium across cell membranes is accompanied by a shift of sodium in the opposite direction, but the *percentage* change in extracellular sodium levels will be much less than that of potassium. A simplified example will demonstrate this. Let us assume extra- and intracellular volumes to be equal, the plasma sodium to be 140 mmol/l and potassium 4 mmol/l, with reversal of these concentrations inside cells. An exchange of 4 mmol/l of sodium for potassium across the cell wall would double the plasma potassium concentration (clinically a very significant change) while only reducing the plasma sodium concentration to 136 mmol/l.

There is *net loss* of potassium from the cell:

if *potassium is lost from the ECF* and replenished from the cell;
if the *sodium pump is inefficient*, as in diabetic ketoacidosis and in hypoxic states;
in acidosis, when potassium is displaced from cells.

There is *net gain* of potassium by the cell:

if the *activity of the sodium pump is increased*, as it is after the administration of glucose and insulin: this effect may be used to treat hyperkalaemia. It is the main cause of the change from hyperkalaemia to hypokalaemia during treatment of diabetic coma;
in alkalosis. Induction of alkalosis can be used to treat hyperkalaemia.

Interrelationship Between Hydrogen and Potassium Ions

Extracellular hydrogen ion concentration affects the entry of potassium into all cells: changing the relative proportions of K^+ and H^+ in distal tubular cells affects urinary loss of potassium. In acidosis increased loss of potassium from the cells into the ECF, coupled with reduced urinary secretion of the ion, causes hyperkalaemia: in alkalosis hypokalaemia is due both to net increased entry of potassium into cells and to increased urinary loss.

As the relationship between K^+ and H^+ is reciprocal, changes in K^+ balance affect the hydrogen ion balance of the body.

In absorptive cells, including those in the renal tubule, the "sodium pump" is probably only on the non-luminal surfaces. In the kidney, the sodium derived from the luminal fluid is pumped through the cell in exchange for K^+ or H^+. If K^+ is lost from the ECF, loss from cells follows, causing a reduction in intracellular K^+, and, unless the loss is due to renal tubular cell damage, sodium is reabsorbed in exchange for less K^+ and more H^+ than usual. An acid urine is passed despite extracellular alkalosis. H^+ within the tubular cell is formed by the carbonate dehydratase (CD) mechanism (p. 97).

$$H_2O+CO_2 \xrightarrow{\text{CD}} H^+ + HCO_3^-$$

As more H^+ is secreted into the urine, the reaction is accelerated and more HCO_3^- is generated and passes into the ECF, accompanied by the reabsorbed sodium. Chronic potassium depletion is therefore accompanied by a *high plasma* $[HCO_3^-]$ *and extracellular alkalosis.* If other causes for a raised plasma bicarbonate (such as respiratory disease) are absent, and especially if factors known to cause potassium depletion (for instance, diuretic therapy) are present, this finding is a sensitive indicator of potassium depletion, even in the absence of hypokalaemia. *The combination of hypokalaemia and a high plasma* $[HCO_3^-]$ *is more likely to be due to* K^+ *depletion, which is common, than primarily to metabolic alkalosis, which is rare.* These are the changes of *chronic* potassium depletion.

In *acute* loss the slight lag in potassium release from cells may result in more severe hypokalaemia for the same degree of depletion than in the chronic state: bicarbonate levels are less likely to be raised, because significant bicarbonate retention by the kidney takes several days.

Theoretically, potassium excess could cause intracellular alkalosis and extracellular acidosis, with a low plasma bicarbonate. However, *the combination of hyperkalaemia and a low plasma* $[HCO_3^-]$ *is more*

TABLE VIII

INTERRELATIONSHIPS OF PLASMA POTASSIUM AND BICARBONATE LEVELS

Plasma [K+]	Plasma [HCO$_3^-$]	Most likely cause	Examples of clinical conditions
Low N or ↓	↑	Chronic K+ depletion	Diuretic therapy *Chronic diarrhoea (chronic purgative takers)
↓ or ↓↓	N or ↓	Acute K+ depletion	Severe acute diarrhoea Fistulae etc.
↓	↓	Respiratory alkalosis	Overtreatment on respirator Hysterical overbreathing
↑	↓	Metabolic acidosis	*Renal failure Diabetic ketoacidosis
↑	↑	Respiratory acidosis	Bronchopneumonia
↑	N	Acute K+ load	Excessive K+ therapy

* It should be stressed that this is *only a guide.* For instance, in severe diarrhoea bicarbonate loss may be so high that $[HCO_3^-]$ is reduced in spite of potassium depletion: similarly, renal tubular lesions cause depletion of both HCO_3^- and K^+.

likely to be due to metabolic acidosis, which is common, than primarily to potassium excess, which is extremely rare. In *respiratory* acidosis the plasma bicarbonate is high (p. 112), but the plasma potassium will also tend to be high.

These various situations are summarised in Table VIII.

ABNORMALITIES OF PLASMA POTASSIUM LEVELS

In any clinical situation no single factor entirely accounts for the changes in plasma potassium concentration. For instance, in conditions associated with intestinal potassium loss, concomitant water loss causes secondary hyperaldosteronism; similarly, most conditions requiring diuretic therapy are associated with hyperaldosteronism. This hyperaldosteronism aggravates urinary loss and may also increase entry of potassium into body cells generally. However, if volume and sodium depletion are very severe, the reduced amount of filtered sodium means that less is available for exchange with potassium in the distal tubule and aldosterone cannot produce maximal effects: hypokalaemia will then be "unmasked" as the patient is rehydrated. This should be anticipated.

Hypokalaemia

This is usually the result of potassium depletion, although, if the rate of loss of potassium from cells equals or exceeds that from extracellular fluid, potassium depletion may not cause hypokalaemia. It can occur without depletion if there is a shift into cells, as in alkalotic states and in the rare condition, familial periodic paralysis.

Misleading and temporary hypokalaemia may occur for a few hours after oral administration of diuretics which cause potassium loss. The finding of a higher value on a later specimen does not usually indicate a "laboratory error".

The causes of hypokalaemia may be classified as follows:

1. predominantly due to **loss of potassium from the body.**

(a) Predominantly due to loss from the ECF in *intestinal secretions*:
  Prolonged vomiting;
  Diarrhoea;
  Loss through intestinal fistulae.

The intestinal loss is usually aggravated by the secondary hyperaldosteronism consequent on water loss. This causes an inappropriately high urinary loss which, in some cases, may contribute more to the depletion than the original pathology. The important points to be noted are:

The concentration of potassium in fluid from a *recent ileostomy* and *diarrhoea stools* may be 5 to 10 times that of plasma, but a prolonged drain of *any* intestinal secretion causes depletion, especially if urinary loss is increased by secondary aldosteronism;

Habitual purgative takers may present with hypokalaemia and are often reluctant to admit to the habit.

The rarely occurring, large *mucus-secreting villous adenoma of the intestine* may cause considerable potassium loss.

(b) Predominantly due to loss from the ECF in *urine*.
   (i) Increased activity of sodium: potassium exchange mechanisms in the distal tubule.

Secondary hyperaldosteronism.—This often aggravates other causes of potassium depletion.

Cushing's syndrome and steroid therapy.—Patients secreting excess of, or on prolonged therapy with glucocorticoids tend to become hypokalaemic, due to the mineralocorticoid effect on the distal tubule.

Primary hyperaldosteronism (p. 58).

Synacthen or ACTH therapy and ectopic ACTH (p. 475).

Carbenoxolone therapy.—Carbenoxolone has been used to accelerate healing of peptic ulcers. It also potentiates the action of aldosterone, and potassium depletion is sometimes a complication of such therapy.

Liquorice contains glycyrrhizinic acid, which also has an aldosterone-like effect. Subjects fond of liquorice-containing sweets may present with hypokalaemia. The same effect has been caused by habitual tobacco chewing.

(ii) Excess available sodium for exchange in the distal tubule.

Diuretics inhibiting the "sodium pump" in the loop of Henle (p. 76). The increased sodium: potassium exchange is aggravated by secondary hyperaldosteronism.

(iii) Decreased renal sodium: hydrogen ion exchange, favouring sodium: potassium exchange.

Carbonate dehydratase inhibitors (p. 111).

Renal tubular acidosis (p. 111).

(iv) Reduced proximal tubular potassium reabsorption.

Renal tubular failure (for example, the recovery phase of acute oliguric renal failure).

"Fanconi syndrome" (p. 19) (hypercalcaemia is a common cause).

2. predominantly due to **reduced potassium intake.**

Chronic starvation.—If water and salt intake are also reduced, secondary hyperaldosteronism may aggravate the hypokalaemia.

3. predominantly due to **redistribution** in the body. Loss into cells.

Glucose and insulin therapy.—This may be used to treat severe hyperkalaemia.

Familial periodic paralysis (very rare).—In this condition episodic paralysis occurs associated with entry of potassium into cells.

4. loss from ECF by **more than one route.**

(a) Into cells and urine
Alkalosis

(b) Into cells, urine and intestine
Pyloric stenosis with alkalosis.—The loss in urine and the loss into cells are probably more important causes of hypokalaemia than the loss in gastric secretion.

HYPERKALAEMIA

This occurs most commonly when the rate of potassium leaving cells is greater than its rate of excretion. The causes of hyperkalaemia may be:

1. predominantly due to **gain of potassium by the body.**

Gain by ECF from the *intestine or by an intravenous route.*
Over-enthusiastic potassium therapy.
Failure to stop potassium therapy when depletion has been corrected.

2. **failure of renal secretion** of potassium.

(a) Decreased activity of sodium: potassium exchange mechanisms in the distal tubule.

Hypoaldosteronism (as in Addison's disease). Prostaglandin inhibitors, such as indomethacin, may produce similar changes if renal function is mildly impaired.
Diuretics acting on the distal tubule by antagonising aldosterone, or direct inhibition of the "sodium pump" (p. 77).

(b) Too little sodium available for exchange in the distal tubule.

Renal glomerular failure.—Hyperkalaemia is usually aggravated by the concomitant acidosis and gain of potassium from cells.
Sodium depletion

3. predominantly due to **redistribution** of potassium in the body.
   Gain of potassium by ECF from cells.
   Severe tissue damage.
4. gain by ECF by **more than one route.**
   Reduced renal excretion in spite of gain by ECF from cells.
   Acidosis
   Hypoxia—Failure of the sodium pump in all cells. Failure in distal tubular cells causes potassium retention. If hypoxia is very severe, lactic acidosis aggravates the hyperkalaemia.

Diabetic ketoacidosis.—In early untreated diabetes potassium leaves the cells, and in spite of a high urinary loss and consequent body depletion, hyperkalaemia is usual. This is due to partial failure of the sodium pump resulting from impaired glucose metabolism because of insulin lack. As the condition becomes more advanced two other factors contribute to hyperkalaemia:

   volume depletion with a low GFR;
   acidosis due to ketone production.

All these factors are reversed during insulin and fluid therapy. As potassium enters cells extracellular levels fall and the depletion is revealed. *Plasma potassium levels should be monitored during therapy,* and potassium should be given as soon as concentrations start to fall.

Measurement of Urinary and Intestinal Losses

Pure urinary or intestinal loss as a cause of hypokalaemia is very rare. Measurement of such losses with a view to quantitative replacement may lead to even more dangerous errors of therapy than in the case of sodium. Exchanges across cell walls cannot be measured. If gain of potassium by the ECF from cells is faster than loss from it into urine and intestine, replacement of measured loss could endanger the patient's life by aggravating hyperkalaemia; if loss from the ECF into cells is predominant, therapy based on urinary excretion may be inadequate. Moreover, high urinary potassium excretion may be appropriate when, for instance, trauma has damaged many cells, reducing cellular capacity and releasing the ion from cells. It would be as rational, and slightly safer, to replace urinary glucose quantitatively in an uncontrolled diabetic as to use urinary potassium measurements in the same way. The student should remember this analogy when considering the value of measuring daily losses of any constituent, especially to control parenteral feeding (p. 229). *It is plasma potassium levels that are important,* and in rapidly changing states frequent estimation of these is the only safe way of assessing therapy. In chronic depletion the plasma bicarbonate level may help to indicate the state of cellular repletion.

The diagnostic use of urinary potassium estimations to determine the primary cause of depletion is also more often misleading than helpful. Most extrarenal potassium loss is associated with volume depletion, and therefore secondary aldosteronism; a high urinary excretion is not proof of a primary renal cause. A urinary potassium excretion of less than about 20 mmol a day can only be expected in the *well-hydrated* hypokalaemic patient with extrarenal losses, in whom aldosterone secretion is inhibited: a low potassium excretion confirms extrarenal loss, but a high one does not exclude it as the primary cause.

Measurement of urinary pH may occasionally be of value to differentiate extracellular alkalosis due to extrarenal potassium depletion and hypokalaemia due to alkalosis. If potassium depletion is primary, more hydrogen than potassium is exchanged for sodium ions and the urine will usually be acid (p. 70): if alkalosis is the primary cause, the urine will usually be appropriately alkaline.

DIURETIC THERAPY

In oedematous states the fluid accumulation is accompanied by an excess of sodium in the body, *even if there is hyponatraemia* (p. 56). Diuretics act by inhibiting sodium reabsorption in the renal tubule and secondarily causing water loss. They may also be used to treat hypertension. All diuretics tend to affect potassium balance, and this effect should be anticipated, especially in patients with secondary aldosteronism.

Diuretics can be divided into two main groups.

1. *Those inhibiting the pump in the loop of Henle*, and therefore water reabsorption: the increased sodium load on the distal tubule and collecting ducts increases sodium: potassium exchange at this site (which is stimulated by the accompanying hyperaldosteronism). In our experience long-term diuretic therapy almost invariably causes significant *potassium depletion*, and sometimes symptomatic hypokalaemia, even if potassium supplements are given (although this has been questioned). *A high plasma bicarbonate* is common in long-continued use of such diuretics. This is because loss of potassium from the ECF results in increased $Na^+:H^+$ exchange in the distal tubule. In this situation high plasma HCO_3^- levels are a more sensitive indication of K^+ depletion than plasma potassium levels.

The *thiazide group* of diuretics act at the junction of the loop and the distal tubule (sometimes called the "cortical diluting segment"). *Frusemide* (furosemide; "Lasix"), *bumetanide* ("Burinex") and *etha-*

crynic acid ("Edecrin") are true "loop diuretics", and inhibit the pump in the ascending limb.

2. Those either directly *inhibiting aldosterone* or inhibiting the exchange mechanisms in the *distal tubule* and *collecting duct*. These cause *potassium retention* and may lead to hyperkalaemia, especially if renal function is impaired: potassium supplements should *not* be used. Potassium-retaining diuretics include:

Spironolactone ("Aldactone")—A competitive aldosterone antagonist.

Amiloride ("Midamor") ⎰ Inhibitors of the $Na^+:K^+$ exchange
Triamterene ("Dytac") ⎱ mechanisms in the renal tubule.

This group of diuretics is often used, together with those causing potassium loss, when hypokalaemia cannot be controlled by potassium therapy, or to potentiate sodium loss by inhibiting reabsorption at more than one site.

In addition carbonate dehydratase inhibitors such as acetazolamide ("Diamox") (p. 111) inhibit sodium reabsorption in the proximal tubule and act as diuretics. They are rarely used for this purpose now because of the danger of acidosis, and because they are relatively ineffective.

CLINICAL FEATURES OF DISTURBANCES OF POTASSIUM METABOLISM

The clinical features of disturbances of potassium metabolism are due to changes in extracellular concentration of the ion.

Hypokalaemia, by interfering with neuromuscular transmission, causes *muscular weakness, hypotonia* and *cardiac arrhythmias*, and may precipitate digitalis toxicity. It may also aggravate paralytic ileus.

Intracellular potassium depletion causes extracellular alkalosis (see p. 71). This reduces ionisation of calcium salts (p. 252) and in long-standing potassium depletion of gradual onset the presenting symptoms may be muscle *cramps* and *tetany*. This syndrome is accompanied by high plasma bicarbonate levels.

Prolonged potassium depletion causes lesions in renal tubular cells, and this may complicate the clinical picture.

Severe hyperkalaemia always carries the danger of cardiac arrest. Both hypokalaemia and hyperkalaemia cause characteristic changes in the electrocardiogram.

TREATMENT OF POTASSIUM DISTURBANCES

Abnormalities of plasma potassium should be corrected whatever the state of the total body potassium. However, an attempt should be made to assess the latter so that sudden changes in plasma potassium (for instance, during treatment of diabetic coma) can be anticipated. Treatment should be controlled by frequent plasma potassium estimations.

Hyperkalaemia.—Treatment of hyperkalaemia is based on three principles. In severe hyperkalaemia the first two principles are used.

Very severe hyperkalaemia can cause cardiac arrest. Calcium and potassium have opposing actions on heart muscle, and the immediate danger can be minimised by infusion of calcium salts (usually as gluconate) (p. 80). This allows time to institute measures to lower plasma potassium.

Plasma potassium can be reduced rapidly (within an hour) by increasing the rate of entry into cells. Glucose and insulin speed up glucose metabolism and the action of the "sodium pump". Induction of alkalosis by infusion of bicarbonate also increases the rate of entry into cells (p. 81). For purely practical reasons this treatment (which involves intravenous infusion) cannot be continued indefinitely, but its use allows long-term treatment to be instituted.

In moderate hyperkalaemia a slower acting method can be used. Potassium can be removed from the body at a rate higher than, or equal to, that at which it is entering the extracellular fluid by using ion exchange resins by the oral or rectal route. These are unabsorbed and exchange potassium for sodium or calcium ions. It will be seen that plasma potassium is lowered at the expense of body depletion. This potassium may have to be replaced later.

Hypokalaemia.—If hypokalaemia is *mild*, potassium supplements should be given *orally* until plasma potassium and bicarbonate levels return to normal. These levels should be monitored regularly. Undertreatment is more common than overtreatment during oral therapy. The normal subject loses about 60 mmol of potassium daily in the urine, much larger amounts being excreted during diuretic therapy. By the time hypokalaemic alkalosis is present the total deficit is probably several hundred mmol. A patient with hypokalaemia should be given *at least* 80 mmol a day: much more may be needed if plasma levels fail to rise.

In *severe* hypokalaemia, particularly if the patient is unable to take oral supplements, *intravenous potassium* should be given cautiously (p. 80). Diarrhoea not only reduces absorption of oral potassium supplements, but may itself be aggravated by them, and may also be an indication for intravenous therapy.

It is sometimes suggested that hypokalaemia could be treated by a high intake of fruit or of fruit juice. Table IX shows that if it were possible to ingest the quantities required, the consequent diarrhoea might well be self-defeating. The relatively high cost is only a minor factor.

TABLE IX

POTASSIUM CONTENT OF FRUIT AND FRUIT JUICE

	Approximate K+ content (mmol)	Approximate quantity containing 50 mmol	Price for 50 mmol (relative to "Slow-K") in U.K. January, 1983
Tomato juice (Heinz)	82 per litre	610 ml	8
Canned orange juice	42 per litre	1200 ml	17
Fresh orange juice	30 per litre	1700 ml	57
Rose's Orange Cordial (*undiluted*)	16 per litre	3100 ml	66
Bananas	8 per banana	6 bananas	20
"Slow-K" (Ciba)	8 per tablet	6 tablets	1

SUMMARY

1. Changes in plasma potassium levels are the net result of changes between ECF and cells, kidney and gut.

2. In any clinical situation many factors are involved, and monitoring of plasma potassium levels is the only safe guide to treatment.

3. As hydrogen and potassium ions compete for exchange with sodium in the renal tubule, disturbances of hydrogen ion homeostasis and potassium balance often coexist. A raised plasma bicarbonate (TCO_2) level may indicate intracellular potassium depletion.

4. Clinical manifestations of disturbances of potassium metabolism are due to its action on neuromuscular transmission and on the heart.

5. Diuretics fall into two main groups:

(*a*) Those inhibiting the "sodium pump" in the loop of Henle, causing potassium depletion.

(*b*) Those antagonising aldosterone either directly, or indirectly by affecting the $Na^+:K^+$ transport mechanism, causing potassium retention.

FURTHER READING

LANT, A. F. (1981). Modern diuretics and the kidney. *J. Clin. Path.* **34**, 1267–1275.
Diuretics. In: *British National Formulary* (1984), No. 7, pp. 68–69. British Medical Association and Pharmaceutical Society of Great Britain.

TREATMENT OF HYPOKALAEMIA

POTASSIUM-CONTAINING PREPARATIONS

One gram of potassium chloride contains 13 mmol of potassium.

FOR INTRAVENOUS USE

These preparations should only be used in serious depletion, or when oral potassium cannot be taken or retained. In most cases oral potassium is preferable. Intravenous potassium should be given with care, especially in the presence of poor renal function and the following rules should be observed:

1. Intravenous potassium should not be given in the presence of oliguria unless the potassium deficit is unequivocal and severe.
2. Potassium in the intravenous fluid should not exceed 40 mmol/l.
3. Intravenous potassium should not usually be given at a rate of more than 20 mmol/hour.

Strong Potassium Chloride Solution B.P.

20 mmol of potassium and chloride in 10 ml.
WARNING. This should *never* be given undiluted. It should be added to a full bottle of other intravenous fluid. (10 ml added to a bottle containing 500 ml of fluid gives a concentration of 40 mmol/l.)

Potassium Chloride and Dextrose Intravenous Infusion (B.P.)

5 per cent dextrose with 40 mmol/l of potassium and chloride. This is hyperosmolal.

FOR ORAL USE

1. Potassium Effervescent Tablets B.P.—6·5 mmol K^+ per tablet (as bicarbonate and acid tartrate).
2. "Slow-K" (Ciba)—8 mmol K^+ per tablet (as chloride).
3. "Sando-K" (Sandoz)—12 mmol K^+ per tablet (as bicarbonate and chloride).
4. "Kloref-S" (Cox-Continental)—20 mmol K^+ per sachet (as chloride).
5. "Kloref" tablets (Cox-Continental)—6–7 mmol K^+ (mostly as chloride and bicarbonate).

TREATMENT OF HYPERKALAEMIA

Emergency Treatment

Calcium chloride (or gluconate). A 10 per cent solution is given intravenously with ECG monitoring. This treatment antagonises the effect of hyperkalaemia on heart muscle, but does not alter potassium levels.
WARNING. Calcium should never be added to bicarbonate solutions, because calcium carbonate is insoluble.
Glucose 50 g with 20 units of soluble insulin by intravenous injection lowers plasma potassium rapidly by increasing entry into cells. If the situation

is less urgent 10 units of soluble insulin may be added to a litre of 10 per cent dextrose.

If acidosis is present, bicarbonate may be used as an alternative to glucose and insulin injection; 40 ml of 8·4 per cent sodium bicarbonate (40 mmol of bicarbonate) may be injected over 5 minutes.

Long-term Treatment

Sodium or calcium polystyrene sulphonate ("Resonium-A"; "Kayexalate" or "Calcium Resonium". Winthrop) 20–60 g a day by mouth in 15 g doses, *or* 10–40 g in a little water as a retention enema every 4–12 hours. This removes potassium from the body.

INVESTIGATION OF RENAL, WATER AND ELECTROLYTE DISORDERS

Protocols for the tests are given on p. 87.

MONITORING FLUID BALANCE

We have pointed out that maintenance and inspection of *accurate* ward fluid balance charts is at least as important as measurement of daily plasma electrolyte levels.

Unconscious patients, or those with abnormal losses, may develop fluid imbalance so gradually that it may not be noticed if individual daily charts are inspected, especially if insensible losses are ignored. Maintenance of a record of cumulative fluid balance is a more sensitive method of demonstrating a trend which may be corrected before a serious deficit or excess develops. This is especially important in patients at risk of predominant water depletion, because this is not as clinically obvious as the more usual isosmolar volume depletion (p. 52); it may not be noticed until hypernatraemia has developed.

The chart below shows how insidiously a serious deficit can develop in a few days. All volumes are in ml, and the minimum of 400 ml has been allowed for insensible loss: calculated losses are therefore more likely to be under- than overestimated, and the danger of fluid overload using this method is minimal.

	Intake	Measured output	Total output (minimum)	Balance	Cumulative balance
Day 1	2000	1900	2300	−300	−300
Day 2	2000	2000	2400	−400	−700
Day 3	2100	1900	2300	−200	−900
Day 4	2200	2000	2400	−200	−1100

The type of fluid infused should be based on the plasma electrolyte results. If the cumulative balance figures are plotted on the same graph as daily plasma sodium and urea values potential hypernatraemia and renal circulatory insufficiency can usually be detected before they become dangerous.

If the ambient temperature is high, or if the patient is pyrexial, hyperventilating, or has large unmeasured intestinal losses (for instance in intestinal obstruction), more should be allowed for insensible loss.

INVESTIGATION OF ACUTE OLIGURIA

The aim is to distinguish between extrarenal and renal causes, and to identify life-threatening abnormalities.

1. Estimate plasma electrolytes (especially potassium) urgently. The level of urea and/or creatinine is of less immediate therapeutic significance. Treat dangerous hyperkalaemia (p. 80).

2. Exclude postrenal causes. If the bladder is palpable the cause in the male is most likely to be prostatic hypertrophy.

3. If the patient is hypotensive and/or dehydrated, collect, if possible, a sample of urine and keep it in the refrigerator. If urinary investigations should later become necessary this ensures that the sample has been taken under appropriate conditions. Examine the specimen for protein and casts: their

presence suggests significant renal damage. Bacteriological examination may reveal urinary tract infection.

4. Rehydrate, watching the patient for signs of incipient overhydration, and measure the urine output accurately. If the output rises rapidly, and the plasma urea and/or creatinine starts to fall within 24 hours, the cause is likely to be prerenal. If, after adequate rehydration, the oliguria persists and the urea continues to rise, renal damage is likely, and fluid balance should be maintained as described on p. 42.

5. In the very rare cases in which doubt remains, estimate the sodium concentration on the urine specimen collected on admission. If it is below 30 mmol/l significant renal damage is unlikely.

6. Plasma potassium levels must be measured at least daily and treated appropriately.

INVESTIGATION OF POLYURIA

A. Confirm polyuria and exclude obvious causes

1. Take a history, and, if possible, distinguish between true polyuria (a high 24-hour urinary volume) and frequency (a normal 24-hour volume but abnormally frequent micturition).

2. Is there a recent history of acute oliguria? If so, the patient has probably entered the polyuric phase and should be treated accordingly.

3. Is the patient receiving intravenous infusion, especially with fluid of low sodium concentration? If so, adjust the infusion accordingly (p. 47).

4. Is there a cause for an osmotic diuresis, for example is there:

gross glycosuria (due to severe diabetes mellitus or infusion of dextrose)?
severe tissue damage leading to a high urea load?
infusion of amino acids?

If so investigate and treat, or change the composition of the infusion.

5. Is there an obvious cause for diabetes insipidus (for example, a head injury)?

6. Is the patient taking diuretics?

B. Distinguish between polyuria appropriate to a high intake (see also 3 above) and failure of homeostatic mechanisms.

1. What is the patient's clinical state of hydration? Dehydration suggests failure of homeostatic mechanisms.

2. If possible, *accurately* monitor fluid intake and output.

   (*a*) A *negative balance* suggests *failure of homeostatic mechanisms*;
   (*b*) A *positive balance* suggests that *polydipsia is primary*.

The patient should be observed closely for secret fluid ingestion ("hysterical polydipsia") or possible addition of fluid to the urine.

3. Estimate plasma urea/creatinine (and electrolytes).

   (*a*) *very high urea/creatinine levels suggest chronic renal failure.* Treat and monitor accordingly (see p. 24);
   (*b*) *low or low normal urea/creatinine levels suggest that polydipsia is the primary cause* and that the polyuria is appropriate. Question the

patient about his intake in relation to true thirst. Remember that patients with "hysterical polydipsia" may give misleading answers;

(c) *High normal or slightly high urea/creatinine levels* (urea up to about 15 mmol/l—90 mg/dl) *suggest failure of normal tubular concentrating ability.*

4. If the plasma urea/creatinine is slightly high, cautiously increase the fluid intake. If the urea level falls in response to rehydration glomerular damage is probably minimal. Seek a cause of tubular damage.

(a) Take a drug history, especially of analgesics;
(b) Send urine for microbiological examination (pyelonephritis);
(c) Estimate plasma calcium and urate concentration;
(d) Consider inborn errors in infants (Chapter XVI);
(e) Exclude other causes listed on p. 19.

5. If it is still not possible to distinguish between polydipsia, tubular impairment and diabetes insipidus, *contact the laboratory* to arrange a urine concentration test, if necessary with Pitressin or DDAVP (see p. 87).

6. Treat mild tubular impairment with a high fluid intake.

7. If diabetes insipidus, without obvious cause, is confirmed, a neurological opinion must be sought.

INVESTIGATION OF THE PATIENT WITH RENAL CALCULI

1. If the stone is available, send it to the laboratory for analysis.

2. Exclude *hypercalcaemia* and *hyperuricaemia*. If either is present it should be treated.

3. If the *plasma calcium is normal*, collect a 24-hour specimen of urine (in a container containing acid to keep calcium in solution) for *urinary calcium estimation*. If hypercalciuria is present it should be treated.

4. *If all these tests are negative*, screen the urine for *cystine*. This is especially important if there is a family history. If the qualitative test is positive, cystine should be estimated quantitatively on a 24-hour specimen of urine (p. 391).

5. If the patient is acidotic and the urine is alkaline, perform an ammonium chloride load test (p. 124).

6. If renal calculi occur in *childhood*, a low *plasma urate* and high *urinary xanthine* suggest xanthinuria. If these are normal the 24-hour excretion of *oxalate* should be estimated to exclude primary hyperoxaluria as a cause.

INVESTIGATION OF ELECTROLYTE DISTURBANCES

In the last two chapters we have outlined conditions in which sodium or potassium concentrations *may* be abnormal. We have pointed out that an abnormal result, particularly of sodium, may not be clinically significant, while "normal" ones do not guarantee "normal" balance. Before making a request, some assessment should be made as to whether the result of an estimation will aid diagnosis or treatment.

Sodium and potassium estimations provide the numerical bulk of the workload of most chemical pathology departments; because of this they are often estimated simultaneously. However the potassium result is more often useful than that of sodium.

The student should bear the following points in mind:

1. *Plasma sodium* should *be estimated regularly*:

   (*a*) *in the unconscious or confused patient and infants losing fluid*, because of the danger of hypernatraemia;
   (*b*) *in the patient in diabetic coma or precoma*, because of the danger of sustained hyperosmolality due to hypernatraemia despite successful control of plasma glucose levels (p. 209);
   (*c*) *in the dehydrated patient*, or those with *abnormal losses*, to help diagnosis and to indicate the type of replacement fluid.

2. *Plasma potassium (and bicarbonate)* should *be estimated regularly* in any patient in whom there is a cause for abnormal levels, because these must be treated:

   (*a*) *in patients with abnormal losses* from the gastro-intestinal tract or kidneys (especially due to *diuretic, steroid or ACTH therapy*);
   (*b*) *in patients on potassium therapy;*
   (*c*) *in patients in renal failure;*
   (*d*) *in patients in diabetic coma or precoma.*

Groups 1 and 2 are numerically small compared to the estimations actually requested.

3. In the fully conscious, normally hydrated patient, with no abnormal losses, plasma sodium estimation rarely helps. Mild hyponatraemia is common (p. 62), but treatment in such subjects is usually contra-indicated. Unless renal failure is present, potassium estimation is also unhelpful in such subjects.

INVESTIGATION OF HYPOKALAEMIA

Diuretic therapy is the commonest cause of hypokalaemic alkalosis in hypertensive patients: primary aldosteronism is very rare. Hypokalaemia due to ectopic ACTH production is even less common.

The following procedure will almost always eliminate the need for expensive and time-consuming hormone assays.

1. Exclude obvious causes of potassium loss, such as diarrhoea.
2. Take a careful drug history, with special reference to potassium-losing diuretics, purgatives, or steroid, ACTH or Synacthen therapy; rare causes of a steroid-like effect are ingestion of carbenoxolone or liquorice.
3. Look at the plasma sodium value. If high, or high normal, the hypokalaemia is likely to be due to a steroid-like effect.
4. If not clinically contra-indicated, stop all therapy known to affect potassium loss. This is, in any case, necessary if hormone assays are later indicated, because the results are affected by treatment.
5. Give adequate oral potassium supplements until plasma potassium *and bicarbonate (TCO$_2$)* are normal.
6. Stop supplements.

Most cases will then remain normokalaemic, and need no further investigation.

If the potassium level falls rapidly again without obvious cause, or if adequate potassium supplementation fails to correct the hypokalaemia, and if there is no evidence of renal tubular damage, investigation of the appropriate hormones may be indicated. Specimens must be taken under carefully controlled conditions, and it is important to contact the laboratory before requesting renin (or angiotensin) and aldosterone assays (for possible primary hyperaldosteronism) or cortisol/ACTH assays (for ectopic hormone production).

ACTH, and therefore, cortisol, secretion is stimulated by stress. Moderately high plasma levels do not necessarily indicate ectopic hormone secretion.

If it is not clear whether hypokalaemic alkalosis is due primarily to potassium depletion or primarily to alkalosis, measurement of urinary pH may be helpful. An acid urine usually suggests primary potassium loss (p. 76).

PROTOCOLS FOR TESTS FOR RENAL, WATER AND ELECTROLYTE DISORDERS

Always contact your laboratory *before* starting any of these tests, both to ensure most efficient and speedy analysis, and to check local variations in protocols.

URINE CONCENTRATION TEST

In the normal subject restriction of water intake for a period of hours results in maximal stimulation of ADH secretion (p. 39). ADH acts on the collecting ducts, water is reabsorbed and a concentrated urine is passed. If the countercurrent multiplication mechanism is impaired maximal water reabsorption cannot take place, and if ADH levels are low the effect is similar.

If the feedback mechanism is intact ADH levels are sufficient to produce maximal effect. Under these circumstances administration of exogenous ADH will not improve renal concentrating power. If, however, the primary disease is diabetes insipidus with normal tubular function administration of ADH will convert this to normal.

This test should not be performed if the patient is already dehydrated. In such cases the demonstration of a low urine to plasma osmolality ratio (see below) is diagnostic without resort to fluid restriction. *The patient should be kept under observation during the test*, which should be terminated if the plasma osmolality rises to high normal or high levels. If the patient becomes distressed, blood and urine should be collected, and the test terminated. If the apparent "distress" does not coincide with a high normal or high plasma osmolality, it is likely to be of psychological origin.

Procedure

The patient is allowed no food or water after 18.00 h on the night before the test.

On the day of the test:

07.00 h—the bladder is emptied. Blood is collected and plasma osmolality measured. If this is low, water depletion is unlikely, and the polyuria is probably an appropriate response to excessive water intake. If it is high the test should be stopped and the osmolality of the urine measured.

08.00 h—The bladder is emptied and blood is collected. The osmolalities of the specimens are measured. If the urine osmolality exceeds 850 mmol/kg the test may be terminated: if it is below this figure and the plasma osmolality is normal the osmolality of urine passed at 09.00 h and, if necessary, at hourly intervals until noon, is measured.

Interpretation

A maximum urinary osmolality of less than 850 mmol/kg in the presence of a normal or increased plasma osmolality indicates impaired renal concentrating power, due either to tubular disease or to diabetes insipidus: the urine to plasma osmolality ratio should be above 3 in normal subjects. In most cases it is clear which of these possibilities is implicated. Where this is still in doubt the DDAVP or Pitressin test may be carried out; this may be done at the end of the concentration test, or on a separate day.

In prolonged overhydration, usually due to hysterical polydipsia, concentrating power may be temporarily reduced due to washing out of medullary hyperosmolality (p. 9). The test should be repeated after several days of relative water restriction. This should not be done if the patient is dehydrated and he must be kept under careful observation, both for signs of genuine distress associated with a rise in plasma osmolality and for surreptitious drinking.

DDAVP OR PITRESSIN TEST

DDAVP (1-Deamino-8-D-Arginine Vasopressin; desmopressin acetate) is a potent synthetic analogue of Pitressin.

The above procedure is followed, but 4 μg DDAVP or 5 units of the oily suspension of vasopressin (Pitressin) tannate is injected intramuscularly at 19·00 h on the evening before the test. If the failure to concentrate is due to tubular disease this will not be improved by the DDAVP or Pitressin: if it is due to diabetes insipidus the urine osmolality will increase.

Limitations of Specific Gravity Measurement

We have recommended measurement of urinary osmolality because of the many limitations of specific gravity measurement listed below. If an osmometer is not available, estimation of the urinary urea concentration may be more useful: values of about 90 to 100 times that of plasma indicate good tubular function.

1. Hydrometers are frequently inaccurate. The specific gravity of distilled water should read 1·000: if it does not, a suitable correction should be made.

2. The urine should be at the temperature to which the hydrometer is calibrated, usually 15°C. Significant errors can be due to measuring specific gravity on freshly passed urine.

3. The hydrometer should be floating freely when the reading is taken. If it touches the wall of the container false values will be obtained. The procedure should be carried out by someone experienced.

4. Sugars, protein and contrast media used in intravenous pyelography contribute to urinary specific gravity, and in their presence high readings are not necessarily indicative of normal renal function. Protein, because of its high molecular weight, contributes very little to osmolality. Glucose does, however, significantly affect it: 150 mmol/l (2·7 g/dl) of glucose in the urine adds 0·001 to the specific gravity and 150 mmol/kg to osmolality.

Chapter IV

HYDROGEN ION HOMEOSTASIS:
BLOOD GAS LEVELS

A NET amount of 50 to 100 $m$mol of hydrogen ions a day is released from cells into the 15 to 20 litres of extracellular fluid. This rate of release fluctuates throughout the day, and rapidly acting homeostatic mechanisms keep the extracellular hydrogen ion concentration almost constant, at about 40 ± 5 $n$mol/l (about pH 7·4). Hydrogen ion balance is finally achieved by secretion, mainly into the urine, and therefore *renal impairment* causes acidosis.

Aerobic metabolism of the carbon skeletons of organic compounds converts the constituent hydrogen, carbon and oxygen to water and carbon dioxide (CO_2). It does not affect hydrogen ion balance directly, but the CO_2 produced is an essential component of the buffering system. Control of CO_2 depends on normal *lung function*. Hydrogen ions are released during *metabolism of amino acids*, or by *incomplete metabolism of the carbon skeletons*.

Conversion of amino nitrogen to urea, or the sulphydryl groups of some amino acids to sulphate, releases equimolar amounts of hydrogen ions. Subjects on a high protein diet pass an acid urine. This source of hydrogen ions is of relatively little importance clinically.

Anaerobic metabolism of carbohydrates (for instance during muscular exercise) yields lactate, while anaerobic metabolism of fatty acids (for instance, during periods of fasting) and of ketogenic amino acids yields acetoacetate; both processes release equimolar amounts of H^+, either directly or indirectly. In pathological states these reactions may be so rapid that they cause a significant fall in pH resulting in lactic acidosis or ketoacidosis.

Many anabolic processes, including gluconeogenesis, utilise hydrogen ions.

Since hydrogen, and not hydroxyl, ions are produced by metabolism the tendency to acidosis is greater than to alkalosis.

DEFINITIONS

An *acid* is a substance which can dissociate to produce hydrogen ions (protons:H^+): a *base* is one which can accept hydrogen ions. Table X includes examples of acids and bases of importance in the body.

TABLE X

Acid		Conjugate Base
Carbonic acid H_2CO_3	$\leftrightharpoons H^+$	$+HCO_3^-$ Bicarbonate ion
Lactic acid $CH_3CHOHCOOH$	$\leftrightharpoons H^+$	$+CH_3CHOHCOO^-$ Lactate ion
Ammonium ion NH_4^+	$\leftrightharpoons H^+$	$+NH_3$ Ammonia
Dihydrogen phosphate $H_2PO_4^-$	$\leftrightharpoons H^+$	$+HPO_4^-$ Monohydrogen phosphate ion
Acetoacetic acid CH_3COCH_2COOH	$\leftrightharpoons H^+$	$+CH_3COCH_2COO^-$ Acetoacetate ion
β-hydroxybutyric acid $CH_3CHOHCH_2COOH$	$\leftrightharpoons H^+$	$+CH_3CHOHCH_2COO^-$ β-hydroxybutyrate ion.

An *alkali* is a substance which dissociates to produce hydroxyl ions (OH^-). Since OH^- is not a primary product of metabolism alkalis are of relatively little importance in the present discussion.

A *strong acid* is highly dissociated in aqueous solution: in other words it produces many hydrogen ions. Hydrochloric acid is a strong acid, and in solution is almost entirely in the form of $H^+ Cl^-$. However, the examples given in the above list are, chemically speaking, *weak acids*, little dissociated in water and yielding relatively few hydrogen ions. In the body even very small changes of pH are important and result in disturbances of physiology.

Buffering is the term used for the process by which a strong acid (or base) is replaced by a weaker one, with a consequent reduction in the number of free hydrogen ions (H^+); the "shock" of the hydrogen ions is taken up by the buffer with a change of pH smaller than that which would occur in the absence of the buffer.

For example:

$$H^+Cl^- \quad + \quad NaHCO_3 \rightleftharpoons H_2CO_3 \quad + \quad NaCl$$

Strong acid   Buffer   Weak acid   Neutral salt

pH is a measure of hydrogen ion activity. It was originally defined as $\log_{10}$ of the reciprocal of the hydrogen ion concentration ($[H^+]$) in mol/l; although it is now known that this definition is not strictly true, it suffices for present purposes. The $\log_{10}$ of a number is the power to which 10 must be raised to produce that number. Thus $\log 100 = \log 10^2 = 2$ and $\log 10^7 = 7$.

Let us suppose $[H^+]$ is 10^{-7} (0·000 000 1) mol/l
Then $\log [H^+] = -7$

But $$\text{pH} = \log \frac{1}{[\text{H}^+]} = -\log [\text{H}^+] = 7$$

For the non-mathematically minded only a few points need be remembered.

Since at pH 6 $[\text{H}^+] = 10^{-6}$ (0·000 001) mol/l (1 000 nmol/l)

and at pH 7 $[\text{H}^+] = 10^{-7}$ (0·000 000 1) mol/l (100 nmol/l)

a change of *one pH unit* represents a *tenfold change in* $[H^+]$. This is a much larger change than is immediately obvious from the change in pH values. Although changes of this magnitude do not occur in the body during life, in pathological conditions changes of 0·3 of a pH unit can take place. 0·3 is the log of 2. Therefore a *decrease of pH by* 0·3 (for example, from 7·4 to 7·1) represents a *doubling of* $[H^+]$ from 40 nmol/l to 80 nmol/l. Here again the use of pH makes a very significant change in $[\text{H}^+]$ appear deceptively small. (Compare the situation if the plasma sodium concentration had changed from 140 to 280 mmol/l). Urinary pH is much more variable than that in the blood: $[\text{H}^+]$ can increase 1000-fold (a fall of 3 pH units).

The Henderson–Hasselbalch equation.—We have already seen that a buffer absorbs the "shock" of the addition of H^+ to a system by replacing a strong acid by a weak one. It will be seen that when the bases in column 2 of Table X buffer H^+, the corresponding acid in column 1 is formed. This weak acid and its conjugate base form a *buffer pair*. In aqueous solution the pH is determined by the ratio of this acid to its conjugate base.

Let us take the bicarbonate pair as an example. Carbonic acid (H_2CO_3) dissociates into H^+ and HCO_3^- until equilibrium is reached (in this case very much in favour of H_2CO_3), and the ratio of the two forms will now remain constant (K). We can therefore write:

$$\text{K}[\text{H}_2\text{CO}_3] = [\text{H}^+] \times [\text{HCO}_3^-]$$

(that is, at equilibrium, the concentration of H_2CO_3 is K times that of the product of $[\text{H}^+]$ and $[\text{HCO}_3^-]$).

Transposing, $$[\text{H}^+] = \text{K} \frac{[\text{H}_2\text{CO}_3]}{[\text{HCO}_3^-]}$$

Although some laboratories now express results in terms of $[\text{H}^+]$, this practice is not yet widespread. We will use the pH notation in this edition.

$$\text{pH} = \log \frac{1}{[\text{H}^+]}$$

Taking reciprocals and logarithms in the equation for $[H^+]$ given above (when taking logs, multiplication becomes addition).

$$\text{Log}\,\frac{1}{[H^+]} = \log\frac{1}{K} + \log\frac{[HCO_3^-]}{[H_2CO_3]}$$

$$\text{Log}\,\frac{1}{K}\ \text{is called pK}$$

Therefore $\qquad$ $$pH = pK + \log\frac{[HCO_3^-]}{[H_2CO_3]}$$

This equation (an example of the Henderson–Hasselbalch equation) is valid for any buffer pair. It is important to notice that the pH depends on the *ratio* of the concentrations of base (in this case $[HCO_3^-]$) to acid (in this case $[H_2CO_3]$).

In practice it is not possible to measure the very low carbonic acid concentration directly. It is in equilibrium with dissolved CO_2, and if the carbon dioxide concentration is inserted into the equation in place of $[H_2CO_3]$, the overall dissociation constant is now that for the sum of those of the two reactions

$$K_1\,[H_2CO_3] = [H^+] \times [HCO_3^-]$$
$$\text{and } K_2\,[CO_2] \times [H_2O] = [H_2CO_3]$$

This combined constant is usually written as K' and the pK' is about 6·1. The Henderson–Hasselbalch equation for the bicarbonate system then becomes:

$$pH = 6·1 + \log\frac{[HCO_3^-]}{[CO_2]}$$

In practice the partial pressure of CO_2 gas (PCO_2) is measured in blood, and its concentration in solution in plasma is derived by multiplying PCO_2 by the solubility constant for carbon dioxide. If the PCO_2 is expressed in kilopascals (kPa) this constant is 0·23: if it is mmHg it is 0·03.

Therefore, if PCO_2 is expressed in kPa, the equation becomes

$$\boxed{pH = 6·1 + \log\frac{[HCO_3^-]}{PCO_2 \times 0·23}}$$

We shall use this form in the rest of the chapter.

HYDROGEN ION HOMEOSTASIS

The following should be noted:

hydrogen ions can be incorporated into water, maintaining normal pH.
This is the normal mechanism during oxidative phosphorylation.
H^+ is also incorporated into water during the conversion of H_2CO_3 to CO_2 and water.

$$H^+ + HCO_3^- \rightleftharpoons H_2CO_3 \rightleftharpoons CO_2 + H_2O$$

As this is a reversible reaction H^+ will only continue to be so inactivated if CO_2 is removed. This results in bicarbonate depletion.

buffering of hydrogen ions is a temporary measure.—The H^+ is still in the body, and the presence of the weak acid of the buffer pair causes a small change in pH (see the Henderson–Hasselbalch equation). If H^+ is not completely neutralised, or eliminated from the body, and if production continues, buffering power will eventually be so depleted that the pH will change significantly.

hydrogen ions can be lost from the body only through the kidney and the intestine. This mechanism is coupled with generation of bicarbonate ion (HCO_3^-). In the kidney this is the method by which secretion of excess H^+ ensures regeneration of buffering capacity.

Control Systems

Carbon dioxide and hydrogen ions are among the potentially toxic products of aerobic and anaerobic metabolism respectively. Although most CO_2 is lost through the lungs some is converted to bicarbonate, thus providing a buffering system: inactivating one toxic product provides a means of minimising the effect of the other.

The Henderson–Hasselbalch equation for any buffer pair is:

$$pH = pK + \log \frac{[base]}{[acid]}$$

A buffer pair is most effective at maintaining a pH near its pK. The optimum pH of extracellular fluid is 7·4, but the pK′ of the bicarbonate system is 6·1. Despite this apparent disadvantage bicarbonate is the most important buffer in the body, and accounts for over 60 per cent of blood buffering capacity. Moreover, the system is central to all the other important homeostatic mechanisms for dealing with hydrogen ions, including buffering by haemoglobin (which provides most of the rest of the blood buffering capacity) and secretion of hydrogen ions by the kidney.

Aerobic metabolism provides a plentiful supply of CO_2, the denominator in the equation:

$$pH = 6{\cdot}1 + \log \frac{[HCO_3{}^-]}{PCO_2 \times 0{\cdot}23}$$

The Control of CO₂ by the Lungs and Respiratory Centre

The partial pressure of CO_2 in plasma is normally about 5·3 kPa (40 mmHg). Maintenance of this level depends on the balance between production by metabolism and loss through the pulmonary alveoli.

The sequence of events is as follows:

inspired oxygen is carried from the lungs to tissues by haemoglobin;

the tissue cells utilise the oxygen for aerobic metabolism, and the carbon in organic compounds is oxidised to CO_2;

CO_2 diffuses along a concentration gradient from the cells into the extracellular fluid and is returned by the blood to the lungs, where it is eliminated in expired air;

the rate of respiration, and therefore the rate of CO_2 elimination, is controlled by chemoreceptors in the hypothalamic respiratory centre, which respond to the $[CO_2]$ (or pH) of the circulating blood. If the PCO_2 rises much above 5·3 kPa (or if the pH falls) the rate of respiration rises. Normal lungs have a very large reserve capacity for eliminating CO_2.

We now see that not only is there a plentiful supply of CO_2, but that the normal respiratory centre and lungs can control its level within narrow limits, so controlling the denominator in the Henderson–Hasselbalch equation.

Disease of the lungs, or abnormality of respiratory control, will primarily affect the PCO_2.

The Control of Bicarbonate by the Kidneys and Erythrocytes

The renal tubular cells and erythrocytes use some of the CO_2 retained by the lungs to form bicarbonate, and so control the numerator in the Henderson–Hasselbalch equation. Under physiological conditions erythrocytes make fine adjustments to the plasma bicarbonate level in response to changes in PCO_2 in the lungs and tissues: the kidney plays the major role in maintaining the circulating bicarbonate concentration.

The Carbonate Dehydratase System

Carbonate dehydratase (carbonic anhydrase; CD) catalyses the first reaction in the chain:

$$CO_2 + H_2O \xrightarrow{CD} H_2CO_3 \rightarrow H^+ + HCO_3^-$$

Not only do erythrocytes and renal tubular cells have a high concentration of CD, but they also have a means of removing one of the products, H^+; thus both reactions continue to the right, and HCO_3^- will be produced. As one of the reactants—water—is freely available, and because one of the products—H^+—is being removed, HCO_3^- production would be accelerated either by a rise in the intracellular concentration of the other reactant—CO_2—or by a fall in the intracellular concentration of the other product—HCO_3^-.

At a plasma PCO_2 of 5·3 kPa (a CO_2 concentration of about 1·2 mmol/l—see p. 92) these two tissues maintain the extracellular bicarbonate concentration at about 25 mmol/l in the normal subject. The extracellular ratio of $[HCO_3^-]:[CO_2]$ (both in mmol/l) is just over 20:1. It can be calculated from the Henderson–Hasselbalch equation that, with a pK' of 6·1, this ratio represents a pH very near 7·4. An increase of intracellular PCO_2, or a decrease in intracellular $[HCO_3^-]$ accelerates production of HCO_3^-, and minimises changes in the *ratio*, and therefore changes in pH.

Bicarbonate Generation by Erythrocytes (Fig. 3)

Haemoglobin is an important blood buffer.

$$pH = pK + \log \frac{[Hb^-]}{[HHb]}$$

However, it only works effectively in the body in co-operation with the bicarbonate system.

Because erythrocytes lack aerobic pathways they produce relatively little CO_2. Plasma CO_2 diffuses into the cell along a concentration gradient, where carbonate dehydratase catalyses its reaction with water to form carbonic acid. As H_2CO_3 dissociates, much of the H^+ is buffered by haemoglobin. The concentration of HCO_3^- in the erythrocyte rises, and it diffuses into the extracellular fluid along a concentration gradient; electrochemical neutrality is maintained by diffusion of chloride into the cell (the "chloride shift").

Under *physiological conditions*, the higher PCO_2 in blood leaving tissues stimulates erythrocyte HCO_3^- production, and the lower PCO_2 in blood leaving the lungs slows it down; the arteriovenous difference in the ratio $[HCO_3^-]:[CO_2]$, and therefore the pH, is kept

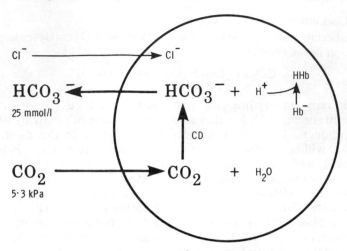

FIG. 3.—Generation of bicarbonate by the erythrocyte.

relatively constant. There is little effect on the overall HCO_3^- "balance".

In *pathological conditions* affecting PCO_2 the same mechanisms operate: a high intracellular PCO_2 tends to cause a net increase, and a low one a net reduction in extracellular $[HCO_3^-]$. However, because the buffering capacity of haemoglobin is limited the erythrocyte can only make a small contribution to homeostasis in chronic disturbances of H^+ balance.

The Kidneys

Carbonate dehydratase is also of central importance in the renal mechanisms involved in hydrogen ion homeostasis. The hydrogen ion is secreted from the renal tubular cell into the tubular lumen, where it is buffered by constituents of the glomerular filtrate. By contrast with the haemoglobin in the erythrocyte, these buffers are constantly being replenished by continuing glomerular filtration. *For this reason, and because excess H^+ can only be eliminated from the body by the renal route, the kidneys are of major importance in compensating for chronic acidosis.* Without them haemoglobin buffering capacity would soon become saturated.

There are two renal mechanisms controlling $[HCO_3^-]$ in the extracellular fluid:

bicarbonate "*reabsorption*", the predominant mechanism in the maintenance of the steady state. As we shall see, the CO_2 driving the carbonate dehydratase mechanism in the renal tubular cell is derived from filtered bicarbonate in the tubular lumen, and there is no *net* loss of hydrogen ions;

bicarbonate generation, a very important mechanism for correcting acidosis, in which the level of CO_2 stimulating the carbonate dehydratase reaction in the renal tubular cell reflects that in the extracellular fluid. There *is* net loss of hydrogen ions. This mechanism is also stimulated by a fall in extracellular $[HCO_3^-]$.

Bicarbonate "reabsorption" (Fig. 4).—Normal urine is almost bicarbonate free.

The luminal surfaces of the renal tubular cells are impermeable to bicarbonate, which therefore cannot be reabsorbed directly. It must first be converted to CO_2 in the tubular lumen, and an equivalent amount of CO_2 is converted to bicarbonate within the tubular cell. The mechanism depends on the action of carbonate dehydratase within the tubular cell, and on H^+ secretion from the cell into the

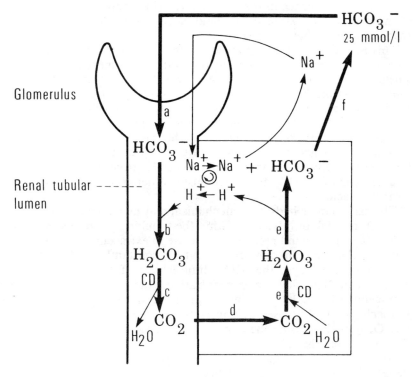

FIG. 4.—"Reabsorption" of filtered bicarbonate by the renal tubular cell.

tubular lumen in exchange for the sodium filtered with the bicarbonate. The sequence of events is shown in Fig. 4.

(a) Bicarbonate passes through the glomerular filter at plasma concentration (about 25 mmol/l).

(b) Filtered bicarbonate combines with H^+ secreted by the tubular cell to form H_2CO_3.

(c) The H_2CO_3 dissociates to CO_2 and water. In the proximal tubule this reaction is catalysed by carbonate dehydratase on the luminal membrane of the tubular cells: in the distal nephron, where the pH is usually lower, it probably dissociates spontaneously.

(d) As the luminal PCO_2 rises, CO_2 diffuses into the tubular cell along a concentration gradient.

(e) As the intracellular $[CO_2]$ rises carbonate dehydratase catalyses its combination with water to reform carbonic acid, which dissociates into H^+ and HCO_3^-.

(f) As H^+ is secreted (and so starts the reactions from (b) again) the intracellular concentration of HCO_3^- rises, and the bicarbonate diffuses into the extracellular fluid accompanied by the sodium reabsorbed in exchange for H^+.

This is a self-perpetuating cycle which reclaims buffering capacity which would otherwise be lost to the body by glomerular filtration. The secreted H^+ is derived from cellular water, and is reincorporated into water in the lumen. Because there is no net change in hydrogen ion balance and no net gain of bicarbonate, this mechanism *cannot correct an acidosis*, but can *maintain a steady state*.

Bicarbonate generation (Fig. 5).—The mechanism in the renal tubular cell for generating bicarbonate is identical to that of bicarbonate reabsorption, but there *is* net loss of H^+ from the body, as well as a net gain of HCO_3^-; this mechanism is therefore ideally suited to correct an acidosis.

The carbonate dehydratase mechanism may be stimulated by a rise in PCO_2 or a fall of $[HCO_3^-]$ within the tubular cell. In this case the rise in $[CO_2]$ is the indirect result of a rise in extracellular PCO_2. The renal tubular cell, unlike the erythrocyte, is constantly producing CO_2 by aerobic pathways: this CO_2 diffuses out of the cell into the extracellular fluid down a concentration gradient. An increase in extracellular PCO_2, by reducing the gradient, slows this diffusion, and intracellular PCO_2 rises. Conversely, reduction of extracellular $[HCO_3^-]$, by increasing the concentration gradient, increases loss of this anion from the cell.

Normally all filtered bicarbonate is "reabsorbed" by the mechanism described above: once all bicarbonate has been "reabsorbed," net secretion of H^+ and net generation of HCO_3^- depend on the presence of other filtered buffers (B^- in Fig. 5). These buffers, unlike bicarbon-

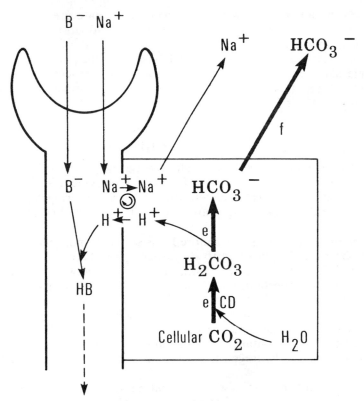

FIG. 5.—Net generation of bicarbonate by the renal tubular cell.
(B^- = non-bicarbonate base.)

ate, do not form compounds capable of diffusing into tubular cells; nor is the hydrogen ion incorporated into water. *The H^+ is lost in the urine* as HB. The bicarbonate formed in the cell is derived from cellular CO_2, not luminal bicarbonate, and therefore represents a *net gain in bicarbonate*. As usual, when a mmol of H^+ is secreted into the urine, a mmol of HCO_3^- passes into the extracellular fluid with sodium.

This mechanism is very similar to that in erythrocytes, but unlike the red cell, the renal tubular cell is not carried between lung and peripheral tissues: it is therefore exposed to a relatively constant PCO_2. Bicarbonate generation coupled with hydrogen ion secretion becomes very important in acidosis, when it is stimulated by a fall in extracellular $[HCO_3^-]$ (metabolic acidosis) or a rise in extracellular PCO_2 (respiratory acidosis).

Urinary buffers.—Buffers other than bicarbonate are involved in bicarbonate generation with H^+ secretion. The two most important of these are phosphate and ammonia.

The phosphate buffer pair.—At pH 7·4 most of the phosphate in the glomerular filtrate is in the form of monohydrogen phosphate ($HPO_4^=$), and this can accept H^+ to become dihydrogen phosphate ($H_2PO_4^-$). The pK of this pair is 6·8.

$$pH = 6·8 + \log \frac{[HPO_4^=]}{[H_2PO_4^-]}$$

Even in mild acidosis bone salts release more free calcium and phosphate ions than at normal pH, and the requirement for increased urinary secretion of H^+ is linked with increased buffering capacity in the glomerular filtrate, due to an increase of phosphate liberated from bone.

The role of ammonia.—As more H^+ is secreted, more monohydrogen phosphate is converted to dihydrogen phosphate until, at a pH below 5·5, most is in this form. In severe acidosis bicarbonate generation is important, but urinary phosphate cannot maintain the essential continued hydrogen ion secretion. The predominant urinary anion, chloride, cannot buffer hydrogen ion (hydrochloric acid is almost entirely in the ionised form). As the urine becomes more acid it can be shown to contain more ammonium ion, and urinary ammonia probably enables hydrogen ion secretion and bicarbonate formation to continue.

The enzyme glutaminase is present in renal tubular cells and catalyses the hydrolysis of the terminal amino group of glutamine ($GluCONH_2$) to form glutamate ($GluCOO^-$) and the ammonium ion.

$$H_2O + GluCONH_2 \rightarrow GluCOO^- + NH_4^+$$

Ammonia and ammonium ion form a buffer pair

$$pH = pK + \log \frac{[NH_3]}{[NH_4^+]}$$

but the pK of the system is about 9·8 and at pH 7·4 the equilibrium is overwhelmingly in favour of NH_4^+. NH_3 can diffuse out of the cell into the urine much more rapidly than NH_4^+, and, if the urine is acid, will be prevented from returning by avid combination with secreted H^+ derived from the carbonate dehydratase mechanism. This allows hydrogen ion to be excreted as ammonium chloride: thus, in severe acidosis, bicarbonate formation can continue even when the phosphate buffering power has been exhausted. Dissociation of NH_4^+ in the cell is maintained by removal of NH_3 into the urine. However, this dissociation liberates H^+ in the cell. On the face of it there seems no advantage in buffering one secreted H^+ in the urine if, at the same time, one is produced in the cell.

A possible explanation is to be found in the fate of the glutamate ($GluCOO^-$) produced at the same time as the ammonium ion. After further deamination to α-oxoglutarate it can be converted to glucose by gluconeogenesis—*a process which utilises an equivalent amount of H^+.* Thus the H^+ liberated into the cell may be incorporated into glucose.

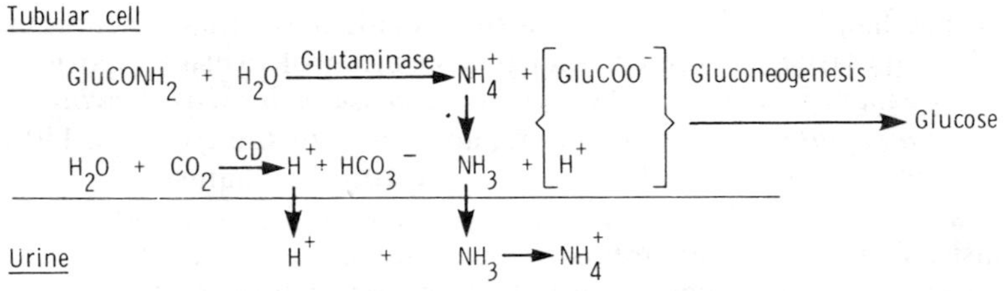

Tubular cell

As usual the net result is a gain of HCO_3^-.

Both glutaminase activity and gluconeogenesis have been shown to be increased in acidosis.

Bicarbonate Formation in the Gastro-intestinal Tract

Intestinal mucosal cells form bicarbonate by the carbonate dehydratase mechanism. The bicarbonate may enter either the extracellular fluid or the intestinal lumen: in either case the mechanism can only continue if H^+ is pumped in the opposite direction. Electrochemical neutrality is maintained by one of two mechanisms:

Na^+ exchange for H^+, by a mechanism which is the reverse of that in the renal tract;
passage of Cl^- with H^+.

Acid secretion by the stomach.—The parietal cells of the stomach secrete H^+ into the lumen together with Cl^-. As $H^+ Cl^-$ enters the gastric lumen, the gain of HCO_3^- by the extracellular fluid accounts for the postprandial "alkaline tide". In the normal subject this is rapidly corrected by bicarbonate secretion as food continues down the intestine. This mechanism is of importance in explaining the *metabolic alkalosis due to pyloric stenosis* (p. 113).

Sodium bicarbonate secretion by pancreatic and biliary cells.—Duodenal fluid is alkaline because of the high bicarbonate content of secretions entering it from the common bile duct. Sodium bicarbonate secretion by pancreatic and biliary cells is the reverse of sodium bicarbonate reabsorption by renal tubular cells (p. 98). The pancreatic and biliary mechanisms are stimulated by the local rise in PCO_2 when H^+ is pumped from the glandular cells into the extracellular fluid and reacts with the HCO_3^- generated by the gastric parietal cells. This is analogous to the stimulation of the renal bicarbonate formation by the rise in luminal PCO_2. *Loss of large amounts of duodenal fluid may cause bicarbonate depletion* (p. 109).

Bicarbonate secretion and chloride reabsorption by intestinal cells.—As the fluid passes down the intestine, bicarbonate enters and chloride leaves the lumen by a reversal of the gastric mucosal mechanism. Thus the gastric loss of chloride and gain of bicarbonate is finally corrected. Preferential reabsorption of urinary chloride by this mechanism explains the *hyperchloraemic acidosis following ureteric transplantation* into the ileum, ileal loops or the colon (p. 110).

Figure 6 summarises these gastro-intestinal mechanisms.

In summary, CO_2 is of central importance in hydrogen ion homeostasis. Despite the apparently unfavourable pK of the bicarbonate buffer system, the tendency of H_2CO_3 to form volatile CO_2, the partial pressure of which can be controlled by the respiratory centre and lungs to about 5·3 kPa, and the ability of the renal tubular cells and erythrocytes to maintain the $[HCO_3^-]$ at 25 mmol/l at this PCO_2, enable the pH to be kept above the pK of the system. Dysfunction of either kidneys or lungs impairs control of extracellular pH. Inhibition of carbonate dehydratase activity (p. 111) decreases bicarbonate formation by erythrocytes and renal tubular cells, and bicarbonate reabsorption from the glomerular filtrate, and causes bicarbonate depletion.

Buffers in the body other than bicarbonate and haemoglobin include plasma and tissue proteins; intracellular proteins have an important local role in buffering, but plasma proteins play a very minor part in hydrogen ion homeostasis. Phosphate, with a pK of 6·8, is a very important buffer in the urine, in which concentrations of 25 mmol/l may be reached: its plasma concentration is only about 1 mmol/l and it does not contribute significantly to blood buffering.

DISTURBANCES OF HYDROGEN ION HOMEOSTASIS

Disturbances of hydrogen ion homeostasis involve the bicarbonate buffer pair. In "respiratory" disturbances abnormalities of CO_2 are primary, while in so-called "metabolic" disturbances $[HCO_3^-]$ is affected early and changes in CO_2 are secondary.

Measurements Used to Assess Hydrogen Ion Balance

Measurement of blood pH indicates only whether there is overt

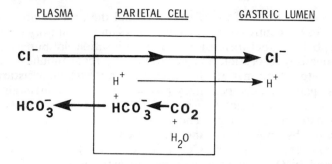

Net effect on plasma – Gain of bicarbonate and loss of chloride

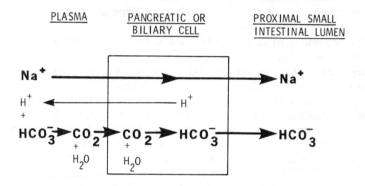

Net effect on plasma – Loss of sodium bicarbonate

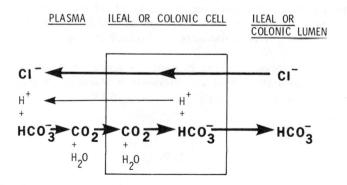

Net effect on plasma – Loss of bicarbonate and gain of chloride

FIG. 6.—Possible mechanisms for gastro-intestinal handling of bicarbonate.

acidosis or alkalosis. If the pH is abnormal the primary abnormality may be in the control of CO_2 by the lungs and respiratory centre, or in the balance between bicarbonate utilisation in buffering and bicarbonate reabsorption and generation by renal tubular cells and erythrocytes. A normal pH, however, does not exclude a disturbance of these pathways: compensatory mechanisms may be maintaining it. Assessment of these factors is made by measuring components of the bicarbonate buffer system. The concentration of dissolved CO_2 is calculated by multiplying the measured PCO_2 by the solubility constant of the gas (0·23 if PCO_2 is in kPa; 0·03 if it is in mmHg).

pH and PCO_2.—There is a significant arteriovenous difference in pH and PCO_2, and measurement of both must be made on arterial blood. Whole blood (heparinised) must be used because the result of PCO_2 estimation by some methods may depend on the presence of erythrocytes.

The technique of sample collection is important, and details of this are given on p. 124.

Measurement of blood [HCO₃⁻].—One of two methods may be used to estimate circulating bicarbonate levels.

Plasma total CO_2 (TCO₂): "plasma bicarbonate concentration".— This is probably the most commonly measured index of hydrogen ion homeostasis and is easily performed on plasma: an adequate amount of sample is needed if the result is to be valid (p. 121). If pH and PCO_2 are not required it has the advantage that venous blood can be used. It is an estimate of the sum of plasma bicarbonate, carbonic acid and dissolved CO_2. At pH 7·4 the ratio of [HCO₃⁻] to the other two components is about 20 to 1 and at pH 7·1 is still 10 to 1. Thus, if the TCO₂ were 21 mmol/l, [HCO₃⁻] would contribute 20 mmol/l at pH 7·4 and just over 19 mmol/l at pH 7·1. Only 1 mmol/l and just under 2 mmol/l respectively would come from $H_2CO_3 + CO_2$. Thus TCO₂ is effectively a measure of plasma bicarbonate concentration.

"Actual" bicarbonate.—This estimate is made on whole arterial blood. It is calculated from the Henderson-Hasselbalch equation, using the measured values of pH and PCO_2 (in kPa).

$$pH = 6·1 + \log \frac{[HCO_3{}^-]}{PCO_2 \times 0·23}$$

It represents whole blood [HCO₃⁻] and for reasons discussed above, usually agrees well with plasma TCO₂. It is the estimate of choice if the other two parameters are being measured.

ACIDOSIS

Acidosis is due to fall in the ratio $[HCO_3^-]$:PCO_2 in the extracellular fluid.

$$pH = 6\cdot1 + \log \frac{[HCO_3^-]}{PCO_2 \times 0\cdot23}$$

In *metabolic (non-respiratory) acidosis* the primary abnormality in the bicarbonate buffer system is a *reduction in* $[HCO_3^-]$.

In *respiratory acidosis* the primary abnormality in the bicarbonate buffer system is a *rise in* PCO_2.

In either metabolic or respiratory acidosis the ratio of $[HCO_3^-]$:PCO_2, and therefore the pH, can be corrected by a change in concentration of the other member of the buffer pair in the same direction as the primary abnormality. This *compensation* may be partial or complete. The compensatory change in a metabolic acidosis is a reduction in PCO_2: in a respiratory acidosis it is a rise in $[HCO_3^-]$.

In a fully compensated acidosis the pH is normal. However, the levels of the other components of the Henderson–Hasselbalch equation are abnormal. Full *correction* of all parameters can only occur if the primary abnormality is corrected.

Metabolic Acidosis

The primary abnormality in the bicarbonate buffer system in a metabolic acidosis is a reduction in $[HCO_3^-]$ which, by reducing the ratio $[HCO_3^-]$:$[CO_2]$ causes a fall in pH. The bicarbonate may be lost in the urine or gastro-intestinal tract, may fail to be generated, or may be utilised in buffering H^+. In the face of a reduction in the number of negatively charged ions electrochemical neutrality might be maintained by replacing it with an equivalent number of other anion(s), or by loss of an equivalent number of cation(s). If we classify metabolic acidosis according to these alternatives many of the associated findings can be explained. The value of chloride estimation can also be assessed.

Because one negative charge balances one positive charge (1 equivalent of cation balances 1 equivalent of anion), and because moles may be multivalent (have more than one charge per mole), milliequivalents (mEq) rather than mmol must be used in any calculation of anion/cation balance. In the normal subject over 80 per cent of extracellular anions is accounted for by chloride and bicarbonate: the remaining 20 per cent or so (sometimes referred to as "unmeasured anion") is made up of proteins, together with the normally low concentrations of, for example, urate, phosphate, sulphate, lactate and other organic anions. The protein concentration remains relatively constant, but the levels of the other unmeasured anions can vary considerably in disease.

Sodium and potassium provide over 90 per cent of plasma cation concentration in the normal subject; the balance includes low concentrations of calcium and magnesium, which are not usually measured with sodium and potassium, and which, as we have seen (p. 35), vary very little even in pathological states.

The difference between the total concentration of *measured* cations, sodium and potassium, and that of *measured* anions, chloride and bicarbonate, is sometimes called the "anion gap": it is normally about 15 to 20 mEq/l. If we represent the unmeasured anion as A^-, we can express the normal situation as:

$$[Na^+]+[K^+] = [HCO_3^-]+[Cl^-]+[A^-]$$
$$140 + 4 = 25 + 100 + 19 \text{ mEq/l}$$

The anion gap, due to $[A^-]$, is 19 mEq/l.

The changes in anion and cation concentrations which coincide with the low $[HCO_3^-]$ of metabolic acidosis are summarised in Table XI.

In the following examples we shall assume a fall in $[HCO_3^-]$ of 10 mmol/l (10 mEq/l).

1. Increase in $[A^-]$.—In **renal glomerular failure** with normal tubular function bicarbonate generation is impaired by a reduction in the amount of sodium available to exchange for H^+, and in the

TABLE XI

CHANGES IN CONCENTRATION OF IONS BALANCING FALL IN $[HCO_3^-]$ IN
METABOLIC ACIDOSIS

Clinical examples	Measured Cation [Na$^+$]	Measured Anion [Cl$^-$]	Unmeasured anion Mixture [A$^-$]	Unmeasured anion Single [X$^-$]	"Anion Gap"
Glomerular failure	N	N	↑	—	↑
Ketoacidosis and lactic acidosis	N	N	—	↑ (Lactate or ketoacid)	↑
Intestinal loss or renal tubular failure	←————————VARIABLE————————→				
Ureteric transplantation, acetazolamide, renal tubular acidosis, NH$_4$Cl	N	↑	N	N	N

The last group is the rarest.

IT MUST BE STRESSED THAT THESE ARE THE CHANGES IN UNCOMPLICATED CASES. IN MANY CLINICAL SITUATIONS THERE IS MORE THAN ONE ABNORMALITY.

amount of buffer anion, B^- (Fig. 5), to accept H^+ (p. 99). These buffer anions are components of the unmeasured anion (A^-). For each mEq retained, 1 mEq fewer H^+ can be secreted, and therefore 1 mEq fewer HCO_3^- is generated. The retained A^- therefore replaces HCO_3^-. There is no change in chloride in the uncomplicated case. (Abnormal figures underlined).

$$[Na^+]+[K^+] = [HCO_3^-]+[Cl^-]+[A^-]$$
$$140 + 4 = \underline{15} + 100 + \underline{29} \text{ mEq/l}$$

In this example, as the $[HCO_3^-]$ has fallen from 25 to 15, $[A^-]$ has risen from 19 to 29 mEq/l. The anion gap (entirely due to $[A^-]$) has risen by the same amount.

In generalised renal failure, direct impairment of tubular mechanisms further reduces bicarbonate generation; electrochemical neutrality is maintained by loss of other ions.

If renal bicarbonate generation is so impaired that it cannot keep pace with its peripheral utilisation the pH will fall.

$$pH\downarrow = 6\cdot1+\log \frac{[HCO_3^-]\downarrow}{PCO_2\times0\cdot23}$$

Compensation occurs as the respiratory centre responds to the acidosis. CO_2 is lost through the pulmonary alveoli and the pH returns towards normal. *In the compensated case the PCO_2 is low.*

The cause of the low $[HCO_3^-]$ is usually obvious if plasma urea levels are estimated.

Correction can only occur when the *GFR increases* (for example, by correction of dehydration). Treatment of the acidosis of irreversible glomerular failure, other than by dialysis, is usually contra-indicated because of the danger of giving sodium salts of alkalis to a patient in whom sodium excretion is impaired; it is, in any case, rarely necessary.

2. Increase in a single anion (X^-) other than chloride.—Such anions are:

acetoacetate and β-hydroxybutyrate in **ketoacidosis**;
lactate in **lactic acidosis.**

In both these syndromes the increase in $[X^-]$ is due to overproduction rather than to underexcretion: in both there is simultaneous production of H^+. Effectively, therefore, acids (H^+X^-) are being added to the body. The reduction in $[HCO_3^-]$ is the result of its utilisation in buffering the H^+ which accompanies X^-.

$$HCO_3^-+H^+\rightarrow H_2CO_3\rightarrow CO_2+H_2O$$

and X^- replaces HCO_3^-.

$$[Na^+]+[K^+] = [HCO_3^-]+[Cl^-]+[A^-]+[X^-]$$
$$140 + 4 = \underline{15} + 100 + \underline{19} + \underline{10} \quad mEq/l$$

"Anion gap"

There is no change in chloride in the uncomplicated case.

In this example, as the $[HCO_3^-]$ has fallen from 25 to 15 mEq/l, it has been replaced by 10 mEq/l of X^-. The anion gap is now the sum of $[A^-]$ and $[X^-]-29$ mEq/l.

In the *uncomplicated* case the renal and erythrocyte mechanisms are functioning normally. Bicarbonate reabsorption from the glomerular filtrate is complete.

The falling $[HCO_3^-]$ in the ECF accelerates its generation in the renal tubular cell and erythrocyte (Figs. 3 and 5), and it is only when this fails to keep pace with utilisation that $[HCO_3^-]$ becomes abnormal. The urine will become acid as the rate of H^+ secretion linked to bicarbonate generation increases.

By this ingenious mechanism the additional, and potentially toxic, H^+ is converted to water after it is buffered by HCO_3^-; it is therefore inactivated near the site of production. Repletion of bicarbonate involves reutilisation of water; this again liberates H^+, but this time in the renal tubular cells—at a site where it is immediately eliminated from the body. Water is being utilised as an inactive carrier for hydrogen from the site of production to the site of secretion (Fig. 7), although it is, of course, not the "same" water nor the "same" CO_2 that is used. Water is freely available and CO_2 is constantly being produced in the tubular cell.

The CO_2 derived from buffering by bicarbonate is lost through the lungs as the respiratory centre is stimulated by the acidosis; because of the large reserve capacity of the lungs, *the blood PCO_2 is never raised if pulmonary function is normal.* The respiratory centre continues to respond to the low pH until the PCO_2 falls low enough to correct

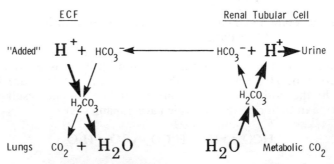

FIG. 7.—Hydrogen ion "shuttle" between site of buffering and kidneys.

the ratio of $[HCO_3^-]$:PCO_2, when the acidosis is *compensated*. Full correction by the kidney is only possible if the rate of H^+ production is reduced to such a level that generation can keep pace with utilisation.

Diagnosis is usually obvious. The commonest cause of *ketoacidosis* is uncontrolled diabetes (p. 207), and the finding of a *high blood glucose concentration* suggests the cause of a low $[HCO_3^-]$; occasionally starvation ketosis can be severe enough to cause acidosis.

The commonest cause of *lactic acidosis* is interference with aerobic metabolism by tissue hypoxia in the *shocked hypotensive patient*; the cause of the acidosis will usually be obvious on clinical grounds. Drugs such as *phenformin* or an overdosage of *salicylates* may also interfere with lactate metabolism and cause lactic acidosis (pp. 201 and 115). A drug history should be sought if a low $[HCO_3^-]$ is found for no immediately obvious reason.

Lactic acidosis can complicate diabetic ketoacidosis if tissue perfusion is poor. In both lactic acidosis and ketoacidosis, volume depletion and hypotension can reduce the GFR so that acidosis is further aggravated by renal factors. It is usually impossible to determine the relative contribution that these different factors make to the fall in $[HCO_3^-]$. However, in practical terms this is not necessary. *Treatment* is directed to improving tissue perfusion by rehydration with the appropriate fluid, and by other measures designed to restore normal blood pressure. Diabetic ketoacidosis should be treated with insulin. If acidosis is very severe intravenous administration of isosmolar, or even hyperosmolar, bicarbonate solution may be indicated: this therapy should be used with caution because of the danger of precipitating plasma hyperosmolality, and hypokalaemia (pp. 59 and 70). Respiratory dysfunction, which impairs compensatory CO_2 elimination, must also be treated.

3. Loss of a mixture of anions and cations.—Loss of intestinal secretions.—Duodenal fluid is alkaline, with a bicarbonate concentration about twice that of plasma. Excessive loss through fistulae may reduce plasma $[HCO_3^-]$ and cause acidosis. Electrochemical neutrality is maintained by loss of anions and cations in equivalent amounts. The plasma electrolyte concentrations depend initially on the composition of the fluid lost, but mainly on the composition of any replacement fluid. Renal generation of HCO_3^- will be accelerated by the falling concentration, and it is only when intestinal loss is very rapid that $[HCO_3^-]$ becomes abnormal. However, if volume depletion is severe enough to reduce the GFR, renal mechanisms may also be impaired and precipitate overt acidosis.

The *diagnosis* is usually obvious on clinical grounds.

Treatment is to restore fluid volume by administering the appropriate solution (p. 65). Usually, if an adequate GFR is maintained, the kidneys will correct the acidosis without bicarbonate administration: in very severe loss it may be necessary to give some bicarbonate.

Renal tubular failure may cause a loss of a similar mixture of ions in the urine. Renal HCO_3^- reabsorption and generation is impaired due to damage to the tubular mechanisms for H^+ secretion. The *cause* of the low $[HCO_3^-]$ is *suggested by polyuria*.

4. Increase in Cl^-.—In groups 1 and 2 the reduction in $[HCO_3^-]$ is compensated for by a rise in unmeasured anion, and $[Cl^-]$ is not affected in uncomplicated cases. In group 3 $[Cl^-]$ is variably affected. The combination of a low $[HCO_3^-]$ and high $[Cl^-]$ is known as "hyperchloraemic acidosis" and is rare; it can usually be predicted on clinical grounds. The "anion gap" in such cases is normal.

$$[Na^+]+[K^+] = [HCO_3^-]+[Cl^-]+[A^-]$$
$$140 + 4 = \underline{15} + \underline{110} + 19 \text{ mEq/l}$$

(*a*) Such a combination would be expected if HCO_3^- were lost in a one-to-one *exchange* for chloride. This occurs if the **ureters are transplanted into the ileum or colon.** The cells of the ileum and colon, like those of the renal tubules, are capable of active transport of ions. Normally reabsorption of water and electrolytes in this part of the intestinal tract is almost complete: however, if fluid which contains chloride enters the lumen, the cells reabsorb some of this chloride in exchange for HCO_3^-. Bicarbonate depletion may therefore occur if urine is delivered into the ileum (or ileal loops) or colon, as after transplantation of the ureters to these sites. This operation may be performed, with total cystectomy, for carcinoma of the bladder. Unless large doses of oral bicarbonate are given, the result is a very low plasma HCO_3^- and very high plasma chloride concentrations.

(*b*) Hydrogen ion secretion, and consequent bicarbonate reabsorption and generation, are affected in tubular disease of any kind. However, if the tubular ability to handle ions other than H^+ is unaffected, hyperchloraemic acidosis results.

Normally most of the filtered sodium is reabsorbed with chloride. Any further sodium returned to the extracellular fluid is exchanged for secreted H^+ or K^+. If H^+ secretion is impaired, and if the same amount of sodium is reabsorbed, then it must either be accompanied by Cl^- or be exchanged for K^+. *This type of hyperchloraemic acidosis is therefore often accompanied by hypokalaemia*—an unusual finding in any other type of acidosis, when hyperkalaemia is the rule.

The two causes of this type of acidosis are:

renal tubular acidosis;
administration of carbonate dehydratase inhibitors.

Renal tubular acidosis.—The group of conditions giving rise to this syndrome is usually the result of acquired tubular lesions; there is a failure to acidify the urine normally, even after ingestion of an acid load such as ammonium chloride (p. 124). There may be either an impairment of the H^+ secreting mechanism itself, or an abnormal permeability of the distal tubular wall to the secreted H^+, allowing its diffusion back into the blood: unlike the situation in generalised tubular failure, the tubular cells retain the ability to form ammonia. Because there is no primary glomerular lesion the plasma urea and creatinine levels are often normal. However, prolonged acidosis increases the release of free ionised calcium from bone (p. 252): this calcium is often precipitated in the renal tubules and the subject may present with uraemia due to nephrocalcinosis and fibrosis. The increased breakdown of bone salts also partially explains the phosphaturia often accompanying renal tubular acidosis.

Acetazolamide therapy.—Acetazolamide is a drug, used in the treatment of glaucoma, which inhibits the action of carbonate dehydratase. Because H^+ secretion by the renal tubular cell and generation of HCO_3^- by the erythrocyte both depend on the action of carbonate dehydratase (Figs. 3, 4 and 5), bicarbonate reabsorption and generation are impaired with resultant metabolic acidosis.

To summarise the findings in metabolic acidosis:

[HCO_3^-] *always low*;
PCO_2 usually low (compensatory change);
pH low (uncompensated or partially compensated) or normal (fully compensated);
chloride concentration unaffected in most cases; raised after transplantation of the ureters, administration of acetazolamide, or in renal tubular acidosis.

Tests which may help to elucidate the cause of a metabolic acidosis are:

plasma urea estimation;
plasma glucose estimation;
urine or blood testing for ketones.

The indications for chloride estimations are discussed on p. 122.

Respiratory Acidosis

The findings in respiratory acidosis are significantly different from those in non-respiratory disturbances.

The primary defect is retention of CO_2, usually due to generalised respiratory impairment (p. 118). The consequent *rise in PCO_2 is the* *constant finding in respiratory acidosis.* As in the metabolic disturbance, the acidosis is due to the fall in the ratio of $[HCO_3^-]$: PCO_2.

$$pH\downarrow = 6{\cdot}1 + \log \frac{[HCO_3^-]}{PCO_2\uparrow \times 0{\cdot}23}$$

Compensatory changes in $[HCO_3^-]$ are initiated by the acceleration of the carbonate dehydratase mechanism in erythrocytes and renal tubular cells by the high PCO_2; $[HCO_3^-]$ generation is speeded up, tending to compensate for the raised PCO_2 (Figs. 3 and 5). The urine becomes acid as more H^+ is secreted.

In **acute respiratory failure** (for instance due to bronchopneumonia or status asthmaticus) the relatively rapid erythrocyte mechanism makes the major contribution to the slight rise in $[HCO_3^-]$. This degree of compensation is often inadequate to prevent a fall in pH.

In **chronic respiratory failure** (for instance due to chronic obstructive airways disease) the renal tubular mechanism is of prime importance. Haemoglobin buffering power is of limited capacity, but, so long as the glomerular filtrate provides an adequate supply of sodium and of buffers to maintain H^+ secretion, tubular cells continue to generate bicarbonate until the ratio $[HCO_3^-]$: PCO_2 is normal. Bicarbonate levels as high as twice normal may sometimes be found with a normal pH in stable chronic respiratory failure.

The findings in the arterial blood in respiratory acidosis are:

    PCO_2 always raised

In *acute* respiratory failure

    pH low

    $[HCO_3^-]$ high normal or slightly raised.

In *chronic* respiratory failure

    pH normal or low depending on severity

    $[HCO_3^-]$ raised.

Mixed Metabolic and Respiratory Acidosis

If there is metabolic acidosis, and CO_2 retention is also present, some of the compensatory increase in blood $[HCO_3^-]$ is used to buffer acid other than H_2CO_3. The rise of $[HCO_3^-]$ is impaired, more CO_2 is produced, and the pH falls to lower levels than during CO_2 retention alone. This situation most commonly occurs in the "respiratory distress syndrome" of the newborn, when there is added lactic acidosis due to hypoxia; it may also be due to coexistence of respiratory failure with, for example, ketoacidosis or renal failure. In such cases the "actual" bicarbonate (and TCO_2) concentration may be high, normal or low, depending on the relative contributions from the respiratory and metabolic components.

ALKALOSIS

Alkalosis is due to a rise in the ratio $[HCO_3^-]$: PCO_2 in the extracellular fluid.

$$pH = 6 \cdot 1 + \log \frac{[HCO_3^-]}{PCO_2 \times 0 \cdot 23}$$

In *metabolic alkalosis* the primary abnormality in the bicarbonate buffer system is a *rise in* $[HCO_3^-]$. There is little, if any, compensatory change in PCO_2.

In *respiratory alkalosis* the primary abnormality in the bicarbonate buffer system is a *fall in* PCO_2. The compensatory change is a fall in $[HCO_3^-]$.

The primary products of metabolism are hydrogen ions and CO_2, not hydroxyl ions and HCO_3^-; alkalosis is therefore less common than acidosis.

Alkalosis may present clinically as tetany in spite of normal total plasma calcium levels: this is due to a reduced amount of free ionised calcium in a relatively alkaline medium (p. 252).

Metabolic Alkalosis

A primary rise in plasma $[HCO_3^-]$ may occur in three situations:

Administration of bicarbonate.—Ingestion of large amounts of alkali, usually sodium bicarbonate, in the treatment of "indigestion", or during intravenous bicarbonate infusion. Usually the cause and treatment are both obvious.

Generation of bicarbonate by the kidney in potassium depletion.— This is discussed more fully on p. 71.

Generation of bicarbonate by the gastric mucosa when hydrogen and chloride ions are lost in *pyloric stenosis or by gastric aspiration*. This causes *hypochloraemic alkalosis*.

Pyloric stenosis.—Secretion of $H^+ Cl^-$ into the gastric lumen is accompanied by generation of equivalent amounts of HCO_3^- by the parietal cells (Fig. 6). Normally this is followed by secretion of HCO_3^- into the duodenum. Therefore if vomiting occurs with free communication between the stomach and duodenum the loss of secreted HCO_3^- counteracts the effect of its generation by the gastric mucosa. Such vomiting causes relatively little disturbance of hydrogen ion balance: the effects are due to volume and electrolyte loss.

Volume depletion also occurs due to the vomiting of pyloric stenosis. However, the obstruction between the stomach and duodenum reduces loss of HCO_3^-, Na^+ and K^+. Loss of gastric H^+ stimulates further production by the carbonate dehydratase of mucosal cells, with generation of equivalent amounts of HCO_3^-. H^+ continues to be lost, and if duodenal secretion and renal loss of bicarbonate are insufficient to correct the rise in plasma $[HCO_3^-]$ alkalosis may result.

$$pH\uparrow = 6{\cdot}1 + \log \frac{[HCO_3^-]\uparrow}{PCO_2 \times 0{\cdot}23}$$

Correction of the alkalosis of pyloric stenosis depends on urinary bicarbonate loss. Although acidosis stimulates the respiratory centre, alkalosis does not seem to depress it significantly: CO_2 is not retained in adequate amounts to compensate for the rise in plasma $[HCO_3^-]$. H^+ secretion into the renal tubular lumen is essential for bicarbonate reabsorption from the glomerular filtrate. The rising extracellular $[HCO_3^-]$, derived from gastric mucosal cells, inhibits the formation of H^+ and HCO_3^- by the CD mechanism in renal tubular cells, and diminished renal secretion of H^+ reduces HCO_3^- reabsorption. The consequent loss of HCO_3^- in the urine may compensate for the increased generation by the gastric mucosa.

Two factors may impair this capacity to lose HCO_3^- in the urine, and so aggravate the alkalosis.

A *reduced GFR* due to volume depletion limits the total amount of HCO_3^- which can be lost.

The *chloride concentration in the glomerular filtrate* may be *reduced*, reflecting hypochloraemia due to gastric loss. Isosmotic sodium reabsorption in the proximal tubule depends on the passive reabsorption of chloride along the electrochemical gradient. A reduction in available chloride limits this isosmotic reabsorption, and more sodium becomes available for exchange with H^+ and K^+. The urine becomes inappropriately acid as H^+ secretion stimulates HCO_3^- reabsorption. Chloride depletion will also increase potassium loss in exchange for sodium, so aggravating the hypokalaemia due to the alkalosis.

Thus vomiting due to pyloric stenosis may cause:

hypochloraemic alkalosis;
hypokalaemia;
haemoconcentration and mild uraemia (due to loss of fluid).

The hypokalaemia may be masked while the patient is dehydrated, but should be anticipated during treatment.

Severe hypochloraemic alkalosis is relatively rare because pyloric stenosis is usually treated before this occurs. Nevertheless, when the typical changes in bicarbonate and chloride occur the diagnosis could (but should not) be made on biochemical grounds alone. The plasma chloride level may be as much as 80 mmol/l lower than that of sodium, rather than the usual 40 mmol/l.

Treatment of the biochemical disorder of pyloric stenosis.—Treatment consists (if renal function is normal) of replacing water and chloride by administering large amounts of saline. The kidney will then correct the alkalosis. The fluid (which, because of the pyloric obstruction, must be given intravenously) should be at least isosmolar saline. Potassium should be added to the saline if the plasma potassium concentration is low normal or low (see p. 80).

Respiratory Alkalosis

A primary fall in PCO_2 is due to abnormally rapid or deep respiration with relatively normal CO_2 transport capacity across the pulmonary alveoli. The causes are:

> **hysterical overbreathing,** which overrides normal respiratory control, with normal lung function;
> stimulation of the respiratory centre by **raised intracranial pressure** or **brain stem lesions,** with normal lung function;
> stimulation of the respiratory centre by **hypoxia,** with relatively normal overall alveolar diffusion of CO_2 (p. 117);
> **pulmonary oedema;**
> **lobar pneumonia;**
> **pulmonary collapse or fibrosis;**
> **excessive artificial ventilation.**

The fall in PCO_2 slows the carbonate dehydratase mechanisms in renal tubular cells and erythrocytes, reducing HCO_3^- generation and reabsorption (Figs. 3, 4 and 5). The *compensatory fall in* $[HCO_3^-]$ tends to correct the pH. It is important not to interpret a low $[HCO_3^-]$ with overbreathing as metabolic acidosis, with stimulation of the respiratory centre. In doubtful cases estimation of arterial pH and PCO_2 are indicated.

The arterial blood findings in respiratory alkalosis are:

> PCO_2 *always reduced*;
> actual $[HCO_3^-]$ low normal or low;
> pH raised (uncompensated or partially compensated) or normal (fully compensated).

Salicylate overdosage, by stimulating the respiratory centre directly, initially causes respiratory alkalosis. However, it also causes uncoupling of oxidative phosphorylation; consequent impairment of aerobic pathways superimposes a metabolic (lactic) acidosis on the respiratory alkalosis. Both respiratory alkalosis and metabolic acidosis result in low blood HCO_3^- levels, but the pH may be high if respiratory

TABLE XII

SUMMARY OF FINDINGS IN ARTERIAL BLOOD IN DISTURBANCES OF
HYDROGEN ION HOMEOSTASIS

		pH	PCO$_2$	Actual HCO$_3^-$	K$^+$ levels (plasma)
Acidosis					
Metabolic	Initial state	↓	N	↓ (underlined)	Usually ↑
	Compensated state	N	↓ (circled)	↓ (underlined)	(↓ in renal tubular acidosis and acetazolamide therapy)
Respiratory	Acute change	↓	↑ (underlined)	N or ↑	↑
	Compensation	N	↑ (underlined)	↑↑ (circled)	
Alkalosis					
Metabolic	Acute state	↑	N	↑ (underlined)	
	Chronic state	↑	N or slightly ↑	↑↑ (underlined)	↓
Respiratory	Acute change	↑	↓ (underlined)	N or ↓	↓
	Compensation	N	↓ (underlined)	↓↓ (circled)	↓

Arrows underlined = Primary change.
Arrows circled   = Compensatory change.

Note:
1. Generalised potassium depletion can cause extracellular alkalosis.
   Generalised alkalosis can cause hypokalaemia.
   Only the clinical history can differentiate the primary cause of the combination of alkalosis and hypokalaemia.
2. Overbreathing *causes* a low [HCO$_3^-$] in respiratory alkalosis.
   Metabolic acidosis with a low [HCO$_3^-$] *causes* overbreathing.
   Only measurement of blood pH and/or PCO$_2$ can differentiate these two.

alkalosis is predominant, normal if the two "cancel each other out", or low if metabolic acidosis is predominant. Only by measurement of blood pH can the true state of hydrogen ion balance be assessed.

The possible findings in disturbances of hydrogen ion homeostasis are summarised in Table XII. *The student is urged to read the section on the investigation of hydrogen ion homeostasis (p. 121) to assess realistically how many of these tests are necessary.*

BLOOD GAS LEVELS

In respiratory disturbances associated with acidosis a knowledge of the partial pressure of oxygen (PO_2) is as important as that of pH, PCO_2 and [HCO_3^-].

Normal gaseous exchange across the pulmonary alveoli involves loss of CO_2 and gain of O_2. However, in pathological conditions a fall in PO_2 and rise in PCO_2 do not always coexist. The reasons for this are as follows:

CO_2 *is much more soluble than O_2 in water*, in which its rate of diffusion is 20 times as high as that of O_2. In *pulmonary oedema*, for example, arterial PO_2 falls because its diffusion across the alveolar wall is hindered by oedema fluid. Respiration is stimulated by the hypoxia and by pulmonary distension, and CO_2 is "washed out". However, the rate of transport of O_2 through the fluid cannot be increased enough to restore normal PO_2. The result is a *low or normal arterial* PCO_2 and a *low* PO_2. Only in very severe cases is the PCO_2 raised;

the haemoglobin of arterial blood is normally 95 per cent saturated with oxygen, and very little oxygen is carried in simple solution in the plasma: the dissolved O_2 is in equilibrium with the oxyhaemoglobin. Overbreathing air with a normal atmospheric PO_2 cannot significantly increase the amount of oxygen carried in the blood leaving normal alveoli: it can, however, reduce the PCO_2. (Breathing pure oxygen by increasing inspired PO_2 can increase arterial PO_2, but not the haemoglobin saturation.)

Let us consider the situation in many pulmonary conditions such as *lobar pneumonia, collapse of the lung*, and *pulmonary fibrosis or infiltration*. In such conditions not all alveoli are affected by the disease to the same extent, or in the same way.

(*a*) Some alveoli will be unaffected. The composition of blood leaving these is initially that of normal arterial blood. Increased rate or depth of respiration can lower the PCO_2 to very low levels, but not alter either the PO_2 or the haemoglobin saturation in this blood.

(*b*) Some alveoli, while having a normal blood supply, may, perhaps because of obstruction of small airways, have little or no air entry. The composition of blood leaving these is near to that of venous blood (it is a right-to-left shunt). Unless increased ventilation can overcome the obstruction it will have no effect on this low PO_2 and high PCO_2.

(*c*) Some alveoli may have normal air entry, but little or no blood supply. These are effectively "dead space": increased ventilation will be "wasted", because however much air enters and leaves these alveoli, there is no gas exchange with blood.

Blood from (*a*) and (*b*) mixes in the pulmonary vein before entering the left atrium: the result is mixed venous and arterial blood. The high PCO_2 and low PO_2 stimulate respiration, and if enough unaffected alveoli (*a*) are present, the very low levels of PCO_2 in blood leaving these may "correct" the high PCO_2 from poorly aerated alveoli. For reasons discussed above, neither the PO_2 nor the haemoglobin saturation will be significantly altered. The final result is therefore *low or normal arterial PCO_2 with a low PO_2.*

If the proportion of type (*b*) and (*c*) alveoli is very high, PCO_2 cannot be adequately corrected by overventilation, and the result is a *high arterial PCO_2 and low PO_2.*

In conditions in which almost all the alveoli have a normal blood supply but poor air entry the result will be a *high arterial PCO_2 and low PO_2.* This may be due to *mechanical or neurological defects in respiratory movement,* or to *obstruction of large or small airways.* However, early in an acute asthmatic attack, stimulation of respiration by alveolar stretching may keep PCO_2 normal, or even low.

The two groups of findings in pulmonary disease are summarised in the list below. The conditions marked with * can fall into either group.

Low arterial PO_2 with low or normal PCO_2 can occur in such conditions as:
  pulmonary oedema (diffusion defect)
  pneumonia*
  collapse of the lung*
  pulmonary fibrosis or infiltration*

Low arterial PO_2 with high PCO_2 can occur in such conditions as:
  chest injury, gross obesity, ankylosing spondylitis (impairment of movement of the respiratory cage)
  poliomyelitis, lesions of the central nervous system affecting the respiratory centre (neurological impairment of respiratory drive)
  laryngeal spasm, severe asthma, chronic bronchitis and emphysema (obstruction to airways)
  pneumonia*
  collapse of the lung*
  pulmonary fibrosis or infiltration*.

SUMMARY

HYDROGEN ION HOMEOSTASIS

1. CO_2 is central to hydrogen ion homeostasis. Arterial blood PCO_2 is controlled by the respiratory centre and lungs to about 5·3 kPa.
2. At a PCO_2 of 5·3 kPa the carbonate dehydratase mechanism in erythrocytes and renal tubular cells maintains the plasma $[HCO_3^-]$ at about 25 mmol/l.
3. In erythrocytes the H^+ produced by the carbonate dehydratase

mechanism is buffered by haemoglobin. This has limited capacity, but is of importance in acute disorders of hydrogen ion homeostasis.
4. The renal tubular cell secretes H^+ into the urine in exchange for Na^+. H^+ secretion is essential for HCO_3^- reabsorption: HCO_3^- generation also depends on secretion of H^+ into the urine.
5. Normal urine is HCO_3^- free. Generation of HCO_3^- to replace utilisation depends on the availability of urinary buffers, particularly $HPO_4^=$.
6. Renal correction of either acidosis or alkalosis is dependent on a normal glomerular filtration rate.
7. Acidosis results from a decrease in the ratio $[HCO_3^-]$:PCO_2. In a compensated acidosis the ratio is normal, but the levels of HCO_3^- and CO_2 are abnormal.
8. Acidosis may be due to excessive production of hydrogen ion, to failure of the lungs or kidneys, or to excessive loss of bicarbonate.
9. Alkalosis results from an increase in the ratio $[HCO_3^-]$:PCO_2. In a compensated alkalosis the ratio is normal, but the levels of HCO_3^- and CO_2 are abnormal.

BLOOD GAS LEVELS

Low PO₂ and Normal or Low PCO₂

1. Carbon dioxide is very much more soluble than oxygen in water. Its level in the blood is therefore less affected than that of oxygen in pulmonary oedema, in which it may even be low due to respiratory stimulation.
2. Arterial blood is 95 per cent saturated with oxygen. Overbreathing therefore cannot increase oxygen carriage in blood from normal alveoli, but can reduce the PCO_2. In ventilation-perfusion defects, when some alveoli have a normal blood supply, but are ventilated poorly, mixture of the "shunted" blood with blood from normal alveoli results in a low PO_2 and normal or low PCO_2 in the peripheral arterial blood.

Low PO₂ and High PCO₂

3. The PO_2 is low and PCO_2 is high in total alveolar hypoventilation when neither gas can be adequately exchanged.

FURTHER READING

ZILVA, J. F. (1983). Disorders involving change in hydrogen ion and blood gas concentrations. In: *Biochemistry in Clinical Practice*, pp. 24–43. (Scientific Foundations of Clinical Biochemistry, Vol. II.) D. L. Williams and V. Marks, eds. London: Heinemann Medical Books.

MORRIS, R. C. and SEBASTIAN, A. (1983). Renal tubular acidosis and Fanconi syndrome. In: *The Metabolic Basis of Inherited Disease*, 5th edit., pp. 1808–1843. J. B. Stanbury, J. B. Wyngaarden, D. S. Fredrickson *et al.*, eds. New York: McGraw-Hill Book Co.

INVESTIGATION OF HYDROGEN ION DISTURBANCES

We have discussed the mechanisms behind the abnormal findings in disorders of hydrogen ion balance. Understanding is an essential preliminary to critical assessment of how much investigation is necessary for diagnosis and management. Without such understanding, this assessment cannot be made, and, more dangerously, the results may be misinterpreted.

INTERPRETATION OF PLASMA TCO_2

As explained on p. 104, this commonly performed estimation is effectively a measure of plasma $[HCO_3{}^-]$. It has the advantage that it can be performed on venous plasma at the same time as, for instance, plasma urea, sodium and potassium.

The measurement of the level of plasma bicarbonate *alone* tells us nothing about the state of hydrogen ion balance. For instance, a low concentration may be associated with compensated or uncompensated non-respiratory ("metabolic") acidosis, or respiratory alkalosis. The pH is determined by the ratio of $[HCO_3{}^-]$ to $[CO_2]$ according to the Henderson-Hasselbalch equation. Nevertheless, with a little thought TCO_2 usually provides adequate information for clinical purposes.

We suggest the following procedure:

If the TCO₂ is Low

(a) **Exclude artefactual causes** due to *in vitro* loss of CO_2 from a specimen that has been standing overnight, or from a small specimen with a large dead space above it.

(b) **Reassess the clinical picture.**—Particularly note the presence of the following:

> diarrhoea or intestinal fistulae;
> any evidence suggesting renal glomerular or tubular dysfunction (Chapter I);
> hypotension, dehydration or other evidence of poor tissue perfusion;
> a history of ureteric transplantation into the ileum or colon;
> a drug history with specific reference to biguanides, such as phenformin, or more rarely, acetazolamide.

(c) **Estimate plasma urea and glucose** levels and test the urine for ketones.

In the great majority of cases the diagnosis will now be obvious. If it is not, among the few remaining possibilities are:

> respiratory alkalosis due to overbreathing;
> renal tubular acidosis.

In such rare cases estimation of arterial blood pH and PCO_2 may help to differentiate respiratory alkalosis from metabolic acidosis. If renal tubular acidosis is suspected, the finding of a high plasma chloride level strengthens the suspicion: an ammonium chloride loading test (p. 124) should be performed.

If the TCO₂ is High

(a) **Reassess the clinical picture.**—Particularly note whether:
there is generalised respiratory disease;
there is a cause for potassium depletion (for instance, is the patient taking potassium-losing diuretics?);
bicarbonate has been ingested or infused;
there is severe vomiting which, especially if there is a history of dyspepsia, might indicate pyloric stenosis.

(b) **Estimate plasma potassium**

INDICATIONS FOR ESTIMATING ARTERIAL pH AND PCO₂

In most cases of metabolic acidosis, in which the fall in $[HCO_3^-]$ is primary, plasma TCO_2 yields adequate information for clinical purposes. Similarly, a subject with chronic bronchitis and a high plasma concentration of bicarbonate undoubtedly has a respiratory acidosis which may or may not be fully compensated. If there is no possibility of improving air entry into the lungs by use of physiotherapy, expectorants and antibiotics, treatment on a respirator is contra-indicated, because a reduction of PCO_2 will lead to a bicarbonate loss (by reversal of the changes described on p. 112): unless the patient continues to be respired artificially for life, removal from the respirator will result in return of the PCO_2 to its initial high levels, but with a delay of the compensatory increase in bicarbonate for some days. Nothing is to be gained, from the therapeutic point of view, by a knowledge of pH and PCO_2 in such a case.

Arterial blood estimation may be indicated:

1. If there is doubt about the cause of an abnormal TCO_2 (for instance to differentiate metabolic acidosis from respiratory alkalosis).

2. In some conditions in which mixed respiratory and non-respiratory conditions may be present (for example cardiac arrest; renal failure complicated by pneumonia; salicylate overdosage). In such conditions, the estimations are indicated only if the results will influence treatment.

3. If there has been an acute exacerbation of the chronic bronchitis, or if the pulmonary disease is of acute onset and potentially reversible. In such cases vigorous therapy or artificial respiration may tide the patient over until the pulmonary condition improves and more precise information than the plasma TCO_2 concentration is required for monitoring adequate control of treatment.

4. If arterial blood is being taken for estimation of PO_2.

INDICATIONS FOR CHLORIDE ESTIMATION

Chloride levels are always affected in only a few conditions. Hyperchloraemia occurs in the metabolic acidosis associated with ureteric transplantation, renal tubular acidosis, and administration of acetazolamide. Hypochloraemia occurs in the metabolic alkalosis of pyloric stenosis. If the approach outlined above is followed the level of chloride can usually be predicted, and estimation adds nothing to diagnostic or therapeutic precision. Chloride estimation may help in two situations:

1. If there is a low TCO_2 of obscure origin. The finding of a high chloride (and therefore a normal "anion gap") strengthens a suspicion of renal tubular acidosis. In acidosis due to most other obscure causes the chloride is normal (and "anion gap" increased).

2. If a patient who is vomiting has a high TCO_2. An equivalently low chloride level favours pyloric stenosis. Similar findings may be due to compensated respiratory acidosis: the finding of severe, chronic lung disease usually makes this cause obvious.

TEST PROTOCOLS

AMMONIUM CHLORIDE LOADING TEST OF URINARY ACIDIFICATION

The ammonium ion (NH_4^+) is potentially acidic because it can dissociate to ammonia and H^+. If ammonium chloride is ingested the kidneys should normally secrete the excess of hydrogen ion.

Procedure

No food or fluid is taken after midnight.

08.00 h. Ammonium chloride 0·1 g/kg body weight is administered orally.

Hourly specimens of urine are collected between 10.00 h and 16.00 h, and the pH of each specimen measured immediately with a pH meter (pH paper is not very accurate). If the pH of any specimen falls to 5·2 or below, the test can be stopped.

Interpretation

In normal subjects the urinary pH falls to 5·2 or below between 2 and 8 hours after the dose. In *renal tubular acidosis* this degree of acidification fails to occur. In generalised renal failure the response of the functioning nephrons may give normal results.

COLLECTION OF SPECIMENS FOR BLOOD GAS ESTIMATION

1. *Arterial* specimens are preferable to *capillary* ones.

2. The syringe should be moistened with *heparin* and the specimen well mixed. Excess heparin may dilute the specimen and cause haemolysis.

   NB—If *sodium* heparin is used, do *not* send an aliquot of the blood for electrolyte estimation: the resultant factitious hypernatraemia, if acted on, may be a danger to the patient.

3. Gas exchange with the atmosphere should be minimised by *leaving the specimen in the syringe* and expelling any air bubbles which may be present. The nozzle should be stoppered.

4. The rate of erythrocyte metabolism should be minimised by keeping the specimen *on ice* or *in the refrigerator* until the estimation is performed.

5. The estimation should be performed *within an hour*.

Capillary Specimens

In newborn infants arterial puncture may be technically difficult, and it is common to request estimations on capillary specimens (usually heel pricks). The following points should be noted.

1. The area from which the specimen is taken should be warm and pink, because the composition of the capillary blood should be as near arterial as possible. In the presence of peripheral cyanosis results may be dangerously misleading.

2. There should be free flow of blood. Squeezing may cause dilution by ECF.

3. The capillary tube(s) must be heparinised.

4. The tube(s) must be completely filled with blood. Air bubbles will invalidate the results.

5. The ends of the tube(s) should be sealed with Plasticine.

6. Mixing with the heparin should be complete. Usually a small splinter of metal is inserted into the capillary tube, and mixing is performed by passing a magnet up and down the tube.

7. The tube(s) should be put on ice, and the estimation performed at once. *Whenever possible an arterial specimen should be obtained.*

Fetal Specimens

In fetal distress, one indication for termination of labour may be fetal acidosis. If the cervix is dilated, and the fetal head presenting, specimens may be taken by puncturing the scalp and allowing blood to flow into long capillary tubes passed up the vagina. Again, air bubbles invalidate the results.

It is preferable to take an arterial specimen from the mother at the same time, and to compare the fetal and maternal results.

Chapter V

THE HYPOTHALAMUS AND PITUITARY GLAND

GENERAL PRINCIPLES OF ENDOCRINE DIAGNOSIS

THE principles of the laboratory diagnosis of endocrine disorders are outlined in this and the following three chapters. A few general points will be mentioned here.

Endocrine glands may secrete *excessive* or *deficient* amounts of hormone. This may be caused by a *primary* abnormality of the gland itself, or may be *secondary* to an abnormality in a controlling mechanism, usually located in the hypothalamus or pituitary gland. In the latter case the endocrine gland itself is initially normal.

The secretion of most hormones is influenced, directly or indirectly, by the final product or the metabolic consequence of its secretion. This. is usually a *negative feedback* whereby a rising final product (hormonal or non-hormonal) inhibits secretion. This concept is used in many tests of endocrine function.

If results of preliminary tests are near the limits of the reference range, it is necessary to determine whether they are abnormal or not: if results are definitely abnormal, the abnormality may be primary, or secondary to a disorder of one of the controlling mechanisms. If either of these points is not clear when the results are considered together with the clinical findings, so-called "dynamic" tests should be carried out. In such tests the response of the gland or feedback mechanism is stimulated or suppressed by administration of exogenous hormones.

Suppression tests are used mainly for the differential diagnosis of *excessive hormone secretion*. The substance (or an analogue) normally suppressing hormone secretion by negative feedback is administered and the response measured. *Failure to suppress* implies secretion that is not under normal feedback control (autonomous secretion).

Stimulation tests are used mainly for the differential diagnosis of *deficient hormone secretion*. The trophic hormone that normally stimulates secretion is administered and the response measured. A normal response excludes an abnormality of the target gland whereas *failure to respond* confirms it.

In this chapter disorders of the pituitary gland and hypothalamus will be discussed. Diseases of their target endocrine organs, the adrenal cortex, gonads and thyroid gland will be considered in Chapters VI, VII and VIII.

HYPOTHALAMUS AND PITUITARY

There is a close relationship between the neural and endocrine systems, most obvious in the interaction between the hypothalamus and the two lobes of the pituitary gland.

The anterior and posterior lobes are developmentally and functionally distinct, but both depend for normal functioning on hormones synthesised in the hypothalamus. The hypothalamus has extensive neural connections with the rest of the brain, and it is not surprising to find that stress and some psychological disorders affect secretion of pituitary hormones, and hence those of many other endocrine glands.

HYPOTHALAMIC HORMONES

The hypothalamus synthesises two main groups of hormones, which are associated with the functions of the posterior and anterior lobes of the pituitary gland respectively.

The Hypothalamus and the Posterior Pituitary Lobe

Three peptides are synthesised in the hypothalamus, and travel down the nerve fibres of the pituitary stalk to be stored in the posterior lobe of the pituitary gland. Their release into the blood stream is under hypothalamic control.

These peptides are:

Antidiuretic hormone (ADH: arginine vasopressin), which is discussed in Chapter II.

Oxytocin, a hormone, of similar structure to ADH, which controls ejection of milk from the lactating breast, and which may have some role in the initiation of uterine contractions during labour (although normal labour can proceed in its absence). It can be used in obstetric practice, to induce labour.

Neurophysin, the function of which is doubtful, but which may help transport and storage of the other two hormones in the posterior pituitary gland.

The Hypothalamus and the Anterior Pituitary Lobe

The hypothalamus has no direct neural connection with the anterior pituitary. Instead it synthesises small molecules (regulating hormones or factors) which are carried to the cells of the anterior lobe by the *hypothalmic portal system.* This network of capillary loops in the median eminence forms veins that, after passing down the pituitary stalk, divide into a second capillary network in the anterior pituitary. The high local concentration of these hypothalamic hormones stimulates or inhibits hormonal secretion from the cells of the anterior lobe into the systemic circulation.

The cells of the anterior pituitary lobe can be classified simply by their staining reactions as acidophils, basophils or chromophobes. More sophisticated techniques can identify specific hormone secreting cells.

Acidophils are of two cell types, one of which secretes growth hormone (GH) and one prolactin. These simple polypeptides, with some amino-acid sequences in common, mainly *affect peripheral tissues directly*. Stimulation and inhibition of the hypothalamus is markedly influenced by neural stimuli.

Basophils secrete hormones that *affect other endocrine glands*. The hypothalamic control is mainly stimulatory. There are three cell types. One secretes thyroid-stimulating hormone (TSH; thyrotrophin) which acts on the thyroid gland. Another secretes the gonadotrophins, follicle-stimulating hormone (FSH) and luteinising hormone (LH), that act on the gonads. These hormones are structurally similar glycoproteins and secretion is influenced more by feedback than by neural mechanisms. The third cell type synthesises a large polypeptide (pro-opiocortin), which is a prohormone for both adrenocorticotrophic hormone (ACTH; corticotrophin) and β-lipotrophin (β-LPH): secretion of these hormones occurs in parallel:

> *ACTH* stimulates secretion of steroids, other than aldosterone, from the adrenal cortex: part of the molecule has melanocyte-stimulating activity and high circulating levels of ACTH are often associated with pigmentation.
>
> β-*LPH* is probably a precursor of endorphins, which have an opiate-like effect. The endorphins will not be considered further.

Chromophobes, once thought to be inactive, do contain secretory granules. Chromophobe adenomas often secrete hormones, particularly prolactin.

Control of Anterior Pituitary Hormone Secretion

Neural and *feedback* control are the two most important physiological factors influencing secretion of the anterior pituitary hormones (Fig. 8).

Extra-hypothalamic neural stimuli modify and, at times, override other controls. Physical or emotional stress and mental illness may give similar findings to, and even precipitate, endocrine disease. The stress of insulin-induced hypoglycaemia is used to test anterior

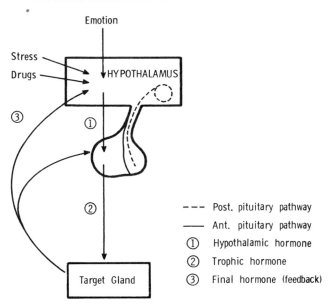

FIG. 8.—Control of pituitary hormone secretion.

pituitary function. Stress may also stimulate secretion of the posterior pituitary hormone, ADH.

Feedback control is mediated by the levels of circulating target-cell hormones; a rising level usually suppresses trophic hormone secretion. This "negative feedback" may directly suppress hypothalamic hormone secretion or may modify its effect on the pituitary cell ("long feedback loop"). The hypothalamic hormone may also be suppressed by rising levels of pituitary hormone in a short feedback loop.

Inherent rhythms.—The release of the hypothalamic (and consequently pituitary) hormones occurs intermittently. There is an underlying regular secretory rhythm of some hormones, and disturbances of such rhythms may be of diagnostic value. This subject will be considered further in the relevant sections.

Effects of drugs.—The hypothalamus translates neural signals into chemical hormone secretion. It is not surprising therefore that drugs stimulating or blocking the action of neurotransmitters (such as catecholamines, acetylcholine and serotonin) influence the secretion of hypothalamic and, consequently, pituitary hormones. For example, drugs such as *chlorpromazine* interfere with the action of dopamine. This results in reduced GH secretion (reduced effect of releasing factor) and increased prolactin secretion (reduced effect of inhibitory factor). *Bromocriptine* (2-bromo-α-ergocryptine) which may stimulate the attachment of dopamine to pituitary receptor sites, and *L-dopa*, which is converted to dopamine, have the opposite effect in normal subjects. *Bromocriptine* causes a *paradoxical suppression of excessive GH secretion in acromegalics.*

These effects, which have been used both in the diagnosis and therapy of hypothalamic-pituitary disorders, will be discussed in later sections.

Assay of pituitary hormones.—All the pituitary hormones are peptides, and are usually measured by radio-immunoassay. The clinician should remember that the specificity and precision of peptide hormone assays is not comparable with that of, say, sodium estimation, and should interpret results with due caution.

DISORDERS OF PITUITARY HORMONE SECRETION

The main clinical syndromes associated with excessive or deficient pituitary hormone secretion are shown in Table XIII. Excessive secretion usually involves a single hormone, but deficiencies are often multiple.

TABLE XIII

CONSEQUENCES OF A PRIMARY ABNORMALITY OF ANTERIOR PITUITARY HORMONE SECRETION

Hormone	Excess	Deficiency
Growth Hormone	Acromegaly Gigantism	Dwarfism
Prolactin	Amenorrhoea Infertility Galactorrhoea	Lactation failure
ACTH (corticotrophin)	Cushing's disease	Secondary adreno- cortical hypofunction
TSH	Hyperthyroidism (very rare cause)	Secondary hypothyroidism
LH/FSH	Precocious puberty	Secondary hypo- gonadism Infertility

GROWTH HORMONE (GH)

Growth hormone secretion from acidophil cells of the anterior pituitary is controlled by the balance between hypothalamic GH-releasing hormone and somatostatin (which inhibits GH release). The relatively infrequent bursts of GH secretion are usually associated with exercise, the onset of deep sleep, or occur in response to the falling plasma glucose level about an hour after meals. At other times levels may be very low or undetectable, especially in children. Somatostatin may also inhibit TSH secretion. It is found not only in the hypothalamus and elsewhere in the brain, but also in gastro-intestinal and pancreatic islet cells. It inhibits secretion of many gastro-intestinal hormones.

The main function of GH is to *promote growth*. This is illustrated by the fact that during childhood, deficiency causes dwarfism and excess causes gigantism. GH *promotes protein synthesis* and, in conjunction with insulin, *stimulates amino acid uptake by cells*. Its effect on growth however is probably mediated by the somatomedins which are polypeptide growth factors synthesised by the liver.

Other factors such as adequate nutrition and thyroxine are also required for normal growth. The growth spurt at the onset of puberty is probably due to androgens.

Growth hormone also affects *carbohydrate and fat metabolism*. It antagonises the insulin-mediated cell uptake of glucose. Although excess may produce carbohydrate intolerance, this tendency is usually counteracted by increased insulin secretion. GH stimulates lipolysis and, as the resultant increase in free fatty acids (FFA) also antagonises insulin release and its action, it is difficult to distinguish between direct and indirect insulin antagonism.

GH secretion is stimulated by hypoglycaemia and suppressed by hyperglycaemia. In adults growth hormone deficiency very rarely produces symptoms.

Control.—In addition to the factors mentioned above, GH secretion may be stimulated by stress, starvation, rapidly falling blood glucose levels and by the administration of certain amino acids, or drugs such as L-dopa. Secretion is suppressed by hyperglycaemia and is affected by the levels of other hormones. An impaired response to stimuli may occur in obese patients, in hypothyroidism and hypogonadism, in some cases of Cushing's syndrome and in patients receiving large doses of steroids. Oestrogens potentiate the GH response.

GH EXCESS: ACROMEGALY AND GIGANTISM

Most patients with GH excess have an acidophil adenoma of the pituitary, but this is probably secondary to excessive hypothalamic stimulation. The condition may therefore recur after removal of the pituitary adenoma. The clinical manifestations depend on whether the condition develops before or after fusion of the bony epiphyses. In childhood GH excess leads to *gigantism*. Epiphyseal fusion may be delayed by accompanying hypogonadism and extreme heights of over 2·4 m (8 feet) have occasionally been reached. Acromegalic features may develop after bony fusion but these giants frequently die in early adult life from infections or progressive tumour growth.

In adults after epiphyseal fusion GH excess causes *acromegaly*. Bone and soft tissues increase in bulk and lead to increasing size of the hands and other parts due to soft-tissue thickening. There may be excessive hair growth and skin secretion. Changes in facial appearance are often marked, due to the increasing size of the jaw and sinuses. The gradual coarsening of the features may, however, pass unnoticed for many years. The patient may first complain that rings or shoes are becoming too small. The viscera, too, enlarge and cardiomegaly is common. Acromegalics are usually euthyroid despite thyroid enlargement. Menstrual disturbances are common. A different group of symptoms may occur due to the encroachment of the pituitary tumour on surrounding structures. For example, compression of the optic chiasma produces visual-field defects.

If destruction of the gland progresses, other anterior pituitary hormones may become deficient. Prolactin levels may, however, be raised (p. 160).

Impaired glucose tolerance may be demonstrated in about 25 per cent of cases. Only about half of these develop symptomatic diabetes. In most cases the pancreas can secrete enough insulin to overcome the antagonistic effect of GH. Probably only those with a diabetic tendency will develop it under these conditions.

Acromegaly is sometimes one of the manifestations of multiple endocrine adenomatosis (p. 470).

Diagnosis

The diagnosis of acromegaly is suggested by the clinical features and radiological findings. GH levels are usually higher than normal and may reach several hundred mU/l, but because of the wide "normal" range, results from ambulant patients may fail to distinguish acromegalics with only moderately raised levels from normal subjects.

The diagnosis is confirmed by the demonstration of a *raised GH level that is not suppressed by a rise in plasma glucose concentration*.

During a glucose tolerance test (GTT) on a normal subject plasma GH falls to very low levels. In acromegaly secretion is autonomous, and GH levels may remain constant, may even increase, or may fall only slightly. Plasma GH levels may also be used to monitor the efficacy of treatment.

GROWTH HORMONE DEFICIENCY

A small percentage of normally proportioned dwarfed children have growth hormone deficiency. In about half of these there is also deficiency of other hormones. In some patients there may be an organic disorder of the anterior pituitary or hypothalamus, and a rare familial autosomal recessive form is described, but in many cases the cause is unknown. Growth failure is progressive and may be noticed during the first two years of life. GH is the only means of treatment, and as supplies of this are limited it is essential to prove GH deficiency.

Emotional deprivation may be associated with growth hormone deficiency indistinguishable by laboratory tests from that due to organic causes.

Diagnosis

Basal plasma GH levels in normal children are usually low and such assays rarely exclude the diagnosis. If blood is taken at a time when physiologically high levels are expected, the more unpleasant stimulation tests may be avoided. Such times are 60 to 90 minutes after the onset of sleep and about 20 minutes after vigorous exercise.

If GH deficiency is not excluded by the above measurements, one or more *stimulation tests* are necessary (p. 171). An unequivocally normal response to these excludes the diagnosis and a clearly impaired one confirms it.

Once GH deficiency is established a cause should be sought by appropriate clinical and radiological means.

HYPOPITUITARISM

Causes of Hypopituitarism

Destruction of, or damage to, the gland or hypothalamus by a primary or secondary tumour. The most common source of secondary tumour is the breast or bronchus;

Infarction, most commonly postpartum (Sheehan's syndrome) or, rarely, other vascular catastrophes;

Pituitary surgery or irradiation.

Rarer causes include:
head injury;
infections or granulomas.

Isolated hormone deficiencies may be idiopathic.

Established panhypopituitarism presents with the characteristic clinical picture described below. It is, however, not common and the question of hypofunction of the anterior pituitary usually arises in patients:

> with clinical and radiological *evidence of a pituitary or localised brain tumour*;
> presenting with *hypogonadism, hypothyroidism* or *adrenocortical insufficiency* which is shown by preliminary testing to be *secondary* in origin;
> with *dwarfism* shown to be *due to growth hormone deficiency*.

While isolated hormone deficiency, particularly of growth hormone or gonadotrophins, may sometimes occur, several hormones are usually involved. If a deficiency of one hormone is demonstrated it is therefore important to establish if the secretion of the others is normal or impaired. This evidence is needed, not only to guide replacement therapy, but also to assess the ability of the pituitary to respond to stress such as surgery.

The anterior pituitary has considerable functional reserve. Clinical features of deficiency are usually absent until about 70 per cent of the gland has been destroyed, unless there is associated hyperprolactinaemia (p. 160), when amenorrhoea and infertility may be early symptoms.

Consequences of Hormone Deficiencies

Progressive pituitary damage usually causes deficiencies of gonadotrophins and GH first. Plasma ACTH and/or TSH levels may remain normal, or may become deficient months or even years later. The clinical and biochemical consequences, usually those of target-gland failure, are summarised below.

> **Secondary hypogonadism** due to gonadotrophin deficiency presents as amenorrhoea, infertility, atrophy of secondary sex characteristics and impotence or loss of libido. Characteristically axillary and pubic hair are lost. In children puberty is delayed.
> **Retarded growth in children** may be due to growth hormone and thyroid insufficiency (TSH).
> **Secondary hypothyroidism** (TSH deficiency) may be clinically indistinguishable from primary hypothyroidism.
> **Secondary adrenocortical hypofunction** (ACTH deficiency) differs from the primary form (Addison's disease) in several respects. Patients are not hyperpigmented because ACTH

secretion is deficient, not excessive. The sodium deficiency so typical of Addison's disease does not occur, as aldosterone secretion (which is controlled by angiotensin, *not* ACTH) is normal. However, there may be marked *dilutional* hyponatraemia due to deficiency of cortisol, which is necessary for normal water excretion: cortisol is also necessary for maintenance of normal blood pressure, and hypotension is associated with ACTH deficiency. Hyperkalaemia is unusual. Because of cortisol and/or growth hormone deficiency, these patients are very sensitive to insulin and fasting hypoglycaemia may be a feature.

Lactation failure due to prolactin deficiency may occur after postpartum pituitary infarction (Sheehan's syndrome). In hypopituitarism due to tumour, however, prolactin levels are often raised and may cause galactorrhoea (secretion of breast fluid).

The patient with hypopituitarism, like the patient with Addison's disease, may die because of an inability to secrete an adequate amount of ACTH, and therefore of cortisol, in response to the stress caused by, for example, infection or surgery. Other life-threatening complications are hypoglycaemia, water intoxication and hypothermia.

Evaluation of Anterior Pituitary Function

Interpretation of basal pituitary hormone results is often difficult, as low levels are not necessarily abnormal and "normal" values do not exclude pituitary disease: the pituitary has a large functional reserve. The diagnosis of suspected hypopituitarism is best confirmed or excluded by *direct measurement of pituitary hormones after stimulation* (see p. 171).

Should direct pituitary hormone determinations not be available, *indirect* evidence of deficiency should be sought by demonstrating target-gland hyposecretion that responds to administration of the relevant trophic hormone (see Chapters VI, VII and VIII); results are often difficult to interpret.

It must be remembered that *prolonged* hypopituitarism may lead to secondary atrophy of the target gland with a consequent diminished response to stimulation.

Note that laboratory tests establish only the presence or absence of hypopituitarism. The *cause* must be sought by clinical and radiological means.

Hypothalamus or Pituitary?

In the above discussion we have not distinguished between

hypothalamic and pituitary causes of pituitary hormone deficiency or, more correctly, between deficient releasing factor and deficient pituitary hormone secretion. Isolated hormone deficiencies are more likely to be of hypothalamic than of pituitary origin. The coexistence of diabetes insipidus suggests a hypothalamic disorder, but symptoms may be masked initially by ACTH, and therefore cortisol, deficiency.

Some tests evaluate both hypothalamic and pituitary function, and some only the latter. It is sometimes possible to distinguish the level of the lesion. The TSH response to TRH (p. 190), for example, differs in hypothalamic and pituitary causes of secondary hypothyroidism. In hypogonadism a hypothalamic cause can be postulated if the basal gonadotrophin levels are low, and there is a delayed, but quantitatively normal, response to LH/FSH-RH.

When other synthetic releasing hormones become available it may be possible to make the distinction in cases of GH and ACTH deficiency by comparing the response to stress (both components) and to releasing factors (pituitary only).

Pituitary Tumours

The clinical presentation of pituitary tumours depends on the types of cell involved and on the size of the tumour.

Tumours of secretory cells produce the clinical effects of excessive hormone secretion:

GH Acromegaly or gigantism
ACTH Cushing's syndrome (p. 145)
Prolactin Infertility, amenorrhoea, and, occasionally, galactorrhoea (p. 160).

Large tumours may present with:

such symptoms as visual disturbances and headache due to pressure on the optic chiasma or to raised intracranial pressure; deficiency of some or all of the anterior pituitary hormones.

Non-secreting tumours are difficult to diagnose by biochemical tests, although the combined pituitary stimulation test may indicate subclinical impairment of function. Many such tumours, however, produce asymptomatic hyperprolactinaemia, either because the tumour secretes prolactin, or because it impairs normal hypothalamic inhibition.

The investigation of suspected pituitary disorders is outlined on p. 166.

SUMMARY

1. The anterior pituitary secretes growth hormone (GH), prolactin, adrenocorticotrophic hormone (ACTH; corticotrophin), β-lipotrophin, thyroid-stimulating hormone (TSH) and the two pituitary gonadotrophins, follicle-stimulating hormone (FSH) and luteinising hormone (LH). The posterior pituitary secretes antidiuretic hormone (ADH) and oxytocin.

2. The secretion of anterior pituitary hormones is controlled by regulating factors from the hypothalamus. These, in turn, are controlled by circulating hormone levels (feedback), or respond to stimuli from higher cerebral centres.

3. GH controls growth and has a number of effects on intermediary metabolism. *Excessive GH* secretion causes gigantism or acromegaly. Laboratory evidence of autonomous GH secretion is obtained by failure of suppression of plasma levels of GH during a glucose tolerance test. *GH deficiency* in childhood leads to dwarfism. Diagnosis of such deficiency is made by demonstrating a subnormal GH response to appropriate stimuli.

4. Anterior pituitary hormone deficiency usually involves several hormones. Less commonly, isolated deficiency occurs. The clinical features of hypopituitarism are those of gonadotrophin and sex hormone deficiencies and of secondary hypofunction of the adrenal cortex and thyroid. Diagnosis is made by demonstrating reduced anterior pituitary reserve after stimulation.

5. Pituitary tumours may produce excess of GH, ACTH, or prolactin, or may cause hypopituitarism. In patients without obvious endocrine disturbance hyperprolactinaemia may be found.

FURTHER READING

General Endocrinology

HALL, R., ANDERSON, J., SMART, G. A. and BESSER, M. (1980). *Fundamentals of Clinical Endocrinology*, 3rd edit. London: Pitman Medical.

Hypothalamus and Pituitary Gland

BESSER, G. M. (ed.) (1977). The hypothalamus and pituitary. *Clinics in Endocr. and Metab.*, **6**, 1–276. (This issue is devoted to the normal and abnormal function of the hypothalamus and pituitary.)

JADRESIC, A., BANKS, L. M., CHILD, D. F. *et al.* (1982). The acromegaly syndrome. Relation between clinical features, growth hormone values and radiological characteristics of the pituitary tumours. *Quart. J. Med.*, **51**, 189–204.

LOCK, W. (1978). Control of anterior pituitary function. *Arch. Intern. Med.*, **138**, 1541–1545. (A good review of the factors, including drugs, that influence anterior pituitary hormone secretion.)

Chapter VI

ADRENAL CORTEX: ACTH

THE adrenal glands are divided into the functionally distinct cortex and medulla. The *cortex* is part of the hypothalamic-pituitary-adrenal endocrine system. Histologically the adult adrenal cortex is divided into three layers. The outer thin layer *(zona glomerulosa)* secretes only aldosterone. The inner two layers, the *zona fasciculata* and the *zona reticularis*, form a functional unit and secrete the bulk of the adrenocortical hormones. The wider fourth layer present in the fetal adrenal gland disappears after birth: during fetal life one of its most important functions is the synthesis of oestriol (p. 458). Although the *medulla* is part of the sympathetic nervous system, glucocorticoids are probably needed for the final stage of adrenaline (epinephrine) synthesis.

CHEMISTRY AND BIOSYNTHESIS OF STEROIDS

Steroid hormones, bile salts, and the D vitamins are all derived from cholesterol, and steroid-producing tissues are rich in cholesterol. Note that if the molecule contains 21 carbon atoms it is referred to as a C_{21} steroid. The carbon atom at position 21 of the molecule is written as C-21. Figure 9 shows the internationally agreed numbering of the 27 carbon atoms, and the lettering of the four rings of the cholesterol molecule: the products of cholesterol are also indicated. The bile acids and D vitamins are dealt with elsewhere in the book (pp. 327 and 254).

As shown in Fig. 9, the side chain on C-17 is the main determinant of the type of hormonal activity, but substitutions in other positions modify activity within a group.

The first hormonal product of cholesterol is pregnenolone. Several important synthetic pathways diverge from it (Fig. 10). The final product depends on the tissue and the enzymes it contains.

Adrenal Cortex

Zonae fasciculata and reticularis.—*Cortisol* (the most important C_{21} steroid) is formed by progressive addition of hydroxyl groups at C-17, C-21 and C-11.

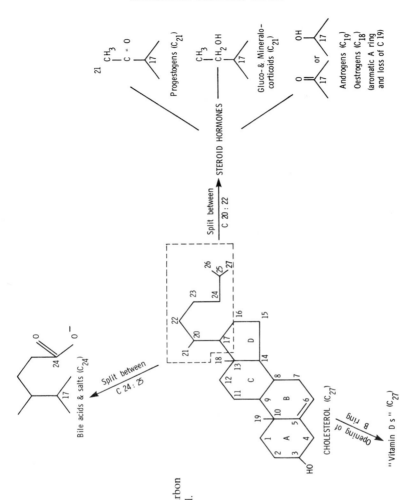

FIG. 9.—Numbering of steroid carbon atoms and products of cholesterol.

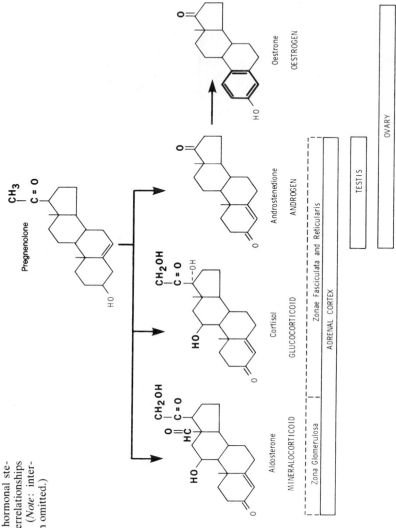

Fig. 10.—Examples of hormonal steroids, showing their interrelationships and sites of synthesis. (*Note*: intermediate steps have been omitted.)

Androgens (for example, androstenedione) are formed after removal of the side chain to produce C_{19} steroids.

Synthesis of these two groups is stimulated by ACTH secreted by the anterior pituitary (p. 128). ACTH secretion is influenced by plasma cortisol levels via negative feedback. Impaired cortisol synthesis, due, for example, to an inherited deficiency of 21 or 11 hydroxylase (congenital adrenal hyperplasia) results in increased ACTH stimulation with increased activity of *both* pathways. The resultant excessive androgen production produces virilisation (p. 152).

Zona glomerulosa.—A different series of hydroxylations produces *aldosterone*. Synthesis of this steroid is controlled by the renin-angiotensin system (p. 38) and not normally by ACTH. In ACTH deficiency, therefore, aldosterone secretion is unaffected.

PHYSIOLOGY

The adrenocortical hormones can be classified into three groups depending on their predominant physiological effects.

Glucocorticoids

Naturally occurring glucocorticoids are *cortisol* and *corticosterone*.

Glucocorticoids stimulate gluconeogenesis and the breakdown of protein and fat and so oppose some of the actions of insulin. In excess they impair glucose tolerance and alter the distribution of adipose tissue.

Cortisol helps to maintain the extracellular fluid volume and normal blood pressure. Most is bound to cortisol-binding globulin (CBG; transcortin) and to albumin. Only the unbound free fraction (about 5 per cent of the total at normal levels) is physiologically active (compare with thyroxine, p. 178). Increased CBG levels, such as those that occur during pregnancy, produce high bound, and therefore high total, cortisol levels with a normal, or only slightly raised, free fraction: low levels, such as occur in the nephrotic syndrome, have the opposite effect (again compare with thyroxine, p. 180).

The liver metabolises glucocorticoids to inactive conjugates: because these conjugates, are more water-soluble than the mainly protein-bound parent hormones, they can be excreted in the urine (compare bilirubin, p. 316).

Cortisone is not secreted in significant amounts by the adrenal cortex, but may be used therapeutically. It is biologically inactive until it has been converted *in vivo* to cortisol (hydrocortisone).

Mineralocorticoids

Aldosterone is discussed on p. 38. It stimulates sodium exchange for potassium and hydrogen ions across cell membranes, and its renal action is of particular homeostatic importance. Like the glucocorticoids it is excreted in the urine after hepatic conjugation.

There is some overlap in the actions of the C_{21} steroids. Cortisol in particular may, at high levels, have a significant mineralocorticoid effect.

Adrenal Androgens

The main adrenal androgens are dehydroepiandrosterone (DHEA) and its sulphate (DHEA-S), and androstenedione. They are only mildly androgenic at physiological levels and they promote protein synthesis. The powerful androgen, testosterone, comes from the testis or ovary, not the adrenal cortex. Most circulating androgens, like cortisol, are protein bound, mainly to the sex-hormone-binding globulin (SHBG) and albumin.

There is extensive peripheral interconversion of adrenal and gonadal androgens. The end-products, *androsterone* and *aetiocholanolone*, together with DHEA, are conjugated in the liver and excreted as glucuronides and sulphates.

CONTROL OF ADRENAL STEROID SECRETION

The hypothalamus, anterior pituitary and adrenal cortex form a functional unit (the hypothalamic-pituitary-adrenal axis—Fig. 11).

Cortisol is secreted in response to *adrenocorticotrophic hormone* (*ACTH*; corticotrophin) from the anterior pituitary. ACTH secretion is in turn dependent on a hypothalamic *releasing hormone* (corticotrophin releasing factor; CRF). At least three mechanisms influence CRF secretion.

Feedback mechanism.—High free cortisol levels suppress and low ones stimulate CRF and therefore ACTH secretion. The melanocyte-stimulating effect of ACTH (p. 128) accounts for the pigmentation of Addison's disease, in which cortisol levels are persistently low and ACTH levels persistently high. Marked pigmentation may also occur if adrenalectomy is performed to treat pituitary-dependent Cushing's syndrome: in this case removal of cortisol feedback causes a further rise in ACTH levels (Nelson's syndrome).

Inherent rhythms.—Sampling at frequent intervals has shown that ACTH is secreted *episodically*, each burst of secretion being followed 5 to 10 minutes later by cortisol secretion. These episodes are most frequent in the early morning (between the 6th and 8th hour of sleep) and least frequent in the few hours before sleep; plasma cortisol levels are usually *highest* between about 07.00 and 09.00 h and *lowest* between about 23.00 and 04.00 h.

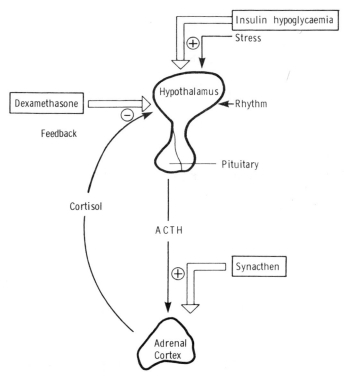

Fig. 11.—Control of cortisol secretion (+ =stimulates; − =inhibits). Dynamic tests shown in boxes.

We have seen that ACTH and cortisol secretion usually vary inversely and the almost parallel circadian rhythm of ACTH and cortisol secretion may be due to a cyclical change in the sensitivity of the hypothalamic feedback centre to cortisol levels. Inappropriately high cortisol levels at any time of day suppress ACTH secretion and inappropriately low levels stimulate it. The first effect can be tested by the *dexamethasone suppression test.*

Loss of the circadian rhythm is one of the earliest features of Cushing's syndrome.

Stress.—Physical or mental stress may override the first two mechanisms and cause sustained ACTH secretion. The consequent diagnostic difficulties are discussed on p. 147. An inadequate stress response may cause acute adrenocortical insufficiency. The stress caused by insulin-induced hypoglycaemia can be used to test the axis.

ACTH (Corticotrophin)

ACTH is a single-chain polypeptide made up of 39 amino acids. Only the 24 N-terminal amino acids are required for biological activity and this peptide has been synthesised (*tetracosactrin*; "Synacthen") and can be used in diagnosis and therapy in place of ACTH. ACTH *stimulates cortisol* synthesis and secretion by the adrenal cortex. It has much less effect on adrenal androgen production and, at physiological levels, virtually no effect on aldosterone production.

ESTIMATIONS USED IN THE DIAGNOSIS OF ADRENOCORTICAL DISORDERS

Cortisol estimation is the most useful. It may be measured in plasma or in urine by competitive protein-binding or radioimmunoassay, or *11-hydroxycorticosteroids* (11-OHCS) may be measured by a fluorimetric method which is, however, more susceptible to non-specific interference.

Whichever method is used, it is important to note that cortisol, like thyroxine, is largely protein bound. Primary changes in cortisol-binding globulin levels will alter values (p. 141) and should not be misinterpreted as indicating adrenocortical dysfunction.

Plasma ACTH assay may help in the differential diagnosis of Cushing's syndrome (p. 147). It is only available in special centres.

Urinary excretion of 17-oxosteroids (17-OS) and total 17-oxogenic steroids (17-OGS).—These substances are metabolites of androgens (adrenal and gonadal) and of cortisol and its precursors respectively. Their excretion is unreliable as a test of hormone secretion, and their estimation for diagnostic purposes has been superseded by plasma and (more rarely) urine assays of the individual hormones.

The selection of tests is discussed in the following pages. Blood is easy to collect but plasma levels reflect adrenocortical activity only at that moment. Because of the episodic nature of cortisol secretion, isolated values may be misleading. The measurement of steroids in a 24-hour urine sample reflects the *overall daily secretion*, but it is difficult to ensure the accuracy of timed collections.

DISORDERS OF THE ADRENAL CORTEX

The main disorders of adrenocortical function are:

Hyperfunction
Excessive secretion of:
cortisol   Cushing's syndrome.
aldosterone   primary aldosteronism (Conn's syndrome).

androgens   adrenocortical carcinoma;
                 congenital adrenal hyperplasia.

Hypofunction

primary   Addison's disease;   (deficient
                 congenital adrenal hyperplasia; } cortisol and
                                                      aldosterone)
secondary   ACTH deficiency (deficient cortisol).

CUSHING'S SYNDROME

Cushing's syndrome is produced by an excess of circulating cortisol
which may be due to:

bilateral adrenal hyperplasia, secondary to high circulating levels of
ACTH from:
   the pituitary (Cushing's disease, the commonest form in adults)
   a non-adrenal tumour (ectopic secretion, p. 475)
   therapeutic administration of ACTH or a synthetic analogue;
an adrenal tumour (adenoma or carcinoma);
administration of cortisol or cortisone. (High-dose therapy with
related steroids may produce a similar clinical picture.)

In the two latter groups plasma ACTH levels are low.

Clinical and Metabolic Features

Cushing's syndrome may occur at any age. The commonest form,
that caused by pituitary-dependent adrenal hyperplasia, usually presents in women of reproductive age. These patients often have a
basophil adenoma of the pituitary which may be very small. It is not
known whether the tumour is the primary lesion or whether it is
secondary to excessive CRF secretion by the hypothalamus.

Consistently high circulating cortisol levels produce truncal obesity
and a round, red, "moon" face. These patients may also develop
psychiatric disorders, especially depression. Although the reason for
these features is not understood, many of the other clinical and
metabolic disturbances can, at least partly, be explained by the known
actions of cortisol.

The most prominent metabolic abnormalities reflect the glucocorticoid action. About two-thirds of the patients have impaired glucose
tolerance and of these many have hyperglycaemia and glycosuria. As
cortisol has an opposite action to insulin the resultant picture in some
ways resembles diabetes mellitus. Protein breakdown is accelerated
and the carbon chains of the liberated amino acids may be converted
to glucose (gluconeogenesis); the nitrogen is lost in the urine leading

to a negative nitrogen balance. This catabolic effect not only causes osteoporosis and muscle wasting and therefore weakness, but also thinning of the skin. The tendency to bruising, and the purple striae, most obvious on the abdominal wall, are probably due to this thinning.

The mineralocorticoid effect of cortisol leads to urinary sodium, and therefore water, retention which produces hypertension. Increased urinary potassium excretion may result in hypokalaemia and aggravate muscle weakness. This is a prominent feature of Cushing's syndrome due to ectopic ACTH secretion.

Adrenal androgen secretion may be increased. In pituitary-dependent Cushing's syndrome this is not marked, but may cause hirsutism and menstrual irregularities in the female. Virilisation is rare except in patients with an adrenocortical carcinoma.

Basis of Investigation of Suspected Cushing's Syndrome

Because untreated Cushing's syndrome has a high morbidity and mortality rate, any suspected case must be adequately investigated. Such patients are usually obese, red-faced, hairy, hypertensive women, most of whom do *not* have Cushing's syndrome. The problem is more often one of exclusion than of confirmation. However, as the tests available are not specific, it is *essential* to apply them in a logical sequence.

What we require to know is:

(a) is there abnormal cortisol secretion?

(b) if so, does the patient have any other condition that may cause it?

(c) if it is Cushing's syndrome, what is the cause?

(a) Is there Abnormal Cortisol Secretion?

Plasma cortisol levels in the morning may be normal or only moderately raised. However, as one of the earliest features of Cushing's syndrome is a reduction in amplitude or loss of the circadian rhythm, high levels in the late evening, when secretion is normally at a minimum are suggestive but not diagnostic of the disorder.

As discussed on p. 141, most of the cortisol in plasma is bound to protein. Only the small free fraction is filtered at the glomerulus and excreted (*urinary "free" cortisol*). Normally, however, CBG is almost fully saturated, and increased cortisol secretion produces a disproportionate rise in the free fraction, with a corresponding increase in urinary free cortisol. Because of the loss of the circadian rhythm, these raised plasma values are present for a longer than normal time, and daily urinary cortisol excretion is further increased. Thus urinary free cortisol excretion will be more markedly raised than plasma levels. Plasma and urinary cortisol levels are much higher when

Cushing's syndrome is due to adrenocortical carcinoma or ectopic ACTH secretion.

In addition, the sensitivity of the feedback centre to circulating cortisol levels is reduced. A small dose of the synthetic steroid *dexamethasone*, which suppresses ACTH, and consequently cortisol secretion in normal subjects, will not do so in those with pituitary-dependent Cushing's syndrome. In cases due to an *adrenal tumour* or to *ectopic ACTH*, ACTH secretion from the pituitary is already suppressed and dexamethasone will have no effect.

The *dexamethasone suppression test* is thus a sensitive, but not completely specific test in evaluating such patients. A normal fall in cortisol levels makes the diagnosis of Cushing's syndrome very unlikely.

(b) Is there Another Cause for the Abnormal Cortisol Secretion?

Abnormal cortisol secretion may be due to causes other than Cushing's syndrome. The following are of particular importance:

> *Stress* overrides the other controlling mechanisms of ACTH secretion and leads to loss of the normal circadian variation of plasma cortisol and a reduced feedback response. Urinary free cortisol values may be markedly increased. This may occur in even relatively minor physical illness as well as in psychiatric disturbances, particularly endogenous depression. Indeed, failure of dexamethasone suppression has been used to distinguish endogenous from other forms of depression.
>
> *Severe alcohol abuse* can produce hypersecretion of cortisol that mimics Cushing's syndrome clinically and biochemically. The abnormal findings revert to normal when alcohol is stopped.
>
> *Simple obesity*, often accompanied by hirsutism and hypertension, may be associated with a shortened half-life, and increased turnover of cortisol. Plasma and urinary cortisol levels are, however, usually normal and are usually suppressed by dexamethasone.

(c) What is the Cause of the Cushing's Syndrome?

Only when Cushing's syndrome has been confirmed should tests to establish the cause be carried out.

The test of choice is *estimation of plasma ACTH*. Levels are moderately raised in pituitary-dependent adrenal hyperplasia, markedly raised in that due to ectopic ACTH secretion and very low in cases of adrenocortical tumour (feedback suppression by high cortisol levels).

If ACTH estimation is not available, a *high-dose dexamethasone test* may help. The principle of the test is as described earlier except that a high enough dose of dexamethasone is given to suppress the relatively insensitive feedback centre in pituitary-dependent disease. In the other two categories, when pituitary ACTH is already suppressed, even this high dose will have no effect.

It must be emphasised that these tests can only be interpreted if the diagnosis of Cushing's syndrome has been established. For example, in an ill patient with high (but appropriate) cortisol levels, plasma ACTH levels are raised but may be suppressed only by a high dose of dexamethasone. This may lead to an erroneous diagnosis of pituitary-dependent Cushing's syndrome.

Some findings may be suggestive of one of the uncommon causes of Cushing's syndrome. Virilisation, rather than mild hirsutism, suggests an adrenocortical carcinoma. Severe hypokalaemic alkalosis may occur with ectopic ACTH-producing tumours and these patients may become increasingly pigmented due to the high ACTH levels.

Table XIV summarises the results of tests in Cushing's syndrome.

ADRENOCORTICAL HYPOFUNCTION

Primary Adrenocortical Hypofunction (Addison's Disease)

Addison's disease is caused by destruction of all zones of the adrenal cortex. Tuberculosis was once the commonest cause but it has now been superseded in many countries by atrophy of the gland, due to an autoimmune process. Rare causes of destruction (which must be bilateral to cause adrenal failure) include amyloidosis, mycotic infections, and secondary malignancy which often originates from the bronchus.

Clinical features.—The presentation of Addison's disease depends on the degree of adrenal destruction. If this is extensive the patient may be shocked or dehydrated (Addisonian crisis): this is a medical emergency requiring immediate treatment. If destruction is less extensive the disease may be diagnosed only after a prolonged period of ill health, or if a crisis is precipitated by stress. The clinical features of this form—tiredness, pigmentation of the skin and buccal mucosa, weight loss and hypotension—are common to many severe chronic diseases.

The most serious consequences are due to the *mineralocorticoid (aldosterone) deficiency. Sodium depletion* is the cardinal biochemical abnormality of the Addisonian state. Loss of sodium through the kidneys is accompanied by loss of water and while these parallel one another the plasma sodium concentration remains in the normal range (see Fig. 2, p. 40). The volume depletion and consequent haemo-

TABLE XIV

Test Results in Cushing's Syndrome

	Adrenocortical hyperplasia		Adrenocortical tumour	
	Pituitary dependent	Ectopic ACTH	Carcinoma	Adenoma
DIAGNOSIS				
Plasma cortisol				
morning	raised, may be N	raised	raised, may be N	
evening	raised			
after 2 mg dexamethasone	no suppression			
Urinary cortisol	raised	usually very high	raised	
AETIOLOGY				
Plasma ACTH	high normal or moderately raised	raised	very low	
Plasma cortisol after 8 mg dexamethasone	suppression	no suppression		
17-oxosteroids or plasma DHEA-S	usually N	often raised	usually very high	normal

concentration are nevertheless evident from the clinical state and usually from the raised haematocrit and total protein concentration. During a crisis, however, acute contraction of the circulating volume may stimulate ADH secretion: water is then reabsorbed in excess of sodium and hyponatraemia develops. As the loss is due to inability of the tubules to reabsorb sodium adequately in the absence of aldosterone the urine will contain an inappropriately high sodium concentration for the degree of volume depletion (p. 12) (Estimation of urinary sodium is, however, *not* a useful diagnostic test.) The Addisonian crisis is therefore the result of massive sodium depletion. Other abnormalities usually include a raised plasma potassium concentration and metabolic acidosis (compare the reverse that occurs with the mineralocorticoid excess of primary hyperaldosteronism p. 59). The fluid depletion leads to a reduced circulating blood volume and renal circulatory insufficiency with a reduced glomerular filtration rate and a moderately raised plasma urea. In the more chronic form of Addison's disease only some of these features may be present.

The absence of glucocorticoids contributes to the hypotension and causes marked sensitivity to insulin; there may even be fasting hypoglycaemia.

Androgen deficiency is not clinically evident because testosterone production by the testis is unimpaired, and because androgen deficiency does not produce obvious effects in females.

The pigmentation that develops in Addison's disease is probably due to high circulating ACTH levels resulting from the lack of cortisol suppression of the feed-back mechanism.

Secondary Adrenal Hypofunction

ACTH release may be impaired by disease of the hypothalamus or of the anterior pituitary, most commonly due to tumour or infarction. Corticosteroid therapy suppresses ACTH release and after such therapy, especially if prolonged, the ACTH-releasing mechanism may be slow to recover. There may be temporary adrenal atrophy after prolonged lack of stimulation.

Complete anterior pituitary destruction results in the picture of panhypopituitarism (p. 134) but if destruction is only partial there may be sufficient ACTH for basal requirements. As in partial adrenal cortical destruction (p. 148) the deficiency may only become evident under conditions of stress, which in these patients, as in those on corticosteroid therapy, may precipitate acute adrenal insufficiency. The most usual causes of stress are infections and surgery.

The adrenal crisis due to ACTH deficiency is due to lack of glucocorticoids only. It differs from an Addisonian crisis in that aldosterone secretion (not under the influence of ACTH) is normal; consequently there is not the characteristic sodium depletion and dehydration, although there may be *dilutional* hyponatraemia. The condition is nevertheless potentially fatal and is ushered in by mental disturbance, nausea, abdominal pain and usually hypotension (cortisol deficiency). As in primary glucocorticoid deficiency there may be hypoglycaemia and marked insulin sensitivity.

Basis of Investigation of Suspected Adrenocortical Hypofunction

If the patient presents with a probable diagnosis of acute adrenal insufficiency, blood should be taken so that cortisol can be measured later, but *treatment must be started at once.*

The performance of further tests can await recovery from the crisis: the underlying abnormality will not be affected by therapy. Cortisol results on the initial sample should distinguish between adrenocortical insufficiency (inappropriately low values) and clinically similar crises when levels are markedly raised due to stress. It will not however, separate primary from secondary adrenocortical failure.

In the less acutely ill patient, a morning plasma cortisol level should be measured, and, if unequivocally high, no further tests are necessary. In most, however, levels will be within the "normal" range. *The essential abnormality in adrenocortical hypofunction is that the patient is unable to respond to stress by increasing cortisol secretion normally.* As this may be due to adrenal (primary) or hypothalamic-pituitary (secondary) pathology, it may be necessary to test the whole axis. Lesions of the adrenal cortex are the commonest cause of adrenal hypofunction, and its integrity should be tested first: furthermore, interpretation of pituitary function tests depends on knowing that the adrenal cortex can respond to ACTH. The response of the cortex may be tested by measuring plasma cortisol levels after stimulation by synthetic ACTH. Tetracosactrin (for example, Synacthen) has the same biological action as ACTH but as it lacks the antigenic part of the molecule there is much less danger of an allergic reaction.

Plasma ACTH estimation is of value in some cases. When inappropriately low cortisol values have been found, a raised ACTH level indicates primary, and a low level secondary, insufficiency. This will distinguish between true adrenocortical disease (when mineralocorticoid replacement is required) and reversible atrophy due to prolonged lack of ACTH.

Urinary cortisol excretion is often low.

Corticosteroid Therapy

There is the risk of adrenocortical hypofunction when long-term corticosteroid therapy is stopped suddenly. This may be due either to secondary adrenal atrophy or to impairment of ACTH-releasing mechanisms.

A simple means of testing the feedback centre and the responsiveness of the pituitary-adrenal axis is to estimate the morning plasma cortisol levels two or three days after stopping steroid therapy. A level within the normal range indicates a functioning adrenal, pituitary and feedback centre. It must be emphasised, however, that this does not test the all-important stress pathway (p. 143).

A suggested sequence of testing is described on p. 167).

CONGENITAL ADRENAL HYPERPLASIA

Rarely there may be an inherited deficiency of one of the enzymes involved in the biosynthesis of cortisol (Fig. 12). As a result, plasma cortisol levels tend to fall, leading to increased secretion of ACTH from the pituitary. This in turn leads to hyperplasia of the adrenal cortex, with increased synthesis of cortisol precursors before the enzyme block. The precursors may then be metabolised by alternative pathways, especially those of androgen synthesis.

C-21 β-hydroxylase, which is involved in the synthesis of both cortisol and aldosterone, is the enzyme most commonly affected.

The increased androgen production may cause:

female pseudohermaphroditism by affecting the development of the female external genitalia *in utero*;

virilisation in childhood with phallic enlargement in either sex, development of pubic hair and rapid growth rate;

milder virilisation in females at or after puberty, with amenorrhoea.

Aldosterone synthesis may be markedly reduced and cause an Addisonian-like picture during the first few weeks of life. Vomiting is severe and there is hyponatraemia and hyperkalaemia with marked volume depletion. This occurs in about half the patients with 21-hydroxylase deficiency but even in those with normal plasma sodium levels, demonstrably increased plasma renin activity may suggest lesser degrees of sodium and volume depletion.

All female babies with ambiguous genitalia should have plasma electrolytes estimated. In male babies with no obvious physical abnormalities the diagnosis may not be suspected.

Diagnosis

The investigation of suspected congenital adrenal hyperplasia is best undertaken in special centres. Only the principles of the diagnosis of 21-hydroxylase deficiency will be outlined here (see Fig. 12).

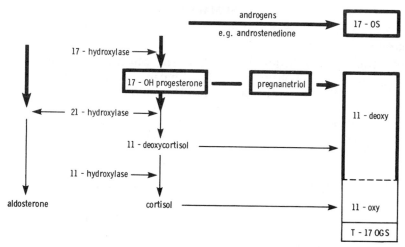

FIG. 12.—The abnormalities occurring in congenital adrenal hyperplasia due to 21-hydroxylase deficiency. Substances of diagnostic importance are shown in boxes.

17-hydroxyprogesterone can only be metabolised by the cortisol pathway in the presence of 21-hydroxylase. Plasma levels of this progestogen are raised and may be measured directly, or its metabolite, *pregnanetriol*, may be measured in the urine.

Excessive androgen synthesis is best shown by demonstrating high *plasma androstenedione* levels. If plasma assays are not available, the less satisfactory *11-oxygenation index* may be measured. Because the final stage of cortisol synthesis is 11-hydroxylation, any block in the pathway (as in congenital adrenal hyperplasia) results in a decreased proportion of total 17-oxogenic steroids with an −OH group at position 11. The ratio of the steroids without an 11-OH to those with 11-OH (the 11-oxygenation index) will usually be greater than normal (Fig. 12).

Detailed evaluation of the pattern of urinary steroid excretion (on a random sample) will indicate which enzyme is deficient.

Congenital adrenal hyperplasia is treated with cortisol or one of the glucocorticoid analogues with, if necessary, a mineralocorticoid. This treatment not only replaces the deficient hormones but, by feedback, suppresses excessive ACTH secretion and therefore excessive androgen production.

SUMMARY

1. The steroids of the adrenal cortex may be classified into three groups.

Glucocorticoids, for example cortisol, which influence intermediary metabolism, body fluid balance and blood pressure. Secretion is controlled by ACTH from the pituitary.

Mineralocorticoids, for example aldosterone, which influence the balance of sodium, and secondarily water, and potassium. Secretion is controlled by the renin-angiotensin system.

Androgens, for example androstenedione, which probably have an anabolic action on protein synthesis.

2. ACTH (and therefore cortisol) secretion is stimulated by CRF from the hypothalamus. The important controlling factors are:

an *inherent rhythm* which produces an overall circadian variation in plasma cortisol levels. These are lowest around midnight and highest in the morning.

negative feedback by circulating *free cortisol* levels;

stress which may override both the above controls.

3. Estimations of plasma and urinary cortisol are the most useful tests. Plasma ACTH measurement may be valuable in the differential diagnosis of Cushing's syndrome and, more rarely, to differentiate primary from secondary adrenal failure.

4. *Cushing's syndrome* is due to excessive circulating cortisol. The causes are hyperplasia or tumour of the adrenal cortex. Hyperplasia is due to excess ACTH either from the pituitary (pituitary-dependent) or from a non-endocrine tumour.

There are three stages in investigation.

    (a) Is there abnormal cortisol secretion?

    (b) If so, can it be accounted for by a condition other than Cushing's syndrome?

    (c) If not, what is the cause of the Cushing's syndrome?

5. Primary adrenocortical hypofunction (*Addison's disease*) is due to destruction of the adrenal cortex with loss of all its hormones. Aldosterone deficiency causes sodium and consequent water depletion.

Diagnosis is made by demonstrating that the adrenal cortex cannot respond to stimulation by exogenous ACTH (or an analogue).

6. *Secondary adrenal insufficiency* is caused by diminished ACTH secretion by the pituitary. This may be due to disease of the hypothalamus or pituitary or it may be a result of corticosteroid therapy.

It is usually diagnosed by demonstrating a definite but impaired response to tetracosactrin (synthetic ACTH) stimulation. In cases without adrenal atrophy the hypothalamic-pituitary-adrenal axis may be assessed by the insulin stress test.

7. *Congenital adrenal hyperplasia* is due to an inherited enzyme deficiency in the biosynthesis of cortisol. Symptoms are due to a deficiency of cortisol and to excessive androgen secretion. Diagnosis of the commonest form is made by estimation of plasma 17-hydroxy-progesterone. Plasma androgen levels are raised.

FURTHER READING

CUSHING, H. (1932). The basophil adenomas of the pituitary body and their clinical manifestations (pituitary basophilism). *Bull. Johns Hopkins Hosp.*, **50**, 137–195.

URBANIC, R. C. and GEORGE, J. M. (1981). Cushing's disease—18 years' experience. *Medicine*, **60**, 14–24.

ARON, D. C., TYRRELL, J. B., FITZGERALD, P. A., FINDLING, J. W. and FORSHAM, P. H. (1981). Cushing's syndrome: problems in diagnosis. *Medicine*, **60**, 25–35.

CRAPO, L. (1979). Cushing's syndrome: a review of diagnostic tests. *Metabolism*, **28**, 955–977.

RUDD, B. T. (1983). Urinary 17 oxogenic and 17 oxosteroids. A case for deletion from the clinical chemistry repertoire. *Ann. Clin. Biochem.*, **20**, 65–71.

(See also the general reference on p. 137.)

Chapter VII

THE GONADS: GONADOTROPHINS: PROLACTIN

THE study of the endocrinology of the male and female reproductive systems is a specialised and rapidly growing field. Nevertheless many disorders can be assessed using the general principles of endocrine diagnosis, based on a knowledge of normal function. This section will deal only with such principles and for more detail the reader should consult the references at the end of the chapter.

THE HYPOTHALAMIC-PITUITARY-GONADAL AXIS

GONADOTROPHINS

The pituitary gonadotrophins, *luteinising hormone* (LH) and *follicle-stimulating hormone* (FSH), control the function of, and secretion of hormones by, the ovary and testis. Gonadotrophin secretion is regulated by a hypothalamic gonadotrophin-releasing hormone (Gn-RH or LH/FSH-RH). Although there is only one releasing hormone, secretions of LH and FSH do not always occur in parallel and may also be modified by feedback by the gonadal oestrogens or androgens. The testis may also produce *inhibin*, which inhibits FSH secretion.

Prolactin

The control of prolactin differs from that of the other pituitary hormones because its secretion is regulated by a hypothalamic *inhibitory* factor, probably *dopamine*; impairment of hypothalamic control, therefore, causes hyperprolactinaemia. Otherwise, prolactin and GH secretion (p. 131) are affected by similar mechanisms; for example, plasma levels of both rise during sleep and in response to physical or psychological stress. There is no obvious feedback control of either prolactin or GH. *Oestrogens*, however, enhance prolactin secretion and plasma concentrations are normally higher in women than in men. Administered thyrotrophin-releasing hormone (TRH, p. 178) stimulates prolactin as well as TSH secretion but does not appear to be a physiological regulatory factor for prolactin.

The function of prolactin in non-pregnant subjects is not known. Its role in pregnancy is discussed later.

THE FEMALE

The hormones secreted by the ovary vary during the menstrual cycle.

Oestrogens are produced from androgens and differ from them in that they have an aromatic A ring and have lost the C-19 methyl group (Figs. 9 and 10, pp. 139 and 140). This minor chemical difference has far-reaching biological effects, and accounts for "la différence". Oestrogens are essential for the development of female secondary sex characteristics and for normal menstruation; levels are usually undetectable in children.

Oestradiol, the most important ovarian oestrogen, is derived from locally synthesised androgens. The liver and subcutaneous fat convert ovarian and adrenal androgens to *oestrone*. Both oestradiol and oestrone are metabolised to the relatively inactive *oestriol*.

The ovary also secretes *androgens*, mainly *androstenedione*, which is converted in extra-ovarian tissues not only to oestrone, but also to the more active androgen *testosterone*. A small amount of testosterone is secreted directly by the ovary. Plasma testosterone levels in women are about a tenth of those in males.

Progesterone is secreted by the corpus luteum and is chemically similar to the adrenocortical progestogens (p. 153). It prepares the endometrium to receive a fertilised ovum, and is necessary for the maintenance of early pregnancy.

The Normal Menstrual Cycle

The menstrual cycle is regulated by changing hormone levels (Fig. 13) and changing sensitivity of ovarian tissue.

Follicular (Pre-ovulatory) Phase

At the beginning of the cycle, during menstruation, ovarian follicles are undeveloped, and oestrogen levels are low. The diminished negative feedback allows FSH and LH secretion to increase.

> *FSH and LH together cause growth of a group of follicles and maturation of the follicular cells. LH also stimulates secretion of hormones* from these cells and circulating oestrogen levels rise steadily. This stimulates regeneration of the previously shed endometrium.

The rising oestrogen levels cause a slight fall in FSH secretion by negative feedback to the pituitary gland.

One of the group of follicles becomes relatively independent of pituitary FSH and continues to grow while the rest atrophy.

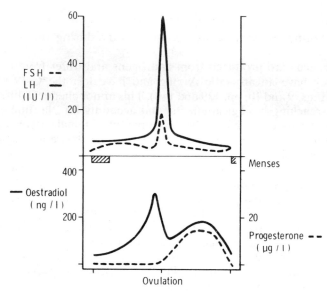

FIG. 13.—Plasma hormone levels during a typical menstrual cycle.

Ovulation

The dominant follicle develops rapidly and secretes a large amount of oestradiol, which triggers a surge of LH release from the pituitary by *positive* feedback. Ovulation occurs about 16 hours after this. The high LH levels inhibit oestrogen and stimulate progesterone secretion, and the follicle gradually develops into the corpus luteum.

Luteal (Postovulatory or Secretory) Phase

This phase is characterised by the rise and fall of the corpus luteum which takes over ovarian hormone secretion.

> *LH stimulates the development of the corpus luteum,* and the secretion of *progesterone and oestrogens* from it.
> *Progesterone prepares the endometrium* to receive a fertilised ovum.

The subsequent events depend on whether the released ovum is fertilised. If it is not, the menstrual cycle takes its course; if it is, pregnancy may supervene.

> Falling levels of ovarian hormones, as the corpus luteum regresses, cause endometrial sloughing and menstrual bleeding.
> As ovarian hormone levels fall, FSH and LH levels begin to rise. The cycle recommences.

Prolactin levels do not change cyclically during the menstrual cycle. Interpretation of sex-hormone levels must be made in relation to the stage of the cycle. For example, it may be important to establish whether a patient complaining of infertility has ovulated, either spontaneously or as a result of therapy to induce ovulation. Plasma progesterone should be measured on a blood sample taken during the second half of the menstrual cycle. A value within the "normal range" for the time of the cycle is good presumptive evidence of ovulation, while a value in the range expected in the follicular phase (Fig. 13) indicates the absence of a corpus luteum, and therefore of ovulation.

Pregnancy and Lactation

If the ovum is fertilised it may implant in the endometrium which has been prepared by progesterone (see "Luteal Phase"). Gonadotrophin secretion is soon taken over by the chorion and developing placenta. This *chorionic gonadotrophin (HCG) is similar in structure and action to LH*, and prevents the involution of the corpus luteum as pituitary gonadotrophin levels fall. *Oestrogen and progesterone levels therefore continue to rise*, and the last two events of the "Luteal Phase" are prevented. After the first trimester these hormones are produced by the placenta (p. 458).

Probably because of the high oestrogen levels, prolactin secretion increases progressively after the eighth week of pregnancy and at term may be 10 to 20 times higher than that of the non-pregnant woman.

Prolactin, oestrogens, progesterone and a placental lactogen (HPL, p. 458) stimulate breast development for lactation. Oestrogens inhibit milk secretion (and were once used clinically for this purpose), and lactation can only start when levels fall after delivery of the placenta.

Early lactation depends on prolactin and in its absence (resulting from, for example, pituitary infarction or administration of bromocriptine, which inhibits prolactin secretion) milk production ceases. Suckling stimulates prolactin secretion; nevertheless, despite continuing successful lactation, plasma concentrations fall progressively postpartum and reach non-pregnant levels after 2 or 3 months. Apart from the effects on the breast, the high concentration of prolactin interferes with gonadotrophin-gonadal function, producing a period of relative infertility.

The Menopause

Postmenopausally, plasma oestrogen levels fall due to ovarian "failure". Removal of normal premenopausal feedback to the pituitary causes increased FSH and, to a lesser extent, LH secretion. The findings are identical with those of primary gonadal failure (p. 160).

Disorders of Gonadal Function in the Female

Women with possible gonadal dysfunction may present with any or all of the following complaints:

    delayed puberty and primary amenorrhoea;
    infertility, with or without amenorrhoea;
    hirsutism;
    virilisation.

The investigation of these disturbances cannot be comprehensively covered here and the attention of the reader is drawn to the references given at the end of the chapter for further details. The following is a guide to the more obvious points, and stresses the underlying biochemical abnormalities.

Amenorrhoea

Amenorrhoea is most commonly due to hormonal abnormalities. If there is ovarian (end organ) failure, pituitary gonadotrophin levels are high ("hypergonadotrophic hypogonadism"); if the cause is in the hypothalamus or pituitary, gonadotrophin secretion is reduced ("hypogonadotrophic hypogonadism").

Primary amenorrhoea is the term used when the patient has *never menstruated*. It is most commonly associated with delayed puberty. The age of the menarche is very variable; unless other signs of endocrine disturbances (such as hirsutism and virilisation) are present extensive investigation should probably be postponed until after the age of 18 years.

Secondary amenorrhoea occurs when the patient who *has previously menstruated* has now stopped. By far the commonest causes are *pregnancy* and the *menopause* (the timing of which is also extremely variable). It may occur during any chronic illness. These causes should always be thought of before extensive, expensive and sometimes uncomfortable and dangerous investigations are started.

(Note that "primary" and "secondary" when used to describe amenorrhoea have different meanings from "primary" and "secondary" endocrine disease).

Hyperprolactinaemia is a common cause of amenorrhoea and infertility. About 30 per cent of patients with high plasma prolactin levels have galactorrhoea. LH and FSH levels are in the low normal range, probably because prolactin inhibits GnRH release.

The most important cause of hyperprolactinaemia is a pituitary tumour (usually a chromophobe adenoma) which may be so small as to be undetectable by other means. *Idiopathic hyperprolactinaemia* may follow the use of oral contraceptive preparations. Patients with no apparent cause for hyperprolactinaemia should be watched for the development of an obvious pituitary adenoma. The higher the prolactin level, the greater the likelihood that a tumour is the cause.

The finding of hyperprolactinaemia should be interpreted with caution. It is particularly important to take samples for prolactin estimation at least 2–3 hours after waking to eliminate misleadingly elevated levels during sleep, and to remember that even the minor stress of venepuncture may cause prolactin secretion. Furthermore, drugs —in particular chlorpromazine, reserpine and methyldopa (p. 129) —may produce this effect, and a drug history should always be taken.

Hyperprolactinaemia is occasionally found, rarely with amenorrhoea or galactorrhoea:

in severe primary hypothyroidism (probably due to stimulation of secretion by excessive TRH). A pituitary pathology should not be diagnosed on the basis of this finding;

after surgical section of the pituitary stalk (lack of inhibitory factor). This is unlikely to pose a diagnostic problem;

in about 20 per cent of cases of chronic renal failure (cause unknown).

Patients with amenorrhoea may also have signs of androgen excess (see below). Amenorrhoea may rarely be the presenting feature of another endocrine disorder such as acromegaly or Cushing's syndrome.

The investigation of amenorrhoea is considered on p. 168.

Hirsutism and Virilism

Increased androgen levels, or increased sensitivity of tissues to androgens, produce effects ranging from increased hair growth to marked masculinisation. As discussed on p. 142, androgens are secreted by both the adrenal cortex and the ovary. The adrenal secretes dehydro-epiandrosterone (DHEA) and DHEA-sulphate (DHEA-S) as well as androstenedione. The ovary secretes androstenedione and testosterone. Although there is extensive peripheral interconversion of androgens, raised levels of DHEA or DHEA-S indicate an adrenocortical source and raised levels of testosterone an ovarian source of hypersecretion. Most of the androgens in plasma are bound to sex-hormone-binding globulin (SHBG) and only the free fraction is active.

Hirsutism and virilism are different and it is essential to distinguish between them.

Hirsutism (excessive growth of facial and body hair) is a common complaint and is often associated with menstrual irregularities. It may be familial or racial. Plasma androgen levels are often within "normal" limits, but may be increased, whether derived from the adrenal cortex, the ovary, or both. In those cases with apparently normal levels there may be a reduced level of SHBG (with increased free androgen levels) or tissues may be abnormally sensitive to androgens.

A common cause is the *polycystic ovary syndrome*, which is characterised by increased plasma LH and testosterone concentrations. By contrast, FSH levels are low and follicle development is impaired. Patients may present with any or all of the symptoms of infertility, menstrual disturbances, hirsutism or obesity, and may be found to have cystic ovaries.

Virilism is uncommon but much more serious and is always associated with increased plasma androgen levels. The patient presents with *other evidence of excessive androgen secretion* such as enlargement of the clitoris, increased hair growth of male distribution, receding temporal hair, deepening of the voice and breast atrophy. The main causes are:

Ovarian tumours (such as arrhenoblastoma and hilus-cell tumour) which secrete androgens, mainly testosterone;

Adrenocortical pathology:
tumours, usually carcinoma;
hyperplasia: pituitary-dependent Cushing's syndrome (rarely);
congenital adrenal hyperplasia.

Plasma DHEA and DHEA-S levels are increased.

Effects of Drugs on the Female Hypothalamic-pituitary-gonadal Axis

Oral contraceptives contain synthetic oestrogens and progestogens. By suppressing pituitary gonadotrophin secretion they *prevent ovulation*. Withdrawal mimics involution of the corpus luteum and results in menstrual bleeding.

Clomiphene, by blocking oestrogen receptors in the hypothalamus, prevents negative feedback and may initiate gonadotrophin release, even when circulating levels of oestrogen and progesterone are high: it may be used therapeutically to *induce ovulation* in subjects with secondary hypogonadism ("fertility drug").

Gonadotrophin therapy may be used if clomiphene fails to induce ovulation in women with secondary hypogonadism. Human menopausal gonadotrophin (predominantly FSH, with some LH) is given to mimic the follicular phase. It carries the risk of "hyperstimulation" and *must be monitored* by frequent plasma or urinary oestrogen estimation until levels reach those of the normal pre-ovulatory peak. HCG is then given to induce ovulation by mimicking the LH peak. Success may be assessed by demonstrating rising progesterone levels.

Bromocriptine rapidly *reduces pathologically high levels of prolactin,* whatever the cause: menstruation restarts and fertility is restored.

THE MALE

LH stimulates the interstitial (Leydig) cells of the testis to secrete the powerful androgen, testosterone. Prepubertal levels of testosterone are low, and the increase at puberty is responsible for the development of male characteristics. *Testosterone inhibits gonadotrophin secretion* by a negative feedback similar to that of the ovarian hormones in the female.

Testosterone is usually estimated in plasma by radioimmunoassay.

FSH stimulates spermatogenesis. Inhibin, probably produced during sperm maturation, inhibits FSH secretion.

Basal prolactin levels are *lower in males* than in females.

GONADAL DISTURBANCES IN THE MALE

Delayed puberty and infertility are the usual presenting features of gonadal disturbances in the male. As in the female, evidence of other endocrine disease and of Klinefelter's syndrome (primary hypogonadism) should be sought (p. 160). Infertility, and very rarely, galactorrhoea may be due to hyperprolactinaemia but *there is no association between hyperprolactinaemia and gynaecomastia.* The method of distinguishing primary from secondary testicular failure is similar to that for distinguishing primary from secondary ovarian failure in the female, but testosterone is estimated instead of oestrogen.

If male infertility is due to *predominant seminiferous tubular failure* (failure of spermatogenesis), Leydig cell function and testosterone secretion may be normal; LH levels are therefore also normal. FSH levels, however, are raised, probably due to deficiency of inhibin production.

INVESTIGATION OF PRECOCIOUS PUBERTY IN EITHER SEX

Hormone estimations may help to distinguish *true precocious puberty*, due to excessive gonadotrophin secretion, from *pseudoprecocious puberty*. In the latter, sex hormone production by an ovarian or testicular tumour will lead to the development of the secondary sexual characteristics, but with undeveloped gonads. Plasma gonadotrophin levels will be suppressed. In boys congenital adrenal hyperplasia must be considered (p. 152).

DETECTION AND FOLLOW-UP OF TROPHOBLASTIC TUMOURS

Trophoblastic tumours (hydatidiform mole, chorionepithelioma), and some teratomas secrete HCG. For diagnostic purposes the ordinary pregnancy tests are usually adequate but they are not sensitive enough for follow-up studies. HCG estimation in plasma or urine by radio-immunoassay is at least 20 times more sensitive and, if performed at regular intervals, allows early detection and treatment of recurrence. It will *not* differentiate between pregnancy and recurrence of a tumour, because in both cases HCG levels are high.

SUMMARY

1. The secretion of gonadal hormones is controlled by the pituitary gonadotrophins LH and FSH. These, in turn, are influenced by hypothalamic-releasing factors and, via negative feedback, by circulating gonadal hormone levels.

2. Cyclical hormonal changes prepare the endometrium for implantation of a fertilised ovum. In the first half of the cycle LH and FSH stimulate ovarian follicle development and oestrogen secretion. A mid-cycle LH surge induces ovulation in one follicle, and converts it into a corpus luteum which secretes oestrogens and progesterone. In the absence of successful implantation the corpus luteum involutes and, as hormone levels fall, endometrial breakdown and shedding occur (menstruation). If implantation occurs the developing placenta produces HCG which maintains the corpus luteum and prevents menstruation.

3. Gonadal hormone levels must be interpreted in relation to the stage of the menstrual cycle.

4. In primary ovarian failure (and at the menopause) oestrogen levels are low and gonadotrophin levels high (negative feedback). In secondary ovarian failure both oestrogen and gonadotrophin levels are low.

5. In the male LH stimulates testosterone secretion from the testicular Leydig cells, and FSH stimulates spermatogenesis. Testicular failure may be confined to failure of spermatogenesis (raised FSH only) or also involve the Leydig cells when levels of testosterone are low and those of both LH and FSH are raised.

6. Hyperprolactinaemia is commonly due to a pituitary adenoma. In the female it produces amenorrhoea and infertility, sometimes accompanied by galactorrhoea. Reducing prolactin secretion by bromocriptine restores fertility.

7. Gonadal hormone and gonadotrophin estimation are of value in:
      detection of ovulation;
      monitoring of therapy designed to induce ovulation;
      assessment of amenorrhoea;
      evaluation of hypogonadism and infertility;
      evaluation of delayed puberty;

evaluation of virilism;
follow-up of trophoblastic tumours.

FURTHER READING

MAROULIS, G. B. (1981). Evaluation of hirsutism and hyperandrogenemia. *Fertil. and Steril.*, **36**, 273–305.

FRANKS, S. (1981). Male reproductive endocrinology. *Clin. Obstet. Gynec.*, **8**, 549–569.

FRANTZ, A. G. (1978). Prolactin. *New Engl. J. Med.*, **298**, 201–207.

KATZ, M. (1981). Polycystic ovaries. *Clin. Obstet. Gynec.*, **8**, 715–731.

McDONOUGH, P. G. (1978). Amenorrhea—etiologic approach to diagnosis. *Fertil. and Steril.*, **30**, 1–15.

INVESTIGATION OF PITUITARY, ADRENAL AND GONADAL DISORDERS

These sections are considered together because pituitary disorders, particularly hypofunction, may present with features of adrenal or gonadal deficiency. Protocols for the tests are given on p. 171.

SUSPECTED HYPOPITUITARISM

Deficiency of pituitary hormones causes hypofunction of the target endocrine glands. Investigation aims to confirm such deficiency, to exclude primary disease of the target gland and then to test pituitary hormone secretion after maximal stimulation.

1. Measure:

   plasma LH, FSH, and oestradiol (female) or testosterone (male);
   plasma thyroxine and TSH;
   plasma prolactin (for hypothalamic or pituitary-stalk involvement);
   plasma cortisol at 09·00 h (to assess the risk of adrenocortical insufficiency during subsequent testing).

2. If low target-gland hormone and raised trophic hormone levels are found, test the target gland affected.
3. If both target gland and trophic hormone levels are low or low normal, proceed to the combined pituitary stimulation test.
4. Investigate the pituitary region using radiological techniques.

SUSPECTED GROWTH HORMONE DEFICIENCY

1. A single unequivocally high plasma GH value probably excludes deficiency. The initial sample should be taken when the highest physiological levels occur—immediately after exercise or during sleep (see p. 133 for details).
2. If GH values under these conditions do not exclude deficiency, proceed to a pituitary stimulation test, either insulin-induced hypoglycaemia or a combined test.

SUSPECTED ACROMEGALY OR GIGANTISM

A raised plasma GH level which fails to suppress adequately with rising blood glucose levels suggests autonomous hormone secretion. Basal levels of GH may be, but are not always, high enough to confirm the diagnosis. It saves time to perform a glucose suppression test initially in a case of suspected acromegaly or gigantism (p. 173).

SUSPECTED CUSHING'S SYNDROME

Cushing's syndrome is a serious condition which may be treatable. It is important to exclude or confirm the diagnosis, even if the clinical features are no more than suggestive. The initial tests may exclude the diagnosis, but may yield some "false" positive results.

A. *Has the patient got Cushing's syndrome?*

The initial tests can be performed without admitting the patient:

1. Perform an overnight dexamethasone suppression test.
2. Estimate urinary free cortisol in several 24 hour collections.

   If these tests are normal, it is unlikely that the patient has Cushing's syndrome. If either is abnormal, or if there is strong clinical suspicion, further testing should be carried out in hospital.
3. Assess the circadian rhythm of plasma cortisol by collecting blood samples at 09.00 h and 23.00 h. This may be followed by repeating the overnight dexamethasone test.

   If all these results are normal, follow up the patient in the outpatient clinic. The manifestations of Cushing's syndrome may be intermittent, and tests may have to be repeated later. If any of the results is still abnormal, exclude causes such as stress, alcoholism or depression. Once the diagnosis of Cushing's syndrome has been made, proceed to the next step.

B. *What is the cause of the Cushing's syndrome?*

1. *Extremely high cortisol levels* are in *favour of adrenocortical carcinoma (especially if the patient is virilised), or of ectopic ACTH production.* Investigate clinically and using other investigations, especially bearing in mind the possibility of carcinoma of the adrenal or bronchus. Severe hypokalaemic alkalosis suggests ectopic ACTH production.
2. It is highly desirable to estimate plasma *ACTH levels*, even if the specimen has to be sent to a special centre.
3. Perform a *high dose dexamethasone* test. *Suppression* suggests either pituitary-dependent Cushing's syndrome or a psychiatric cause. The differentiation can be very difficult and depends on careful clinical assessment and follow-up. *Failure to suppress* suggests either an adrenocortical tumour (low ACTH levels) or ectopic ACTH production (very high ACTH levels).
4. If there is virilisation estimate plasma androgens, especially DHEA-sulphate.

Interpretation of these tests is summarised in Table XIV, p. 149.

SUSPECTED ADRENAL HYPOFUNCTION

A. Suspected Addisonian Crisis

1. Take blood *first*, for immediate electrolyte and later cortisol estimation.
2. Start steroid therapy *at once*. Do not wait for laboratory results.
3. Hyponatraemia, hyperkalaemia and uraemia, although compatible with Addisonian crisis, are common in many clinically similar acute conditions. Treat appropriately (see Chapters II and III).
4. *Plasma cortisol* may be estimated later:

(a) If it is *very high an Addisonian crisis is excluded*;
(b) If it is *very low or undetectable*, and if there is no reason to suspect CBG deficiency, an *Addisonian crisis is confirmed*.

5. *Plasma cortisol levels which would be "normal" under basal conditions may be inappropriately low for a stressed patient.* Perform a *short tetracosactrin test* (p. 173).

B. **Suspected Chronic Adrenal Hypofunction**

1. Measure plasma cortisol levels. A high level *at any time of day* excludes Addison's disease.
2. If the plasma cortisol results are equivocal, perform a *short tetracosactrin test. A normal result excludes Addison's disease* and makes long-standing secondary adrenocortical insufficiency unlikely (prolonged ACTH deficiency causes reversible adrenal insensitivity to trophic stimulation).
3. If the results of the short tetracosactrin test are equivocal, take blood and send to the laboratory before proceeding further, in case ACTH assay seems indicated later (see 7).
4. Admit the patient, and perform a *5-hour tetracosactrin test. A normal result* excludes primary adrenal hypofunction: *if there was a subnormal response to the short tetracosactrin test, it suggests secondary adrenal hypofunction.*
5. If doubt remains, perform a *3-day tetracosactrin test.* The same impaired response to the short, 5-hour and 3-day tetracosactrin tests would confirm *primary adrenal hypofunction*.
6. An *increasing response* to the short, 5-hour, and 3-day tests indicates gradual recovery of the adrenal following prolonged lack of ACTH, and suggests *hypothalamic or pituitary hypofunction*. Perform a combined pituitary function test (p. 171).
7. If doubt remains, and if the facilities are available, assay ACTH on the plasma specimen taken earlier. *If the cortisol level is low or low normal*:
   (a) *a high ACTH level* confirms *primary adrenal hypofunction*;
   (b) *a low ACTH level* suggests *secondary adrenal hypofunction*, and a combined pituitary function test should be performed.

C. **Patients on Steroid Treatment with Suspected Adrenal Hypofunction**

It is not uncommon for patients to be treated before adrenal hypofunction has been proved. Before investigation of such patients is started any steroids, such as cortisone or hydrocortisone (cortisol), that might interfere with the tests, must be stopped. If there is any danger that this may precipitate an adrenal crisis, stimulation tests may be performed while dexamethasone is given. This compound is biologically active, but does not interfere with plasma cortisol assays. Consult your laboratory to find out which steroid analogues interfere with the estimations they perform.

AMENORRHOEA

A full clinical assessment should be made, and the patient should be questioned carefully to determine whether amenorrhoea is primary or secondary. Amenorrhoea may accompany any severe disease.

The aim of laboratory investigations is to detect hypogonadism and distinguish between pituitary and ovarian causes. *Before embarking on such investigation it is essential to test for pregnancy.*

If the pregnancy test is negative proceed as follows:

1. Measure plasma LH, FSH and oestradiol. If oestradiol assay is not available, a vaginal smear should be taken to assess oestrogen effects.
2. Measure plasma prolactin.

Raised gonadotrophin (especially FSH) with *low oestrogen* levels indicate ovarian failure. If the amenorrhoea is primary, chromosome studies are indicated.

Low oestrogen with *low or low normal gonadotrophin* levels suggest a hypothalamic or pituitary cause. This would be supported by high *prolactin* levels, and is further investigated as follows:

3. Perform a Gn-RH test.
   If there is a *normal* gonadotrophin response to Gn-RH the lesion is probably hypothalamic. This may involve only gonadotrophin releasing hormone secretion. If there is a *subnormal* response, a pituitary lesion is likely.
4. Proceed to a combined pituitary stimulation test.

Other endocrine disorders such as thyrotoxicosis and Cushing's syndrome must be considered and, if necessary, excluded by appropriate tests. If there is hirsutism or virilism, proceed as outlined below.

HIRSUTISM AND VIRILISM

The aim of investigation, after full clinical assessment, is to detect cases with significant elevation of plasma androgen levels and to identify the source as the ovary or adrenal cortex.

Measure:

1. plasma testosterone;
2. plasma DHEA or DHEA-S;
3. plasma 17-OH progesterone if congenital adrenal hyperplasia is suspected (p. 152);
4. plasma gonadotrophins.

TABLE XV
Plasma Findings

Testosterone	DHEA-S	LH	FSH	17-OH Progesterone	
N or ± ↑	N or ± ↑	N	N	N	Simple hirsutism
↑	N	↑	N or ↓	N	Polycystic ovaries
↑↑	N	N or ↓	N or ↓	N	Ovarian tumour
N or ± ↑	↑↑	N or ↓	N or ↓	N	Adrenocortical tumour
± ↑	↑	N or ↓	N or ↓	↑	Congenital adrenal hyperplasia

If there is evidence of other endocrine disease such as Cushing's syndrome, (p. 166) investigate accordingly.

The diagnosis of primary aldosteronism (Conn's syndrome) is discussed on p. 85.

PROTOCOLS FOR TESTS OF THE PITUITARY-ADRENAL-GONADAL AXIS

Always contact your laboratory *before* starting a test, both to ensure the most efficient and speedy analysis, and to check local variations in protocols.

Procedure for Repeated Sampling

Many "dynamic" tests of endocrine gland function require several blood samples over a short period of time. Repeated venepuncture is unpleasant for the patient and is also undesirable because it introduces an element of stress which may interfere with the results. An indwelling needle helps to avoid these problems. The following is a suitable procedure:

(i) An indwelling needle (or cannula) of at least 19G is introduced into a suitable forearm vein. It is secured in position with adhesive strapping.

(ii) Isotonic saline is infused *slowly* to keep the needle open. Heparin may interfere with some assays.

(iii) *At least* 30 minutes is allowed between insertion of the needle and sampling to allow any stress-induced elevation of hormones to return to normal.

(iv) All samples may be taken through this needle:
    (a) disconnect the saline infusion (this can be avoided by the use of a disposable 3-way tap between the infusion set and the needle);
    (b) aspirate and *discard* 1–2 ml of saline-blood mixture;
    (c) *using a separate syringe*, aspirate the required volume of blood:
    (d) re-establish the saline flow.

If there is an untoward reaction such as severe hypoglycaemia after insulin administration, intravenous therapy can be given without delay through the indwelling needle.

COMBINED PITUITARY STIMULATION TEST

This test measures the level of anterior pituitary hormones in blood after stimulation by stress, TRH and Gn-RH. Plasma cortisol is usually measured as an index of ACTH secretion: the entire hypothalamic-pituitary-adreno-cortical axis is therefore tested.

After an overnight fast:
1. Insert an indwelling intravenous cannula (see above)
2. *Wait at least 30 minutes* for hormone levels to return to basal levels.
3. Take *basal blood* samples.
4. Inject soluble insulin intravenously (for dose see below) followed by 200 μg of TRH and 100 μg of Gn-RH.
5. Take *further blood samples* at 30, 45, 60, 90 and 120 minutes after insulin administration.
N.B. The test is potentially dangerous and should be done only under *direct medical supervision*. It is contra-indicated in patients with ischaemic heart disease or epilepsy. *Glucose* for intravenous administration should be *immediately available* in case severe hypoglycaemia

develops. At the conclusion of the test the patient should be given something to eat.

Notes

1. Glucose, cortisol, GH, LH and FSH, TSH and prolactin are estimated on all blood samples. The samples for hormone estimation should be kept at 4° C until delivered to the laboratory.

2. The dose of soluble insulin must be sufficient to lower plasma glucose levels to less than 2·5 mmol/l (45 mg/dl) and produce symptoms of hypoglycaemia. The usual dose is 0·15 U/kg body weight. If pituitary or adrenocortical hypofunction is probable, or if a low fasting plasma glucose has previously been found, the dose should be reduced to 0·1 or 0·05 U/kg. Conversely, if there is resistance to the action of insulin (Cushing's syndrome, acromegaly or obesity) 0·2 or 0·4 U/kg may be required.

3. If it is necessary to administer glucose, *continue with the blood sampling.* The stress has certainly been adequate.

Interpretation

Check with your laboratory for their reference values. Methods of assay vary and results should not be compared with values issued by other departments.

As a guide, provided the plasma glucose has fallen below 2·5 mmol/l (45 mg/dl) with clinical evidence of hypoglycaemia, plasma cortisol should rise by more than 200 nmol/l (7 μg/dl) and exceed 550 nmol/l (20 μg/dl); GH levels should exceed 20 mU/l.

TSH values should increase by at least 2 mU/l and exceed the upper limit of the basal normal range. Prolactin and gonadotrophin levels should also rise significantly.

Primary adrenocortical hypofunction as a cause of failure of plasma cortisol to rise must be excluded by a tetracosactrin stimulation test (p. 173). In Cushing's syndrome neither cortisol nor GH levels rise.

GONADOTROPHIN-RELEASING HORMONE TEST

The synthetic Gn-releasing hormone stimulates the release of gonadotrophins from the normal anterior pituitary gland. 100 μg of Gn-RH is given by rapid intravenous injection. *Plasma LH and FSH levels are measured in blood drawn before and at 20 and 60 minutes after* the injection. No side-effects have been described.

In normal subjects plasma LH rises by at least 5 U/l. This rise is absent in patients with pituitary failure.

INSULIN STRESS TEST

If only ACTH and/or GH reserve need to be assessed, TRH and Gn-RH may be omitted from the combined pituitary stimulation test.

GLUCOSE SUPPRESSION TEST FOR ACROMEGALY

A rising plasma glucose level suppresses GH secretion when this is under normal physiological control. If GH secretion is autonomous, as it is in patients with acromegaly or gigantism, this does not happen.
After an *overnight fast*:

1. Insert an indwelling intravenous cannula (see p. 171).
2. *Wait at least 30 minutes.*
3. Take basal samples of blood for glucose and GH estimation.
4. The patient takes 50 g of glucose, dissolved in water, by mouth.
5. Take further blood samples for glucose and GH measurement at 30, 60, 90 and 120 minutes after glucose ingestion.

Interpretation

In normal subjects an elevated plasma GH level will fall to less than 4 mU/l. Failure to suppress suggests acromegaly or gigantism but may also be found in severe liver or renal disease, in heroin addicts or in patients taking levodopa.

TETRACOSACTRIN STIMULATION TESTS OF ADRENAL CORTICAL FUNCTION

Tetracosactrin is marketed as Synacthen (Ciba) or Cortrosyn (Organon).

1. **30-minute stimulation test**

   The patient should be resting quietly;
   (a) Blood is taken for basal cortisol (or 11-OHCS) levels;
   (b) 250 μg of tetracosactrin, dissolved in about 1 ml of sterile water or isotonic saline, is given by intramuscular injection;
   (c) 30 minutes later blood is taken for cortisol estimation.

   Normally plasma cortisol increases by at least 200 nmol/l (7 μg/dl), to a level of at least 550 nmol/l (20 μg/dl).

2. **5-hour stimulation test**

   (a) Blood is taken for basal cortisol;
   (b) 1 mg depot tetracosactrin is injected intramuscularly;
   (c) Further blood samples are taken 1 hour and 5 hours after (b).

   Normally plasma cortisol rises to a level of between 600 and 1300 nmol/l (22 and 46 μg/dl) at 1 hour, and to between 1000 and 1800 nmol/l (37 and 66 μg/dl) at 5 hours. Such a response excludes primary (but not secondary) adrenocortical hypofunction. The cause of an impaired response should be investigated by prolonged stimulation.

3. **3-day stimulation test**

   (a) 1 mg depot tetracosactrin is given daily by intramuscular injection;
   (b) Plasma cortisol is estimated in blood drawn 5 hours after each injection.

   No further response after 3 days indicates *primary adrenocortical hypofunc-*

tion. An increasing response over this period indicates *secondary adrenal atrophy*.

If there is a danger of an adrenal crisis if steroids are withdrawn, dexamethasone, which does not contribute significantly to the cortisol estimation, may be given. Repeated injections of depot tetracosactrin are painful and may, if there is an adrenocortical response, lead to sodium and water retention: *this test is therefore contra-indicated in patients in whom such retention may be dangerous*.

DEXAMETHASONE SUPPRESSION TEST IN SUSPECTED CUSHING'S SYNDROME

1. **Overnight Test** (for diagnosing Cushing's syndrome)

   Dexamethasone (2 mg) is given as a single oral dose at 23·00 h. Plasma cortisol levels are measured at 09·00 h the next morning. Suppression is defined as a plasma cortisol level of less than 190 nmol/l.

2. **High-Dose Test** (for deciding the cause of diagnosed Cushing's syndrome).

   Dexamethasone (2 mg) is given orally every six hours for two days, starting at 09.00 h. At 09.00 h on the third day plasma cortisol is measured. Suppression is defined as levels less than 50 per cent of previously measured basal values.

   Interpretation is discussed on p. 167.

   Anti-epileptic drugs, particularly *phenytoin*, may interfere with dexamethasone suppression tests. These drugs, by inducing liver enzymes, increase the rate of metabolism of dexamethasone, and plasma levels may be inadequate to suppress the feedback centre.

SPECIMEN COLLECTION

Your laboratory should be consulted for details of specimen collection and handling.

Many hormones are stable in drawn blood for several hours or even days. Specimens for others, such as insulin, renin, ACTH, and 11-OHCS need to be kept cool and sent to the laboratory without delay.

The *time* of sampling should be recorded on the request form.

FACTORS AFFECTING RESULTS OF ASSAY

Plasma cortisol is usually measured by a radioimmunoassay or similar method. Some laboratories use a fluorescence assay for 11-hydroxycorticosteroids (11-OHCS).

Many drugs may affect the results of hormone assays. The following should be especially noted:

  Hydrocortisone (cortisol) and *cortisone* (after conversion) will of course, be measured by both methods.

  Some *oestrogens and oral contraceptives* increase CBG (and consequently bound cortisol) levels.

Prednisolone will contribute to the "cortisol" value of most RIA methods but is not measured by the fluorescence 11-OHCS assay.

Spironolactone produces falsely high 11-OHCS values in blood and urine.

Dexamethasone in plasma does not contribute to either assay but its urinary metabolites may interfere with some RIA methods.

NOMENCLATURE OF PITUITARY HORMONES

An alternative nomenclature to the one used in this book should be mentioned for reference. The pituitary hormone is usually identified by the suffix "-*tropin*", and the corresponding hypothalamic hormone by "-*liberin*" (releasing) or "-*statin*" (inhibiting). For example, the release of *somatotropin* (growth hormone) may be promoted by *somatoliberin* or inhibited by *somatostatin*. The other pituitary hormones are *prolactin* (unchanged), *corticotropin* (ACTH), *thyrotropin* (TSH), *lutropin* (LH), and *follitropin* (FSH).

Chapter VIII

THYROID FUNCTION: TSH

THE three hormones known to be produced by the thyroid are thyroxine, tri-iodothyronine and calcitonin. Thyroxine and tri-iodothyronine are products of the thyroid follicular cell and influence metabolism throughout the body. Calcitonin is produced by a specialised cell (the C cell) and influences calcium metabolism. This is an anatomical rather than a functional relationship and in lower animals the calcitonin-secreting cells may be completely separated from the thyroid. Calcitonin is considered briefly on p. 256.

PHYSIOLOGY

The thyroid hormones are synthesised in the thyroid gland by iodination and coupling of two molecules of the amino acid tyrosine, a process dependent on an adequate supply of iodide.

Iodide in the diet is absorbed rapidly from the small bowel. Most natural foods contain adequate amounts of iodide except in regions where the content of the soil is very low. In these areas there used to be a high incidence of goitre, but general use of artificially iodised salt has made this a less common occurrence. Sea foods have a high iodide content and fish and iodised salt are the main dietary sources of the element.

Normally about one-third of absorbed iodide is taken up by the thyroid and the other two-thirds is excreted via the kidneys.

Metabolism of Thyroid Hormones

The steps in the biosynthesis and secretion of the thyroid hormones are outlined below. A knowledge of this pathway is essential for the understanding of the mode of action of drugs—whether used to diagnose or treat thyroid disorders—and of the effects of congenital enzyme deficiences (p. 185).

Iodide is actively taken up by the thyroid gland. The concentration in the gland is normally about twenty times that in plasma but may exceed it by a hundred times or more. The salivary glands, gastric mucosa and mammary glands are also capable of concentrating iodide.

Uptake is *blocked by thiocyanate and perchlorate*.

Trapped *iodide is rapidly converted to iodine*.

Tyrosine residues in a large glycoprotein, thyroglobulin, are *iodinated* to form mono- and di-iodotyrosine (MIT and DIT). This step is *inhibited by carbimazole, propylthiouracil and related drugs.*
The iodotyrosines are coupled to form thyroxine (T_4) (DIT + DIT) and tri-iodothyronine (T_3) (DIT + MIT) (Fig. 14). Normally much more T_4 than T_3 is synthesised but if there is an inadequate supply of iodine the ratio of T_3 to T_4 in the gland increases. The thyroid hormones, still incorporated in thyroglobulin, are stored in the colloid of the thyroid follicle.

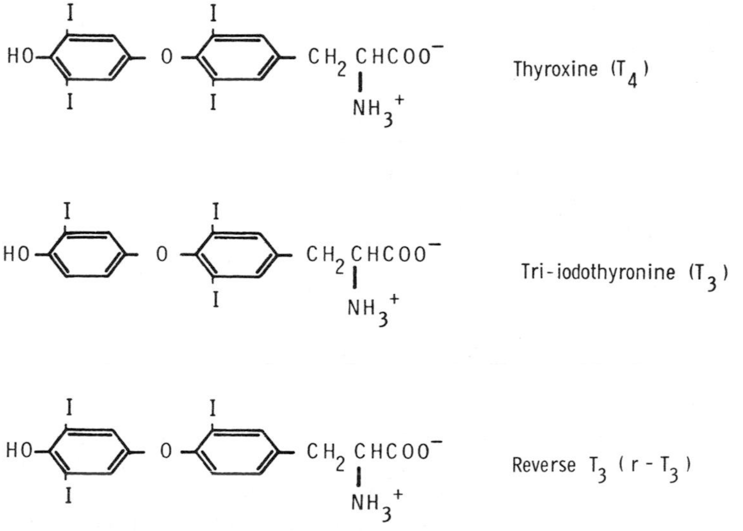

FIG. 14.—Chemical structure of the thyroid hormones and reverse T_3.

Before secretion of the thyroid hormones, thyroglobulin is taken up by the follicular cell. T_4 and T_3 are released from it by proteolytic enzymes and pass into the bloodstream where they are immediately bound to plasma proteins (see next page). Mono- and di-iodotyrosine released at the same time are de-iodinated by another enzyme and the iodine re-utilised.
    Each step is controlled by specific enzymes and congenital deficiency of any of these enzymes can lead to a goitre and, if severe, hypothyroidism. The uptake of iodide as well as the synthesis and secretion of thyroid hormones is regulated by thyroid-stimulating hormone (TSH) from the pituitary.

Circulating Thyroid Hormones

More than 99 per cent of plasma thyroxine and tri-iodothyronine is protein-bound, mainly to an α-globulin, *thyroxine-binding globulin (TBG)*, and, to a lesser extent, to albumin and thyroxine-binding pre-albumin. It is the small free fractions which are physiologically active, and which regulate pituitary TSH secretion (compare calcium, p. 252, and cortisol, p. 141).

Some of the secreted T_4 is de-iodinated by enzymes in peripheral tissues, particularly the liver. Removal of an iodine atom from the outer (β) ring produces the active hormone T_3. Normally most of the plasma T_3 is derived in this way: direct thyroid secretion contributes only a small part. Drugs such as phenytoin, which induce hepatic enzymes, increase the rate of this conversion and result in low normal plasma T_4 levels.

Removal of an iodine atom from the inner (α) ring produces reverse-T_3 (r-T_3) which is probably inactive. In stressed subjects, including those who are chronically ill, and after administration of propranolol, this pathway predominates and T_3 production falls. The advantage of this may be to reduce energy consumption.

The level of plasma T_4 is usually about 100 nmol/l (8 μg/dl) but is higher in the first month of life. Plasma tri-iodothyronine levels are much lower (about 1·5 nmol/l) and tend to be lower in old age. Despite this difference in total concentrations, the free fractions of T_4 and T_3 are almost equal and T_3 is metabolically the more active.

Control

TSH secretion from the pituitary is regulated primarily by negative feedback by circulating free thyroid hormone (particularly free T_4) levels. It is also stimulated by the hypothalamic *thyrotrophin-réleasing hormone* (TRH) but this effect can be overridden by abnormally high circulating free T_4 levels: this is the reason for the lack of effect of exogenous TRH on TSH secretion during the test for hyperthyroidism.

The metabolism and control of thyroid hormones are summarised in Fig. 15.

Actions of Thyroid Hormones

Thyroid hormones influence and speed up many metabolic processes in the body. They are essential for normal growth, mental development and sexual maturation. They also increase the sensitivity of the cardio-vascular and central nervous systems to catecholamines and so influence cardiac output and heart rate.

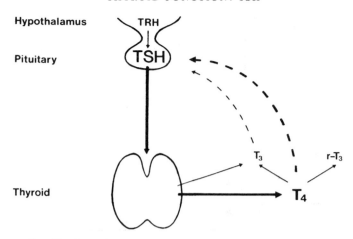

FIG. 15.—Secretion, metabolism and control of thyroid hormones.
Solid lines—secretion and metabolism of hormones;
Dotted lines—negative feedback mechanisms.

THYROID FUNCTION TESTS

Treatment of thyroid disease may be prolonged and, in the case of hypothyroidism, life-long, and it is essential to confirm the clinical diagnosis by laboratory tests. During treatment the original clinical picture disappears and it is not an uncommon problem to be faced with a patient who has taken thyroxine for many years for probable "thyroid trouble". Without an adequately established diagnosis it may be difficult to assess the past history.

The tests described in this section determine only the *functional state* of the thyroid gland. The underlying disease must be established by other means, some of which will be mentioned, briefly.

Plasma Thyroxine and "Corrected" Thyroxine (Refer to Fig. 16)

Plasma thyroxine.—Most techniques in common use assess levels of protein-bound plus free thyroxine: more than 99 per cent of that circulating is protein-bound, and it is predominantly this fraction that is being measured. Clinical effects of changes in thyroxine secretion by the thyroid gland are, on the other hand, due to changes in the *free* fraction. Despite this, the protein-bound and free fractions parallel each other *provided that the level of the binding protein stays nearly constant.* Note that the number of unoccupied binding sites on TBG varies inversely with the T_4 level in hyperthyroidism (a) and in hypothyroidism (b).

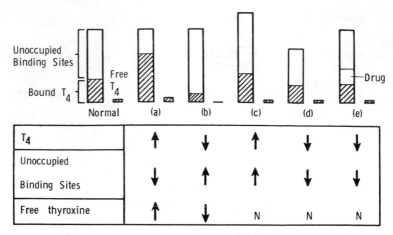

Fig. 16.—Interpretation of tests of circulating hormone (see text).

There may, on the other hand, be a primary change in the level of binding protein. In the euthyroid subject the *proportion* of TBG occupied by the hormone remains at about 30 per cent of the total binding sites, but the free T_4 level is normal.

A primary increase in TBG (c) results in a proportional increase in both bound T_4 and unoccupied binding sites. Such an increase may occur:

> due to high oestrogen levels in pregnancy or in the newborn infant;
> during oestrogen therapy, or while taking oral contraceptives;
> rarely, with inherited TBG excess.

Families have been described who have albumin or pre-albumin variants with a higher than normal avidity for T_4. The findings are similar to those of TBG excess.

A primary decrease in TBG (d) decreases both bound T_4 levels and unoccupied binding sites. Such changes may occur:

> temporarily in severely ill patients;
> due to loss of low-molecular-weight proteins, usually in the urine (nephrotic syndrome);
> in patients taking androgens or danazol;
> rarely, in inherited TBG deficiency.

These changes might be misinterpreted as diagnostic of hyperthyroidism and hypothyroidism respectively if only T_4 were measured. Usually, if there is doubt, results of the tests discussed later in the chapter, together with the clinical picture, will elucidate the true thyroid status. Occasionally it may be desirable to assess the TBG level. We have seen that:

in primary disorders of thyroid function, if TBG levels are unchanged, the measured T_4 and unoccupied binding sites vary inversely;

in a primary alteration of binding protein the measured T_4, TBG concentration *and* unoccupied binding sites change in the same direction.

If, therefore, we could assess the unoccupied binding sites or the TBG level, it should clarify whether measured T_4 truly reflects free T_4 concentrations.

Correction for Altered TBG Levels

Several methods have been used in an attempt to assess or correct for changes due to altered T_4-binding protein levels. It should be noted that whichever method is used, "correction" is often incomplete when TBG levels are very abnormal.

Assessment of unoccupied binding sites.—Methods of assessing the level of unoccupied binding sites all depend on *in vitro* addition of radioactive thyroid hormone to the patient's plasma in amounts which exceed the capacity of TBG to bind it. The more unoccupied sites there are on the TBG in the specimen, the more the added radioactive hormone will be bound by the protein. A resin may now be added to take up the *unbound* hormone: the more unoccupied sites there were on the protein the less radioactive hormone will be left to bind to the resin (Fig. 17). If the resin is separated from the plasma, the amount of radioactivity *either* left on the TBG, *or* bound to the resin can be determined. The more left in the plasma, *or*, the less on the resin, the more unoccupied sites there were on the TBG.

There are many modifications of this technique, and many formulae have been devised, using these two measurements, to "correct" the T_4 and therefore to give an assessment of free T_4 (free thyroxine index or FTI). The student should consult his own laboratory for details.

Plasma TBG may be measured directly and a T_4:TBG ratio calculated. Of these two approaches, the resin uptake test (used for calculation of the FTI) is simple and more generally available than TBG measurement. It also has the theoretical advantage of assessing all T_4 binding proteins, not only TBG. It is worth noting that the use of any formula which incorporates the results of two assays, both with some degree of analytical imprecision, increases the likelihood that the answer may be misleading. It may be better to inspect the results of the two assays separately. There is an additional problem in patients taking drugs which bind to TBG and displace thyroxine (e): examples

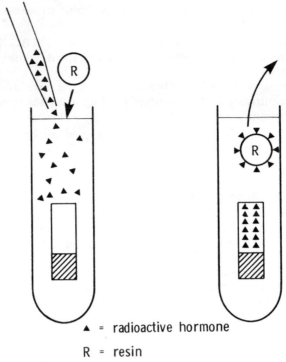

▲ = radioactive hormone

R = resin

Fig. 17.—Assessment of free binding sites on thyroxine-binding globulin.

of such drugs are salicylates and fenclofenac*. The resultant low T_4 may or may not be accompanied by a raised resin uptake. Low T_4 values can only be interpreted with a knowledge of what drugs the patient is taking (p. 192).

"Normalised" thyroxine levels (for example, *effective thyroxine ratio*). A number of tests are available in which the T_4 value is corrected for abnormalities of protein binding in a single procedure and is expressed as a ratio of a standard serum. This is analogous to the FTI and may be used as a screening test.

A disadvantage of this approach is that abnormal TBG levels are not recognised and so are not considered in interpretation. Borderline results, especially if low normal, should be confirmed by other tests.

Free Thyroxine Levels

Free thyroxine assays are now fairly widely available. Although they may sometimes be more informative than the FTI, they do not

---

* This drug has recently been withdrawn from the U.K. market.

always correctly diagnose the thyroid status of patients with abnormal TBG levels. Results must be interpreted with this in mind.

Plasma Tri-iodothyronine (T₃)

Plasma T_3 radioimmunoassay may be helpful as a test of hyperthyroidism, but should not be used to diagnose hypothyroidism. T_3, like T_4, is bound to protein, and it may be necessary to assess binding sites to interpret the results.

Plasma TSH may be measured by radioimmunoassay. Estimation is of particular value in the diagnosis of primary hypothyroidism. Unfortunately, many currently available techniques are unable to distinguish between low normal and subnormal levels. Dynamic tests, however, overcome this problem (p. 190).

Radioiodine Uptake Tests

Tracer doses of radioiodine are metabolised in the same way as the natural element. The fact that iodine is taken up by the gland more rapidly than normal in hyperthyroidism, and less rapidly than normal in hypothyroidism has been used to test thyroid function. The radioiodine uptake test has now been almost entirely replaced by measurement of circulating hormones.

DISORDERS OF THE THYROID GLAND

The commonest presenting features in patients with thyroid disease are:

> hyperthyroidism—due to excessive thyroid hormone secretion;
> hypothyroidism—due to deficient thyroid hormone secretion;
> enlargement of the thyroid gland, either diffuse (goitre), or due to one or more nodules within the gland. There may or may not be a disturbance of secretion.

Excess or deficiency of circulating thyroid hormone produces characteristic clinical changes. Disease of the thyroid gland may, however, be present without hyper- or hypofunction.

HYPERTHYROIDISM (THYROTOXICOSIS)

The syndrome produced by sustained excess of thyroid hormone may be easily recognised or may remain unsuspected for a long time. The key feature is a speeding up of metabolism. Clinical features include tremor, tachycardia (and sometimes arrhythmias), weight loss, tiredness, sweating and diarrhoea. Anxiety and emotional symptoms may be prominent. In some cases a single feature predominates, such as weight loss, diarrhoea, tachycardia or atrial fibrillation.

The *common causes* of hyperthyroidism are:

Graves' disease (commonest form);
toxic multinodular goitre or single functioning nodule (occasionally an adenoma);
ingestion of thyroid hormones.

Rare causes are:

secretion of a thyroid-stimulating hormone by tumours of trophoblastic origin;
struma ovarii (thyroid tissue in ovarian teratoma);
administration of iodine to a subject with iodine-deficiency goitre;
TSH-secreting tumour of the pituitary.

Graves' disease occurs at any age and is commoner in females. It is characterised by one or more of the following:

hyperthyroidism due to diffuse hyperplasia of the thyroid;
exophthalmos;
localised (pretibial) myxoedema.

Graves' disease is one of the autoimmune thyroid diseases. Thyroid antibodies are detectable in some cases, in addition to *thyroid-stimulating immunoglobulins* which probably cause hyperfunction.

Toxic nodules, single or multiple, in a nodular goitre may secrete thyroid hormones *autonomously. TSH is suppressed* by negative feedback (as it is in Graves' disease) and the thyroid tissue not in the nodules is inactive. The "hot" areas can be detected by a thyroid scan. Toxic nodules are found more commonly in the older age groups, and the patient may present with only one of the features of hyperactivity, most commonly cardiovascular symptoms.

The *feature common to all forms of hyperthyroidism* (with the exception of the *very* rare TSH-secreting pituitary tumour) is that *secretion is not dependent on TSH from the pituitary.*

Pathophysiology

In most patients with hyperthyroidism the levels of both T_3 and T_4 in plasma are raised. Much of the T_3 is secreted directly by the thyroid, and plasma levels are increased more above "normal", and are usually evident earlier, than those of T_4. Sometimes only T_3 levels are elevated (T_3 thyrotoxicosis). The higher free hormone levels suppress TSH secretion, which is not even stimulated by giving TRH.

The diagnostic approach to hyperthyroidism is considered on p. 189.

HYPOTHYROIDISM

Hypothyroidism is due to suboptimal circulating levels of one or both thyroid hormones. The condition develops insidiously and, in its early stages vague symptoms, such as tiredness, may not be recognised as being due to hypothyroidism. Many of the features of frank hypothyroidism are, not unexpectedly, the opposite of those in hyperthyroidism. There is a generalised slowing down of metabolism, with mental dullness, physical slowness, weight gain and menstrual disturbances. The skin is dry, hair falls out and the voice becomes hoarse. The face is puffy and the subcutaneous tissues thickened; this pseudo-oedema, with a histologically myxoid appearance, accounts for the term myxoedema, used to describe severe hypothyroidism.

In the most severe cases myxoedema coma, with profound hypothermia, may develop. Congenital thyroid hormone deficiency leads to cretinism, and in children growth may be impaired.

The *common* causes of *primary* hypothyroidism are:

autoimmune thyroiditis;
Hashimoto's disease.
   In these there is progressive destruction of thyroid tissue.
   Circulating thyroid antibodies are present, often in high
   concentration.
other inflammatory diseases of the gland;
following treatment of hyperthyroidism:
   post-thyroidectomy;
   post ^{131}I therapy;
"idiopathic" hypothyroidism.

Rare causes of *primary* hypothyroidism are:

exogenous goitrogens and drugs (including lithium);
dyshormonogenesis.
   The term "dyshormonogenesis" includes inherited deficiencies
   of any of the enzymes involved in thyroxine synthesis.
   Although the biochemical and clinical features differ, the
   end result is hypothyroidism due to reduced thyroxine
   synthesis. In most cases prolonged TSH stimulation causes
   goitre. The commonest form is due to failure to incorporate
   iodine into tyrosine.

Hypothyroidism *secondary* to anterior pituitary or hypothalamic deficiency is much less common than the primary form. In long-standing secondary hypothyroidism the thyroid gland may atrophy irreversibly.

The essential difference in findings between the primary and secondary forms is in the level of plasma TSH; this is high in primary and low in secondary hypothyroidism.

Pathophysiology

In primary hypothyroidism plasma T_4 levels are usually low. However, in the early stages they may be within the reference range, although low for the individual patient. TSH secretion is increased, and a high plasma TSH level may be the only abnormal finding. Because plasma T_3 levels may be normal in hypothyroidism the assay is of no help in making the diagnosis.

Secondary hypothyroidism is due to impaired TSH secretion. This may be caused by pituitary insufficiency, in which case TRH fails to increase TSH secretion, or by failure of TRH secretion from the hypothalamus.

The diagnosis of hypothyroidism is considered on p. 190.

Special problems in the diagnosis of hypothyroidism.—Two situations warrant special consideration.

Neonatal hypothyroidism.—The incidence of neonatal hypothyroidism varies in different populations from 1/5 000 to 1/10 000. It is therefore commoner than many inborn errors for which routine screening is advocated (Chapter XVI). The clinical signs of hypothyroidism in the newborn are minimal but if treatment is not started within the first few months permanent brain damage results. Because of the high but variable plasma T_4 values in the first month of life (due to the high maternal oestrogen levels in late pregnancy, which increase TBG levels in both mother and infant) interpretation of this test is difficult. Plasma TSH levels are usually raised in hypothyroidism, and in some countries all newborn infants are screened by plasma TSH assay about a week after delivery: the test should, in any case, be done on any infant thought, on clinical evidence, to be hypothyroid.

Interpretation of results in very ill patients.—The interpretation of plasma thyroid hormone levels in very ill patients is difficult. T_3 levels usually fall because more T_4 than usual is converted to reverse-T_3 (p. 178). Sometimes decreased binding to TBG may lead to a fall in total T_4 levels. Free T_4 concentrations are probably normal and the diagnosis must be made mainly on clinical evidence. Interested readers may consult the reference at the end of the chapter for further discussion of this topic (Utiger, 1980).

EUTHYROID GOITRE

Thyroxine synthesis may be impaired by iodine deficiency, by drugs such as para-aminosalicylic acid, or possibly by minor degrees of enzyme deficiency. The tendency for circulating T_4 to fall, increases

TSH secretion by feedback. This stimulates synthesis of T_4 and T_3 and maintains adequate plasma levels of these hormones. The thyroid therefore enlarges (goitre), but hypothyroidism is avoided. In areas with low iodine content of the soil, iodine deficiency used to be common (endemic goitre).

Inflammation of the thyroid (thyroiditis), whether acute or sub-acute, may produce marked but temporary aberrations of thyroid function tests. These conditions are relatively uncommon.

Other Biochemical Findings in Thyroid Disease

Cholesterol level.—In *hypothyroidism* the synthesis of cholesterol is impaired but its catabolism is even more impaired and plasma cholesterol levels are high.

Hypercalcaemia is very rarely found with severe thyrotoxicosis. There is an increased turnover of bone, probably due to direct action of thyroid hormone (p. 256).

Plasma creatine kinase (CK) levels are often raised in hypothyroidism but the estimation does not help diagnosis. CK levels may also be raised in thyrotoxic myopathy.

SUMMARY

1. There are two circulating thyroid hormones (apart from calcitonin): thyroxine and tri-iodothyronine. Their synthesis depends on an adequate supply of iodine, and is controlled by circulating thyroid hormone levels via TSH from the anterior pituitary gland.

2. Circulating thyroid hormone levels may be assessed by estimating plasma T_4 and, if hyperthyroidism is suspected, T_3. Changes in the level of TBG may give misleading T_4 and T_3 results, and occasionally this possibility should be investigated by assessing free binding sites, and calculating the free thyroxine index. The value of free T_4 assays is still to be proven.

3. An excess of circulating thyroid hormone produces the syndrome of hyperthyroidism. This may occur in Graves' disease or may be due to a hyperfunctioning nodule of the thyroid.

The diagnosis of hyperthyroidism is made by the finding of an unequivocally high plasma T_4 level (after, if necessary, correcting for raised TBG levels) and, if the result is equivocal, a clearly raised plasma T_3 level. In doubtful cases a TRH test in which there is a subnormal response of plasma TSH *in the presence of high levels of thyroid hormones* supports the diagnosis.

4. A decreased circulating thyroid hormone level produces the syndrome of hypothyroidism. This may be primary, due to disease of

the thyroid gland, or be secondary to pituitary or hypothalamic disease. TSH levels are raised in the first group and low in the second.

The diagnosis of primary hypothyroidism is made by the finding of a low plasma T_4 and high plasma TSH levels. If both T_4 and TSH levels are low the TSH response to TRH may confirm secondary hypothyroidism of pituitary origin, or distinguish it from that due to hypothalamic causes.

5. Euthyroid goitre represents compensated thyroid disease. Thyroid function tests may be normal.

FURTHER READING

DEGROOT, L. J. and NIEPOMNISZCZE, H. (1977). Biosynthesis of thyroid hormone: basic and clinical aspects. *Metabolism*, **26**, 665–718.

BURGER, H. G. and PATEL, Y. C. (1977). Thyrotrophin releasing hormone—TSH. *Clin. Endocr. Metab.*, **6**, 83–100.

LARSEN, P. R. (1982). Thyroid-pituitary interaction. Feedback regulation of thyrotropin secretion by thyroid hormones. *New Engl. J. Med.*, **306**, 23–32.

STERLING, K. (1979). Thyroid hormone action at the cell level. *New Engl. J. Med.*, **300**, 117–123 and 173–177.

UTIGER, R. D. (1980). Decreased extrathyroidal triiodothyronine production in non-thyroidal illness: benefit or harm? *Amer. J. Med.*, **69**, 807–810.

WENZEL, K. W. (1981). Pharmacological interference with *in vitro* tests of thyroid function. *Metabolism*, **30**, 717–732.

(See also the general endocrine reference on p. 137.)

INVESTIGATION OF THYROID FUNCTION (Table XVI)

T_4 assay is available in most laboratories, and should usually be performed first. It may sometimes be necessary to assess abnormalities of binding proteins.

INVESTIGATION OF SUSPECTED HYPERTHYROIDISM

Measure Plasma T₄

If the value is clearly high, and high TBG levels are not suspected, in a clinically thyrotoxic patient no further tests are necessary.

Measure Plasma T₃

If the plasma T_4 is normal or only slightly elevated, measure plasma T_3. Any binding protein abnormality will also affect the T_3 value and must be allowed for. A clearly elevated T_3 concentration confirms the diagnosis of hyperthyroidism.

Rarely, both T_4 and T_3 results are equivocal. If thyrotoxicosis is still considered likely on clinical grounds, it is possible that the levels are abnormally high for that particular patient. If so, TSH secretion from the anterior pituitary gland will be suppressed. It may not be possible to assess this directly because some TSH assays do not distinguish "low normal" from "abnormally low" levels. We can however assess the TSH response to administered TRH.

Perform a TRH Test

A normal rise in plasma TSH levels after administration of TRH rules out the diagnosis of hyperthyroidism. A subnormal response is suggestive but not diagnostic, and if there is clinical doubt it is best to observe and reassess the patient at a later date.

TABLE XVI

RESULTS OF THYROID FUNCTION TESTS

	True T₄ abnormalities			TBG abnormalities	
	Hyperthyroidism	Hypothyroidism		Raised levels	Low levels
		Primary	Secondary		
Plasma T₄	↑	↓	↓	↑	↓
Plasma TSH levels	↓	↑	↓	N	N
Response to TRH	Absent	Exaggerated	See text	Normal	Normal
Free-thyroxine index	↑	↓	↓	N*	N*

*Correction may not be complete if levels are very abnormal.

Monitoring Treatment

The progress of a patient being treated for hyperthyroidism is monitored by estimating plasma T_4 with, if clinically indicated, T_3 levels. Overtreatment may induce hypothyroidism: plasma TSH levels should be monitored after the plasma T_4 has fallen to normal.

INVESTIGATION OF SUSPECTED HYPOTHYROIDISM

Measure Plasma T₄ and TSH Levels

The lower limit of "normal" for T_4 is poorly defined. Levels within the "normal" range may in fact be suboptimal for that particular patient. Because of the T_4-pituitary feedback, however, this fact will be established by the resultant increased TSH secretion.

A *very low* T_4 value indicates hypothyroidism unless there is gross TBG deficiency. In most cases the TSH level will be high (primary hypothyroidism). If it is low, proceed to a TRH test.

A *borderline low* T_4 value with

(a) *a slightly elevated TSH* may indicate early hypothyroidism. Measure circulating thyroid antibodies to assess the presence of autoimmune thyroiditis. As replacement therapy is probably not indicated at this stage, repeat the tests after a period of 3 to 6 months.

(b) *a normal TSH* value suggests either that there is competition by a drug for binding sites on TBG, or that the hypothyroidism is *secondary* (p. 185). A slightly low plasma T_4 with a normal TSH level may also be due to severe disease. Reassess the drug history and, if secondary hypothyroidism is suspected, proceed to a TRH test.

Perform a TRH Test (for protocol see p. 192)

(a) A subnormal rise of TSH confirms the diagnosis of *secondary hypothyroidism* of *pituitary* origin.

(b) A normal, or exaggerated but delayed rise, with TSH levels higher at 60 than at 20 minutes suggests *secondary hypothyroidism* due to hypothalamic dysfunction.

If clinically indicated, pituitary and hypothalamic function should be investigated (p. 166).

Monitoring Treatment

The progress of a patient with primary hypothyroidism on thyroid hormone replacement should be monitored by estimating plasma TSH and T_4 levels.

In monitoring secondary hypothyroidism TSH assays are of no value.

INVESTIGATION OF A PATIENT ALREADY ON TREATMENT

No patient should be given thyroid replacement therapy until the clinical diagnosis has been confirmed and documented by laboratory tests. If it is necessary to confirm the diagnosis in a patient already on treatment, but without adequate laboratory confirmation of the disease, plasma T_4 and TSH should be measured and, unless hypothyroidism is confirmed by a low T_4 and raised TSH, treatment should be stopped. The tests should then be repeated at intervals, but the diagnosis cannot be excluded until at least 6 weeks after stopping therapy.

PROTOCOL FOR TRH TEST

It is best to contact your laboratory *before* starting the test.

(a) A basal blood sample is taken;

(b) 200 μg of TRH in 2 ml saline is injected intravenously over about a minute;

(c) Further blood samples are taken 20 and 60 minutes after the TRH injection;

TSH is measured on all samples.

In normal subjects plasma TSH levels increase by at least 2 mU/l and exceed the upper limit of the reference range. The maximum response occurs at 20 minutes.

DRUG INTERFERENCE WITH THYROID HORMONE ASSAYS

Drugs may alter plasma T_4 and T_3 values as measured by radioimmunoassay or competitive protein binding techniques. The commoner effects are summarised below in Table XVII. If the primary change is in binding protein levels, free T_4 (F-T_4) values are usually normal but correction of measured T_4 by, for example, the FTI, may *not* be complete.

TABLE XVII

Drug	T_4	F-T_4	T_3	Remarks
Oestrogens	↑	N	↑	↑TBG
Oral contraceptives	↑	N	↑	↑TBG
Some radio-contrast media (e.g. ioponoate)	↑	N	↓	Blocking $T_4 \rightarrow T_3$ (transient)
Propranolol	N	N	↓	Blocking $T_4 \rightarrow T_3$
Carbimazole	↓	↓	↓	Therapeutic effect
Propylthiouracil	↓	↓	↓	Therapeutic effect
Androgens	↓	N	↓	↓TBG
Danazol	↓	N		Reduced binding to TBG
Salicylates	↓	N		Reduced binding to TBG
*†Fenclofenac	↓	N	↓N	Reduced binding to TBG
Phenytoin	↓	↓	N	Increased $T_4 \rightarrow T_3$
Carbamazepine	↓		N	

*T_4 may be elevated in some methods.
†This drug has recently been withdrawn from the U.K. market.

Chapter IX

CARBOHYDRATE METABOLISM
AND ITS INTERRELATIONSHIPS

IN most parts of the world carbohydrate is the major source of energy intake. Under normal circumstances starch is the main dietary carbohydrate, disaccharides contribute significantly and monosaccharides are a minor component of the diet.

CHEMISTRY

The main physiologically important *monosaccharide hexoses* are all reducing sugars; they therefore react with Clinitest tablets, which contain a copper compound that changes colour when reduced. They are:

glucose;
fructose;
galactose.

The common *disaccharides* are:

sucrose (fructose + glucose);
lactose (galactose + glucose);
maltose (glucose + glucose).

Lactose and maltose are reducing sugars, sucrose is not.

Naturally occurring *polysaccharides* are long-chain carbohydrates composed of glucose subunits:

starch, found in plants, is a mixture of amylose (straight chains) and amylopectin (branched chains);
glycogen, found in animal tissue, is a highly branched polysaccharide.

PHYSIOLOGY

THE IMPORTANCE OF EXTRACELLULAR GLUCOSE LEVELS

The brain cells are most vulnerable to hypoglycaemia. They derive their energy from aerobic metabolism of glucose and they *cannot*:

store glucose in significant amounts;
synthesise glucose;

metabolise substrates other than glucose and ketones. The latter usually provide very little of the energy requirements of the brain because normal plasma ketone levels are very low;

extract enough glucose for their needs from the extracellular fluid at low concentrations, because entry of glucose into the brain is not facilitated by insulin.

It is therefore clear that the brain is very dependent on extracellular concentrations of glucose for its energy supply, and that hypoglycaemia is likely to impair cerebral function. As we saw on p. 35, hyperglycaemia—especially of rapid onset—can also cause cerebral dysfunction by its effect on extracellular osmolality. In normal subjects the plasma (extracellular) glucose concentration usually remains between 4·5 and 11 mmol/l, despite the intermittent load entering the body from the gastro-intestinal tract.

The maintenance of plasma glucose levels below about 11 mmol/l also minimises loss of this energy source from the body. The renal tubule reabsorbs almost all glucose from the glomerular filtrate up to this concentration (the "renal threshold"), so that normal urine is nearly glucose free, even after a carbohydrate meal. As we shall see, this retained glucose can be stored until required.

Maintenance of Extracellular Glucose Levels

Plasma glucose concentrations depend on the balance between glucose entering and leaving the extracellular compartment. Because little is normally lost unchanged from the body, maintenance of the relatively narrow range of 4·5 to 11 mmol/l in the face of widely varying input from the gastro-intestinal tract is most likely to depend on exchange with cells. If we understand the interaction between tissues which effects this control we can also explain pathological disturbances of carbohydrate metabolism (including ketosis and lactic acidosis).

The Liver

The liver is the most important single organ in ensuring a constant energy supply for other tissues, including the brain, under a wide variety of conditions. It is also of importance in helping to control the plasma glucose concentration post-prandially. It is well-adapted to these roles for many reasons.

Portal venous blood leaving the absorptive area of the intestinal wall reaches the liver first. The hepatic cells are in a key position to buffer the hyperglycaemic effect of a high carbohydrate meal (Fig. 18).

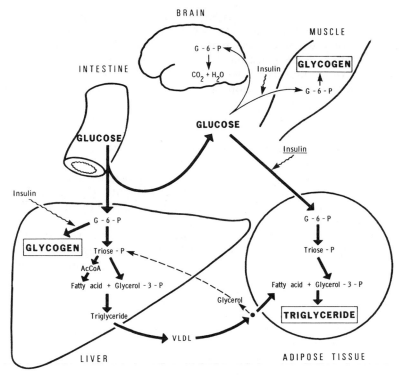

FIG. 18.—Postprandial storage of glucose.

The liver cell can store some of a temporary excess of glucose as glycogen. The rate of glycogen synthesis (glycogenesis) from glucose-6-phosphate (G-6-P) may be increased by insulin (Fig. 18) which is secreted by the β-cells of the pancreas in response to systemic hyperglycaemia.

The liver can convert some of a temporary excess of glucose to fatty acids, which are ultimately stored as triglyceride in adipose tissue (Fig. 18 and p. 197).

The entry of glucose into hepatic cells (as into cerebral cells) is not affected directly by insulin, but depends on the extracellular concentration. The conversion of glucose to G-6-P—the first step in glucose metabolism in all cells—is catalysed in the liver by the enzyme glucokinase, which has a low affinity for glucose compared with that of hexokinase found in most tissues. Glucokinase activity is induced by the insulin secreted in response to systemic hyperglycaemia. For these reasons proportionately less glucose is extracted by hepatic cells during fasting, when levels in portal venous blood are low, than after carbohydrate ingestion. This helps to maintain a fasting supply to vulnerable tissues such as brain.

Under aerobic conditions the liver can synthesise glucose by *gluconeogenesis*, using glycerol, lactate or the carbon chains resulting from deamination of most amino acids (mainly alanine), which are products of other tissues.

The liver contains the enzyme, glucose-6-phosphatase, which, by hydrolysing the G-6-P yielded by glycogen breakdown (glycogenolysis) or by gluconeogenesis, releases glucose and helps to maintain extracellular fasting levels. Hepatic glycogenolysis is stimulated by the hormone glucagon, secreted by the α-cells of the pancreas.

During fasting the liver can convert fatty acids released from adipose tissue to ketones which can be used by other tissues, including brain, as an energy source when glucose is in short supply.

This combination of properties is unique to the liver. The renal cortex is the only other tissue capable of gluconeogenesis, and of converting G-6-P to glucose. The gluconeogenic capacity of the kidney is probably mainly of importance in hydrogen ion homeostasis (p. 101).

Other tissues can store glycogen to a greater or lesser extent, but can only use it locally as they do not contain glucose-6-phosphatase; this glycogen plays no part in maintaining plasma glucose levels.

Systemic Effects of a Glucose Load (Fig. 18)

We have seen that the liver modifies the potential hyperglycaemic effect of a high carbohydrate meal by extracting relatively more glucose from the portal blood than in the fasting state. Some glucose, however, passes the liver unchanged and the rise in systemic concentration stimulates the β-cells of the pancreas to secrete insulin, which may further stimulate hepatic and muscle glycogenesis. More importantly, *entry of glucose into adipose tissue and muscle cells*, unlike that into liver and brain, *is stimulated by insulin*, and plasma glucose falls rapidly to near fasting levels. This does not happen if there is relative or absolute insulin deficiency (diabetes mellitus). Conversion of intracellular glucose into G-6-P in adipose and muscle cells is catalysed by the enzyme *hexokinase which, because its affinity for glucose is greater than that of hepatic glucokinase*, ensures that glucose enters the metabolic pathways in these tissues at lower concentrations than in the liver.

Both muscle and adipose tissue store the excess post-prandial glucose, but the mode of storage and the function of the two types of cell is very different: by examining the interrelationship of each of these tissues with the liver many of the disturbances of carbohydrate metabolism can be explained.

KETOSIS

Adipose Tissue and the Liver

Adipose tissue triglyceride is the most important long-term energy store in the body. A much increased utilisation of fat stores is associated with ketosis.

Adipose tissue, in conjunction with the liver, converts excess glucose to triglyceride and stores it in this form rather than as glycogen. The component fatty acids are derived from glucose entering the liver, and the component glycerol from glucose entering adipose tissue cells.

In the *liver* triglycerides are formed from:
glycerol-3-phosphate (from triose phosphate),
fatty acids (from acetyl CoA).

This triglyceride is transported to adipose tissue in VLDL, where it is hydrolysed by lipoprotein lipase (p. 236). The released fatty acids (of hepatic origin) condense with glycerol-3-phosphate derived from glucose entering *adipose tissue* under the influence of insulin, and the resultant triglyceride is stored. Far more energy can be stored as triglyceride than as glycogen.

During *fasting*, when exogenous glucose is unavailable, endogenous adipose tissue triglyceride is reconverted to free fatty acids (FFA) and glycerol by lipolysis (Fig. 19). Both are transported to the liver where glycerol enters the gluconeogenic pathway at the triose phosphate stage; the glucose synthesised can be released into the blood stream at a time when plasma glucose concentration would otherwise tend to fall. Most tissues other than the brain use the FFA as an energy source after conversion to acetyl CoA. In addition the liver can also form *acetoacetic acid* by enzymatic conversion of two moles of acetyl CoA; acetoacetic acid can be reduced to β-*hydroxybutyric acid* and decarboxylated to *acetone*. These "ketones" can be used as an energy source by brain and other tissues at a time when glucose is in relatively short supply.

Ketosis therefore occurs when fat stores are the main energy source and may result from *fasting*, or from reduced nutrient absorption due to vomiting. Mild ketosis may occur after fasting for as little as 12 hours (and should not then be misinterpreted as diabetic ketosis). After short fasts acidosis is not usually detectable, but after longer

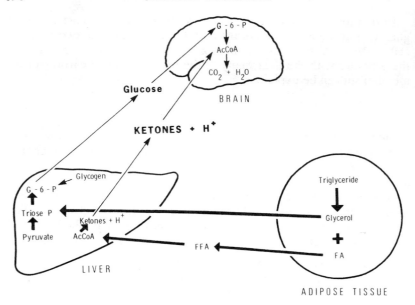

Fig. 19.—Fasting pathways: ketosis.

periods more hydrogen ions may be produced than can be dealt with by homeostatic mechanisms and plasma bicarbonate levels fall. For many weeks the plasma glucose concentration is maintained by hepatic mechanisms, but in prolonged starvation (as in anorexia nervosa) or in infancy (p. 217) ketotic hypoglycaemia may occur. The brain may suffer less from ketotic than from the same degree of insulin-induced hypoglycaemia: in the former the brain adapts to ketone metabolism, while in the latter ketone levels are low, depriving the brain of its only non-glucose energy source.

Diabetic ketoacidosis is more severe. Hyperglycaemia differentiates it from the ketosis of fasting, but the mechanism of the ketosis is the same. In starvation ketosis the supply of glucose to cells of adipose tissue is insufficient for normal glycolysis and lipogenesis. In insulin deficiency the intracellular glucose deficiency is due to impaired entry of high extracellular concentrations of glucose into these cells: the high extracellular levels are misleading as an index of intracellular events. Ketosis reflects the fact that lipolysis is the predominant pathway.

LACTATE PRODUCTION AND LACTIC ACIDOSIS

Striated Muscle and the Liver

Glucose enters the muscle postprandially under the influence of insulin and is stored as glycogen. Because of the absence of glucose-6-phosphatase this glycogen cannot be reconverted to glucose and can only supply local needs. Quantitatively muscle glycogen stores are second only to those in liver.

Muscular contraction (Fig. 20).—During muscular activity glycogenolysis is stimulated by adrenaline, and the resultant G-6-P is drawn upon by rapid glycolysis and by oxidation in the tricarboxylic acid cycle to supply the necessary energy. Under these conditions the rate of glycolysis may outstrip the availability of oxygen, and glycolytic products then exceed the immediate aerobic capacity to oxidise them.

The overall reaction for anaerobic glycolysis is

$$\text{Glucose} \rightarrow 2 \text{ Lactate}^- + 2\text{H}^+$$

The lactate is carried in the blood stream to the liver where it can be used for gluconeogenesis, providing further glucose for the muscle

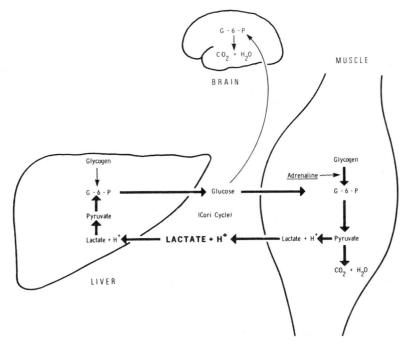

FIG. 20.—Pathways during muscular contraction.

(Cori cycle). During gluconeogenesis H⁺ is also reutilised. Under aerobic conditions the liver consumes much more lactate than it produces.

This physiological accumulation of lactic acid during muscular contraction is a temporary phenomenon, and rapidly disappears at rest, when slowing of glycolysis allows aerobic processes to "catch up".

Pathological Lactic Acidosis

Lactic acid produced by anaerobic glycolysis may be oxidised to CO_2 and water in the TCA cycle, or be reconverted to glucose by gluconeogenesis in the liver. Both the TCA cycle and gluconeogenesis require oxygen: *anaerobic glycolysis is the only pathway that does not need oxygen*.

Pathological accumulation of lactate might be due to increased production, or to decreased utilisation.

Production may be increased by:
increased rate of anaerobic glycolysis.
Utilisation may be decreased by:
impairment of the TCA cycle;
impairment of gluconeogenesis.

The *clinical syndromes* associated with lactic acidosis usually involve more than one of these factors.

Tissue hypoxia, due to the poor tissue perfusion of the "shock syndrome", is *the commonest and most important cause of lactic acidosis* (Fig. 21). Under these circumstances tissue hypoxia increases plasma lactate levels because:

the TCA cycle cannot function anaerobically and oxidation of pyruvate and lactate to CO_2 and water is impaired;
hepatic and renal gluconeogenesis from lactate cannot occur anaerobically;
anaerobic glycolysis is stimulated because the falling ATP levels cannot be regenerated by the TCA cycle, as they are in aerobic conditions.

The combination of impaired gluconeogenesis and increased anaerobic glycolysis converts the liver from a lactate and H⁺ consuming organ to one generating large amounts of lactic acid.

In severe hypoxia (such as after cardiac arrest) acidosis is severe. This hypoxic syndrome may also complicate diabetic ketoacidosis when there is volume depletion.

Some other causes of lactic acidosis are outlined below. In all of them there is imbalance between pyruvate production and utilisation.

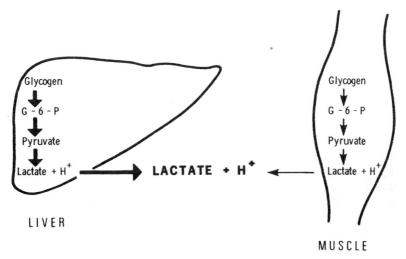

Fig. 21.—Hypoxic pathways.

Metformin or phenformin, drugs used to treat diabetes, can cause severe lactic acidosis. They inhibit both the TCA cycle and gluconeogenesis. They are now rarely used because of the danger of lactic acidosis, and phenformin is banned in some countries.

Severe illnesses such as leukaemias may be associated with lactic acidosis. A variety of factors may be involved, including poor tissue perfusion and increased anaerobic glycolysis by malignant tissue.

Infusion of fructose may cause lactic acidosis. Unlike glucose it enters hepatic cells at low plasma concentration where it is converted to glucose: the rate of anaerobic glycolysis may exceed the capacity of the TCA cycle in even a mildly hypoxic liver. For this reason fructose is now rarely used as an energy source for parenteral nutrition.

In *glucose-6-phosphatase deficiency* (von Gierke's disease, p. 215) the rate of glycolysis increases when G-6-P cannot be converted to glucose in liver and kidney.

The treatment of lactic acidosis is that of the cause, and of the acidosis. *Measurement of blood lactate levels is rarely necessary, because it is the acidosis that is dangerous: lactate itself is harmless.*

Hormones Concerned with Glucose Homeostasis

Some of the important effects of these hormones have already been described. The more detailed actions are summarised in Table XVIII.

Insulin is the most important hormone controlling the metabolic

TABLE XVIII
ACTIONS OF HORMONES ON INTERMEDIARY METABOLISM

	Insulin	Glucagon	Growth Hormone	Glucocorticoids	Adrenaline
Carbohydrate metabolism					
(a) in liver					
—glycolysis	+ +				
—glycogenesis					
—glycogenolysis	—	+ +			+
—gluconeogenesis				+	
(b) in muscle					
—glucose uptake	+ +		—		
—glycogenesis					
—glycogenolysis					+
Protein —synthesis	+		+		
—breakdown				+	
Fat —synthesis	+				
—lipolysis	—		+	+	+
Secretion—stimulated by	Hyperglycaemia Amino acids Glucagon GIT hormones	Hypoglycaemia Amino acids Fasting	Hypoglycaemia Stress and sleep	Hypoglycaemia Stress	Stress
—inhibited by	Adrenaline Fasting	Insulin			
Results	Uses and stores available glucose	Provides glucose	Spares glucose	Provides glucose	
			Provide FFA as alternative fuel →		
Plasma FFA levels	Fall		Rise		
Plasma glucose levels	Fall		Rise		

+ stimulates   — inhibits

pathways described earlier. The β-cells of pancreatic islets produce proinsulin, which incorporates the 51-amino-acid polypeptide insulin, and a 33-amino-acid linking peptide which joins one end of the A chain to the other end of the B chain of insulin. Proteolysis of proinsulin releases insulin, two amino acids, (one from each end of the connecting peptide), and the rest of the connecting peptide (called *C-peptide*). Both insulin and C-peptide are stored in islet cells and are released into plasma in equimolar amounts, mainly in response to hyperglycaemia.

Insulin binds to specific receptors on the surface of insulin-sensitive cells of adipose tissue and muscle: the most important effect is stimulation of glucose entry into these cells with a resultant fall in plasma levels.

The normal response to hyperglycaemia therefore depends on normal:

> insulin secretion;
> insulin receptors;
> intracellular responses to receptor binding of insulin ("post-receptor events").

C-peptide is probably of little physiological importance, but its measurement may help in the differential diagnosis of the causes of hypoglycaemia (p. 222).

When insulin levels are low, for example during fasting, the hyperglycaemic actions of growth hormone, glucocorticoids, adrenaline and glucagon become apparent even if there is no increase in secretion rates. Secretion of these hormones may actually increase in stress and in acromegaly (GH) (p. 132), Cushing's syndrome (glucocorticoids) (p. 145) and phaeochromocytoma (adrenaline and noradrenaline) (p. 465).

Glucagon is a single-chain polypeptide synthesised by the α-cells of the pancreatic islets and secretion is *stimulated by hypoglycaemia*.

Some of the principles that we have discussed are applied in parenteral feeding. A brief discussion of this subject will be found on p. 229.

URINARY GLUCOSE

Glycosuria is, by definition, a concentration of urinary glucose detectable using relatively insensitive, but specific screening tests (for example Clinistix, which contains glucose oxidase, (p. 225)). Usually most filtered glucose is reabsorbed by proximal tubular cells. Although very low urinary concentrations may be detectable, even in normal subjects, by more sensitive methods, glycosuria as defined above occurs only when the plasma, and therefore glomerular filtrate levels, greatly exceed the tubular reabsorptive capacity. This may be because:

the plasma and glomerular filtrate concentrations are more than about 11 mmol/l, and therefore the normal tubular reabsorptive capacity is significantly exceeded;

the tubular reabsorptive capacity is reduced, so that glycosuria occurs at a lower filtrate concentration (*"renal glycosuria"*). This is usually a harmless condition.

Occasionally glycosuria does not occur despite plasma glucose levels above 11 mmol/l; if the volume of glomerular filtrate is low the *total amount* of glucose delivered to tubular cells may be less than normal, even if the concentration is high. Urine testing may then be invalidated as an index of the required dose of antidiabetic drugs in patients with impaired renal function.

A test on a urine specimen passed about an hour after a meal, when plasma glucose levels are likely to be highest, is most sensitive for screening for glycosuria. Specimens collected after a period of fasting will yield positive results only when fasting plasma glucose levels are above 11 mmol/l (severe diabetes mellitus, or during infusion of glucose), or if there is gross renal glycosuria.

Reducing substances in the urine (including glucose) can be detected using Clinitest tablets. This test is important in the neonatal period, when the presence of reducing substances, other than glucose may suggest an inborn error of metabolism. It may also be used by diabetics to control drug dosage (p. 225).

HYPERGLYCAEMIA AND DIABETES MELLITUS

Hyperglycaemia occurs:

in the syndrome of diabetes mellitus;
in patients receiving glucose-containing fluids intravenously;
temporarily in severe stress;
sometimes after cerebrovascular accidents.

DIABETES MELLITUS

Diabetes mellitus is due to absolute or relative insulin deficiency. It has been defined by the World Health Organisation, on the basis of laboratory findings, as a fasting plasma glucose level of 8 mmol/l or more, or, even if the fasting concentration is normal, a level of 11 mmol/l or more 2 hours after oral ingestion of the equivalent of 75 g glucose (p. 227). Severe cases have persistent hyperglycaemia.

The provisional WHO classification divides diabetes into the following categories:

Insulin-dependent diabetes mellitus (IDDM, Type 1) is the term used to describe patients for whom, because they are prone to develop ketoacidosis, insulin therapy is essential. Onset is most common during childhood, and subjects with HLA tissue types DR3 and DR4 are most at risk. It has been suggested that many cases follow a viral infection which has caused destruction of the β-cells of the pancreatic islets.

Non-insulin-dependent diabetes mellitus (NIDDM, Type 2) is the commonest variety. Patients are much less likely to develop ketoacidosis, and although insulin may sometimes be needed it is not essential for survival. Onset is most common in adult life. It is further subdivided into obese and non-obese NIDDM. A variety of inherited disorders may be responsible for the syndrome, either by reducing insulin secretion, or by causing relative insulin deficiency, despite high plasma hormone levels, because of resistance to its action, or to receptor or post-receptor defects.

Diabetes associated with other conditions including:

absolute insulin deficiency due to pancreatic disease (chronic pancreatitis, haemochromatosis, cystic fibrosis);

relative insulin deficiency due to excessive growth hormone (acromegaly) or glucocorticoid secretion (Cushing's syndrome), or increased glucocorticoid levels due to administration of steroids;

relative insulin deficiency due to some other drugs (for example, thiazide diuretics).

Gestational diabetes is first diagnosed during pregnancy, and is classified separately because of potential danger to the fetus. If diabetes mellitus is confirmed after delivery, these cases should then be allocated to the appropriate group.

Impaired Glucose Tolerance

The WHO definition of impaired glucose tolerance includes patients with a plasma glucose concentration between 8 and 11 mmol/l two hours after a standard glucose load. Only a small proportion of such cases later develop diabetes mellitus, and patients are *not* now diagnosed as such on the basis of impaired glucose tolerance because of the serious psychological, social and economic implications of the diagnosis. It is not possible to predict the outcome at the time of discovery of impaired tolerance.

Subjects at Risk of Developing Diabetes Mellitus

People with a normal glucose tolerance may be thought to be at risk of developing diabetes because, for example, of a strong family history or because they have given birth to abnormally large babies.

Subjects with impaired tolerance, or those at risk, used to be called pre-diabetics, latent diabetics or potential diabetics. These terms should no longer be used.

THE CLINICAL AND METABOLIC FEATURES OF INSULIN-DEPENDENT DIABETES MELLITUS

Most of the metabolic changes of diabetes mellitus are a consequence of insulin deficiency.

Hyperglycaemia at some time is, by definition, an invariable finding. If plasma glucose levels exceed about 11 mmol/l and renal function is normal there is usually *glycosuria*. In severe cases high urinary glucose concentrations produce an osmotic diuresis and therefore *polyuria:* severe hyperglycaemia, because of its osmotic effect, causes *thirst* (p. 39). Prolonged osmotic diuresis may cause excessive urinary electrolyte loss. *These "classical" symptoms of diabetes are only present in advanced cases.*

Hyperglycaemia also increases the rate of glucose binding (glycosylation) to proteins such as haemoglobin. The measurement of glycosylated haemoglobin (HbA_{1C}) has been used as a measure of the *mean* plasma glucose concentration during the life of the red cell. Although it reflects long-term glucose control, its use as an index of the adequacy of antidiabetic therapy is limited by technical problems of the assay; moreover, although it may be satisfactory as an index of the mean plasma glucose concentration, short-term swings between hyper- and hypoglycaemia can only be detected by serial glucose measurements. Glycosylation of proteins such as those in the lens of the eye or in the glomerular basement membrane may possibly have a role in the development of the long-term complications of diabetes. If this is so, strict control of plasma levels is important.

Other Metabolic Changes due to Insulin Deficiency

Abnormalities in *lipid* metabolism secondary to insulin deficiency also occur. Lipolysis is stimulated and plasma FFA levels rise. In the liver FFA are converted to acetyl CoA and ketones, or are re-esterified into endogenous triglycerides and incorporated into VLDL. Cholesterol synthesis is also increased with an increase in LDL and if insulin deficiency is very severe, chylomicrons may accumulate in the blood.

Increased breakdown of *protein* may cause muscle wasting.

Long-term Effects

Vascular disease is common and may present as cerebrovascular or peripheral vascular insufficiency. Abnormalities of small blood vessels particularly affect the retina and kidney. In the latter, diffuse, and more rarely nodular glomerulosclerosis (Kimmelstiel-Wilson kidney) may produce the nephrotic syndrome. Infections are common and may aggravate renal and peripheral vascular disease. Diabetics tend to give birth to large babies.

Principles of Management of Diabetes Mellitus

The management of diabetes mellitus will be only briefly considered.

The long-term complications of diabetes mellitus may be due to hyperglycaemia. Treatment should aim to keep plasma glucose concentrations "normal". Some patients are being taught to measure their blood glucose levels throughout the day, using reagent strips: the colour change of the strip may be quantitated using a portable reflectance meter. This procedure involves the discomfort of several skin punctures, but an intelligent patient may be able to adjust his insulin dose more accurately than by testing his urine. It is *essential* that both reagent strips and meters are regularly checked by the laboratory. The patient should still be under regular medical supervision.

Insulin requirements vary in patients with IDDM. For example, the dose may need to be increased during any illness or during pregnancy.

Glucose levels in patients with NIDDM are frequently controlled by *diet* with *weight reduction*. In this group insulin secretion may be stimulated by the *sulphonylurea* drugs such as tolbutamide or glibenclamide. *Biguanides*, such as metformin or phenformin, have been used to lower blood glucose; the mechanism of their action is not completely elucidated, but they inhibit gluconeogenesis either directly or indirectly. The danger of lactic acidosis has been mentioned on p. 201. Patients with NIDDM may require insulin during periods of stress.

Acute Metabolic Complications of Diabetes

The diabetic patient may develop one of several complications requiring emergency therapy. The main ones are:

diabetic ketoacidosis;
hyperosmolal non-ketotic coma;
hypoglycaemia due to insulin in excess of needs (p. 213).

Diabetic Ketoacidosis

Diabetic ketoacidosis is an extension of the metabolic events outlined on p. 206. It may be precipitated by infections or gastro-intestinal upsets with vomiting. Insulin may mistakenly be withheld by the patient who reasons "no food, therefore no insulin". The consequences are due principally to three factors—*glycosuria, plasma hyperosmolality* and *ketosis.*

Plasma glucose levels are usually in the range of 20–40 mmol/l (400–700 mg/dl) but may be considerably higher. This results in marked *glycosuria* which produces an *osmotic diuresis* leading to loss of water and depletion of body electrolytes. Vomiting is common in ketoacidosis and adds to the fluid and electrolyte depletion. Because of the plasma hyperosmolality, fluid loss involves both intra- and extracellular compartments with reduction of circulating blood volume, reduced renal blood flow and glomerular filtration rate, and severe cellular dehydration. The plasma urea level is usually raised and there is often evidence of haemoconcentration such as elevated haematocrit and total protein levels. Both these findings are the result of *volume depletion.*

As a result of intracellular glucose deficiency *lipolysis* is accelerated. More FFA are produced than can be metabolised by peripheral tissues and they are converted in the liver to ketones, or incorporated into endogenous triglycerides. Severe *hyperlipaemia* may be present. The hydrogen ions produced with ketones (other than acetone) are buffered by plasma bicarbonate. Urinary pH drops as the hydrogen ions are secreted by the mechanism described on p. 98: although H^+ secretion is, as always, linked with equimolar generation of bicarbonate, this cannot keep pace with utilisation in buffering and plasma bicarbonate levels fall. The developing metabolic acidosis stimulates the respiratory centre and breathing becomes deeper with a lowering of PCO_2. This respiratory compensation may maintain blood pH within the normal range for a while, but at the cost of further lowering of the plasma bicarbonate to very low levels. The deep sighing respiration (*Kussmaul respiration*) with the odour of acetone on the breath is a classical feature of diabetic ketoacidosis.

Plasma *potassium levels may be raised* before treatment is started, due probably to the acidosis, to reduced entry of glucose into cells and to the low GFR. It must be realised, however, that urinary potassium loss has been high and that there is a *total body deficiency.* This will be revealed during treatment, as potassium re-enters the cells, with resultant, and possibly severe, hypokalaemia.

Plasma *sodium* concentrations on presentation may be low or low normal despite clinically obvious dehydration. This may be due to the osmotic effect of the high extracellular glucose concentrations which, by drawing water from cells, dilutes plasma sodium. Severe hyper-lipidaemia may cause spurious hyponatraemia (p. 37). When insulin is given glucose enters cells and water follows along the osmotic gradient. Despite a satisfactory fall in plasma glucose levels, patients may remain confused or even comatose due to persistence of plasma hyperosmolality if plasma sodium levels rise rapidly. This is especially common if isosmolar or stronger saline is given.

The losses in a typical case of diabetic ketoacidosis may amount to 6 or 7 litres of water and 300–500 mmol each of sodium and potassium. The most usual findings in diabetic ketoacidosis are therefore:

Clinical

> confusion and later coma (hyperosmolality);
> overbreathing (acidosis);
> volume depletion (osmotic diuresis).

Plasma

> hyperglycaemia;
> acidosis with low total CO_2 (bicarbonate);
> normal or raised potassium concentration;
> variable sodium concentration;
> haemoconcentration and mild uraemia.

Urine

> glycosuria ⎫
> ketonuria ⎭ while the urinary flow is adequate;
> low pH (unless there is renal failure).

Associated findings.—Changes in plasma phosphate levels parallel those of potassium. Low levels may be found for several days after recovery from diabetic coma. Amylase activities may be markedly elevated in both urine and plasma, and should not necessarily be interpreted as indicating acute pancreatitis. These may be found coincidentally, but are of no importance in diagnosis, prognosis or treatment. They are mentioned to avoid the dangers of misinterpretation.

Hyperosmolal Non-ketotic Coma

In diabetic ketoacidosis there is always some hyperosmolality due to the hyperglycaemia, and many of the symptoms, including the coma, are probably related to it; the term "hyperosmolal" coma (or "precoma") is, however, usually confined to a condition in which there is marked hyperglycaemia, but no detectable ketonaemia or acidosis. (The reason for these different presentations is not clear.) It is more common in older subjects. Plasma glucose levels are very high, often greater than 50 mmol/l (900 mg/dl). The resulting glycosuria produces an osmotic diuresis, with severe water and electrolyte depletion. Hypernatraemia and uraemia are often present due to predominant water loss, and these aggravate extracellular hyperosmolality. Coma is thought to be the result of consequent cerebral cellular dehydration, which may also lead to hyperventilation: the consequent respiratory alkalosis may cause a slight fall in plasma bicarbonate concentration which should not be confused with that of metabolic acidosis.

Osmolarity can be calculated with sufficient accuracy for clinical purposes as described on p. 35.

Other Causes of Coma in Patients with Diabetes Mellitus

In addition to the metabolic complications described above, a known diabetic may present in coma, or a confused state (precoma) due to hypoglycaemia, a cerebrovascular accident or to an unrelated cause.

> *Hypoglycaemia* is most commonly due to accidental overadministration of insulin or may result from a failure to eat normally, or participation in excessive exercise after the usual dose of insulin or oral antidiabetic drugs.
>
> *Cerebrovascular accidents* are common in diabetics because of the increased incidence of vascular disease. Such an event may be associated with hyperglycaemia and glycosuria ("piqûre diabetes"). Stimulation of the respiratory centre may cause overbreathing and consequent respiratory alkalosis, with a low plasma bicarbonate level. This should not be confused with metabolic acidosis. Diagnosis depends on clinical findings.

The assessment of a diabetic patient presenting in coma or precoma is considered on p. 221.

Principles of Treatment of Diabetic Coma

Only the outline is discussed here. For details consult the references given at the end of the chapter. The treatment of hypoglycaemia is considered on p. 217.

Ketoacidosis.—*Repletion of fluid and electrolytes* should be vigorous. If the plasma sodium concentration is normal or low, isotonic saline is used initially and is continued if the sodium concentration remains normal. A careful watch should be kept for the development of hypernatraemia, and if it is found, hypotonic saline should be used. If acidosis is very severe, bicarbonate may be given, but it is unnecessary to correct the bicarbonate level entirely, as it rapidly returns to normal following adequate fluid and insulin therapy. It should be remembered that 8 per cent sodium bicarbonate is grossly hyperosmolar (Table IV p. 65), and that a rapid rise in blood pH may aggravate hypokalaemia. *Potassium* should be administered as soon as plasma levels start to fall. Urine output should be monitored:

if it falls despite adequate rehydration, fluid and potassium should be given with care.

Insulin is given immediately by continuous intravenous infusion, or frequently repeated small doses.

The factor which precipitated the coma (such as infection) should be sought and treated.

Frequent monitoring of plasma glucose, potassium and sodium levels is necessary to assess progress and to detect developing hypoglycaemia, hypokalaemia or hypernatraemia. The latter may delay the fall of plasma osmolality to normal, with persisting confusion or coma.

Hyperosmolal coma.—Treatment of hyperosmolal coma is similar to that of ketoacidosis; a sudden reduction of extracellular osmolality may do more harm than good (p. 35), and it is especially important to give *small doses of insulin* to reduce plasma glucose levels slowly. These patients are often very sensitive to the action of insulin. Fluid should be replaced with hypotonic solutions but, if severe hypernatraemia is present, this, too, should be done cautiously.

HYPOGLYCAEMIA

Hypoglycaemia is defined as a plasma glucose level of below 2·5 mmol/l (45 mg/dl). This criterion applies *only if a specific glucose assay is used* and the sample is collected into a tube containing an *inhibitor of glycolysis* (see p. 224). Low values obtained on an improperly collected sample cannot be interpreted. The symptoms of hypoglycaemia may develop at levels greater than this when there has been a rapid fall from a previously elevated, or even normal, value; by contrast, some people may show no symptoms at levels much below 2·5 mmol/l, especially if these have fallen gradually. As discussed earlier, cerebral metabolism is dependent on an adequate supply of glucose from the blood and the symptoms of hypoglycaemia resemble those of cerebral hypoxia. Faintness, dizziness or lethargy may progress rapidly to coma and, if untreated, death or permanent cerebral damage may result. If the fall of plasma glucose is rapid there may be a phase of sweating, tachycardia and agitation due to adrenaline secretion, but this phase may be absent if the fall is gradual, or if the autonomic nervous system is unresponsive because the patient is on β-blocking agents or has severe peripheral neuropathy. Existing cerebral or cerebrovascular disease may aggravate the condition. Rapid restoration of plasma glucose concentration is essential.

CAUSES OF HYPOGLYCAEMIA

There is no completely satisfactory classification of the causes of hypoglycaemia as overlap between different groups occurs. A practical approach is based on the history. Particular attention is paid to drugs used and to the relation of symptoms to meals or a particular food. The patient can usually be provisionally allocated to one of two main groups.

Fasting hypoglycaemia.—Symptoms typically occur at night or in the early morning, or an episode may be precipitated by a prolonged fast or strenuous exercise. This implies excessive utilisation of glucose or an abnormality of the glucose-sparing or glucose-forming mechanisms (p. 196).

The main causes are:

> inappropriately high insulin levels due to tumour or hyperplasia of the pancreatic islet cells;
> deficiency of glucocorticoids;
> severe liver disease;
> non-pancreatic tumours.

Non-fasting hypoglycaemia.—Symptoms typically occur within 5 to 6 hours of a meal and may be related to a particular type of food, or are associated with medication. Substances that may provoke hypoglycaemia include:

> drugs (especially insulin);
> alcohol;
> glucose (reactive hypoglycaemia);
> galactose (in milk);
> fructose (in sucrose-containing foods);
> leucine (an amino acid in casein).

It is not always possible to distinguish between provoked and early fasting hypoglycaemia on the basis of history alone.

Because hypoglycaemia presents a particular problem in the early years of life this will be considered separately after a more general discussion.

HYPOGLYCAEMIA, PARTICULARLY IN ADULTS

Symptoms can only be attributed to hypoglycaemia if hypoglycaemia is demonstrated. This statement may seem superfluous but it is not unknown for patients to be submitted to multiple tests for the differential diagnosis of hypoglycaemia that does not exist, or worse, to undergo treatment for it.

Once demonstrated the following main causes must be considered:

Insulin- or other drug-induced hypoglycaemia.—This is probably the commonest cause and unless deliberately concealed by the patient the offending drug should be easily identified. Hypoglycaemia in a diabetic may follow accidental overdosage, be due to changing *insulin* requirements, or to a failure to take food after insulin has been given. Self-administration for suicidal purposes or to gain attention is not unknown, and homicidal use is a remote possibility. *Sulphonylureas* may also induce hypoglycaemia, especially in the elderly, and salicylate poisoning may be complicated by hypoglycaemia. Other drugs such as antihistamines have been suspected in some cases and a drug history is an important part of the investigation.

Hypoglycaemia due to exogenous insulin suppresses C-peptide secretion by feedback. Measurement may help to differentiate insulin administration from endogenous insulin secretion—whether the latter is from an insulinoma, follows pancreatic islet stimulation by antidiabetic drugs, or is from non-pancreatic insulin-secreting tumours.

Insulinoma.—An insulinoma is a tumour of the islet cells of the pancreas. As with the other endocrine tumours, hormone secretion is inappropriate and usually excessive. It may occur at any age and is usually single and benign but may rarely be malignant. Multiple tumours may be present and it may be part of the syndrome of multiple endocrine adenomatosis (p. 470). The tumour involves β-cells, although other islet cells are usually present in it. C-peptide is released in parallel with insulin, and plasma levels are therefore inappropriately high.

Symptoms may be sporadic, with trouble-free intervals. Hypoglycaemic attacks occur typically at night and before breakfast or may be precipitated by strenuous exercise. Personality or behavioural changes may be the first feature and many of these patients present to psychiatrists.

Alcohol-induced hypoglycaemia.—Hypoglycaemia may develop between 2 and 10 hours after ingestion of alcohol, usually in considerable amounts. It is described most frequently in chronic alcoholics when starvation or malnutrition is present, but may occur in young subjects after first exposure. The probable mechanism is reduced hepatic output of glucose due to suppression of gluconeogenesis during metabolism of the alcohol. This is potentiated in the fasting or starved state when gluconeogenesis is the main source of plasma glucose. Clinical differentiation from alcoholic stupor may be impossible without plasma glucose estimation. Frequent infusions of glucose may be needed in treatment. It is difficult to reproduce the hypoglycaemia after infusion of alcohol unless the patient fasts for 12 to 72 hours.

Non-pancreatic tumours.—Although many types of malignant tumours—carcinoma (especially hepatocellular) as well as sarcoma—

have been incriminated in the production of hypoglycaemia, it occurs most commonly in association with mesodermal tumours resembling fibrosarcomata, that occur mainly retroperitoneally. They are slow-growing and may become very large. Hypoglycaemia may be the presenting feature. The mechanism is uncertain and may be secretion of an insulin-like substance or excessive utilisation of glucose (p. 476).

Functional (reactive) hypoglycaemia (sensitivity to glucose).—Some people develop symptomatic hypoglycaemia 2–4 hours after a meal or after a glucose load. Unconsciousness is very rare. A similar "reactive" hypoglycaemia may be seen after *gastrectomy*, when rapid passage of glucose into the intestine and rapid absorption may stimulate excessive insulin secretion (one type of "dumping syndrome". See p. 298).

True reactive hypoglycaemia, except after gastrectomy, is probably rare, and is diagnosed too often.

Endocrine causes.—Hypoglycaemia may occur in adrenal or pituitary insufficiency. Rarely it is the sole manifestation of these conditions, which are considered in Chapters V and VI.

Impaired liver function.—The functional reserve of the liver is so great that, despite its central role in the maintenance of plasma glucose levels, hypoglycaemia is uncommon in liver disease, except in very severe *hepatitis* or liver necrosis—conditions in which the whole liver is affected. Its importance lies more in the recognition of hypoglycaemia as a cause of coma in such patients than as a problem of differential diagnosis of hypoglycaemia.

HYPOGLYCAEMIA IN CHILDREN

Hypoglycaemia in infancy is not uncommon and is important because permanent brain damage can result, especially in the first few months of life. Only the main causes will be outlined.

Neonatal Period

Plasma glucose levels in the neonatal period as low as 2·2 mmol/l (about 40 mg/dl) or, in the first 72 hours, 1·7 mmol/l (about 30 mg/dl) may be considered "normal" and are often asymptomatic. Such hypoglycaemia often lasts only 24 hours. In premature (less than 2·5 kg) infants, levels below 1·1 mmol/l (about 20 mg/dl) may occur without clinical evidence of hypoglycaemia. Signs of hypoglycaemia at this age are convulsions, tremors and attacks of apnoea with cyanosis; treatment in such cases is urgent, and such symptomatic hypoglycaemia may last for up to a week. Neonatal hypoglycaemia occurs particularly:

in babies of diabetic mothers. If the fetus is exposed to hyperglycaemia during pregnancy, islet-cell hyperplasia occurs and the consequent hyperinsulinism may produce hypoglycaemia when the supply of excess glucose from the mother is removed after parturition. Severe hypoglycaemia has been noted in babies whose mothers have been treated with *sulphonylurea drugs* during pregnancy. In some series the incidence of hypoglycaemia in babies of diabetic mothers is 50 per cent or higher. Most of these infants show no clinical signs and the significance of these low levels is controversial. Similar islet-cell hyperplasia and neonatal hypoglycaemia may be found with severe erythroblastosis fetalis;

in babies suffering from intra-uterine malnutrition (for instance, infants of toxaemic mothers, or the smaller of twins). These babies are usually "small for dates" and may show a tendency to hypoglycaemia during the first week, possibly due to poor liver glycogen stores. Prematurity is an aggravating factor as most of the liver glycogen is laid down after 36 weeks. Low fat stores also limit the availability of ketones as an alternative energy source for the brain, and these babies usually show clinical signs of hypoglycaemia.

Early Infancy

Soon after birth, or after the introduction of milk or sucrose, hypoglycaemia may be due to:

Glycogenoses.—A deficiency of one of the several enzymes involved in glycogenesis or glycogenolysis results in the accumulation of normal or abnormal glycogen. In the commonest of these diseases (von Gierke's) *glucose-6-phosphatase* is deficient. As this enzyme is essential for the conversion of glucose-6-phosphate to glucose, *fasting hypoglycaemia* occurs. The further metabolic consequences include *ketosis* and *endogenous hypertriglyceridaemia* (excessive lipolysis due to intracellular glucose deficiency), *lactic acidosis* (excessive glycolysis) and *hyperuricaemia* (p. 409).

Hepatomegaly occurs due to the accumulated glycogen.

The *diagnosis* is made directly by demonstrating the absence of the enzyme in a liver biopsy specimen, or indirectly by the failure of blood glucose levels to rise after glucagon administration (which stimulates glycogenolysis). Infusion of galactose or fructose, normally converted to glucose, also fails to raise plasma glucose levels in these patients as the G-6-P formed cannot be converted to glucose.

Treatment is to take frequent meals to maintain plasma glucose levels and prevent cerebral damage.

Galactosaemia.—Galactose is necessary for the formation of cerebrosides, some glycoproteins and, during lactation, of milk. Any excess is rapidly converted to glucose.

The commonest form of galactosaemia is due to deficiency of *hexose-1-phosphate uridylyltransferase* (previously known as *galactose-1-phosphate uridyltransferase*).

The condition becomes apparent only after milk has been added to the infant's diet. The main features are:

vomiting and diarrhoea with failure to thrive;
hepatomegaly leading to jaundice and cirrhosis;
cataract formation;
mental retardation;
renal tubular damage ("Fanconi syndrome");
hypoglycaemia.

Galactose in the urine is a cause of a positive reaction with Clinitest tablets. This feature is dependent on ingestion of galactose and may be absent if the subject is not receiving milk. Tubular damage may result in generalised aminoaciduria.

The *diagnosis* is made by identifying the urinary sugar as galactose by chromatography and by demonstrating a deficiency of the relevant enzyme in the erythrocytes. The latter test should be done on cord blood in all newborn infants with affected siblings.

Treatment.—Galactose (in milk and milk products) should be eliminated from the diet. Sufficient galactose (as UDP-galactose) is synthesised endogenously for the body's needs.

Hereditary fructose intolerance.—This is a *rare* cause of hypoglycaemia. Symptoms start after the introduction of sucrose or fruit juice to the diet. Fructose administration leads to symptoms of hypoglycaemia in half to one hour, accompanied by nausea or vomiting and abdominal pain. There is failure to thrive and progressive liver damage with hepatomegaly, jaundice and ascites. Cirrhosis may develop. Diagnosis is based on the demonstration of fructosuria and hypoglycaemia after oral or intravenous administration of fructose (the basic defect is a deficiency of the enzyme *fructose-1-phosphate aldolase* and accumulation of *fructose-1-phosphate*). The hypoglycaemia is probably due to inhibition of glycogenolysis and gluconeogenesis by fructose-1-phosphate.

Later Infancy

Idiopathic hypoglycaemia of infancy may be of multiple aetiology.

Symptoms usually develop after fasting or a febrile illness. The diagnosis is made by excluding other causes. There is a high incidence of brain damage.

In some cases there is excessive insulin secretion and differentiation from insulinoma is not possible.

Leucine sensitivity.—In the first six months of life *casein* may precipitate severe hypoglycaemia. This is due to its content of the amino acid, leucine. Leucine sensitivity is probably due to excessive stimulation of insulin secretion and there is often a familial incidence. The condition appears to be self-limiting and does not usually persist above the age of 6 years. Diagnosis is confirmed by the demonstration of hypoglycaemia within half-an-hour after an oral dose of leucine or casein. Treatment is a low-leucine diet.

Normal individuals do not show a significant drop in plasma glucose after leucine, but a number of patients with insulinoma do demonstrate leucine sensitivity.

Ketotic hypoglycaemia is the commonest cause in the second year of life and develops after fasting or a febrile illness. These children were usually "small-for-dates" babies. Ketonuria precedes the hypoglycaemia (as it does in starvation) and the diagnosis can be established by feeding a ketogenic diet (high fat, low calorie) for 48 hours, during which clinical hypoglycaemia occurs.

"Adult" causes, including insulinoma, must be considered at all times.

TREATMENT OF HYPOGLYCAEMIA

Hypoglycaemia should be treated by urgent intravenous administration of at least 10 per cent, and in the adult 50 per cent, glucose solution *after withdrawal of a blood sample for diagnosis* (p. 222). Some cases may need to be maintained on a glucose infusion until the cause has been established and treated.

The treatment of *insulinoma* is by surgical removal if possible. If not, a combination of diazoxide and chlorothiazide may maintain normoglycaemia.

SUMMARY

1. Glucose is the main product of dietary carbohydrate.

2. The brain is almost entirely dependent on extracellular glucose as an energy source, and maintenance of plasma glucose levels is important for normal cerebral function.

3. After a carbohydrate-containing meal, excess glucose is:
   stored as glycogen in liver and muscle;
   converted to fat and stored in adipose tissue.
   Insulin stimulates these processes.

4. During fasting:
   glycogen breakdown in the liver (and kidney) releases the glucose into the blood;
   triglyceride breakdown in adipose tissue releases glycerol, which can be converted to glucose, and fatty acids, which can be metabolised by most tissues *other than the brain*.

5. The liver converts excess fatty acids to ketoacids. Ketones can be used as an energy source by the brain and other tissues.

6. If ketoacid formation exceeds the capacity of homeostatic mechanisms, ketoacidosis may develop.

7. Anaerobic glycolysis produces lactic acid:
   lactic acid production occurs temporarily in contracting muscles;
   lactic acid is produced in hypoxic tissues. The hypoxic liver becomes a major lactic-acid-producing rather than a lactic-acid-consuming organ, and lactic acidosis results.

Other factors increasing glycolysis or reducing utilisation of lactic acid may also cause lactic acidosis. Some drugs, such as biguanides, reduce lactic acid utilisation.

8. Diabetes mellitus is the result of relative or absolute insulin deficiency. It is characterised by hyperglycaemia, which may be intermittent. In severe diabetes mellitus excessive lipolysis may cause ketosis and later acidosis. Coma is probably due to hyperosmolality, and in both ketoacidosis and hyperosmolal coma there is severe water and electrolyte depletion.

9. Hypoglycaemia may occur during fasting, or be provoked by drugs or by some foods.

10. In adults *insulinoma* is the *most important* cause of fasting hypoglycaemia; it is diagnosed by demonstrating high plasma insulin levels despite hypoglycaemia. *Insulin administration* is the *commonest* cause of hypoglycaemia.

11. Factors causing hypoglycaemia in childhood vary with age. Many inborn errors of carbohydrate metabolism may cause hypoglycaemia.

12. It may be necessary to supply energy substrate intravenously over long periods. Glucose minimises lipolysis and ketosis, but may need to be supplemented by fat solutions to provide enough energy in a manageable volume of fluid (See p. 229).

FURTHER READING

WHO EXPERT COMMITTEE ON DIABETES MELLITUS. (1980). Second Report. *WHO Tech. Rep. Ser.* No. **646**.

CZECH, M. P. (1981). Insulin action. *Amer. J. Med.*, **70**, 142–150.

STANLEY, J. C. (1981). The glucose-fatty acid cycle. *Brit. J. Anaesth.*, **53**, 123–129.

STANLEY, J. C. (1981). The glucose-fatty acid-ketone body cycle. *Brit. J. Anaesth.*, **53**, 131–136.

STANLEY, J. C. (1981). The regulation of glucose production. *Brit. J. Anaesth.*, **53**, 137–146.

CAHILL, G. F. (Jr.) and McDEVITT, H. O. (1981). Insulin-dependent diabetes mellitus: the initial lesion. *New Engl. J. Med.*, **304**, 1454–1465.

ROTTER, J. I. and RIMOIN, D. L. (1981). The genetics of the glucose intolerance disorders. *Amer. J. Med.*, **70**, 116–126.

OLEFSKY, J. M. (1981). Insulin resistance and insulin action. An *in vitro* and *in vivo* perspective. *Diabetes*, **30**, 148–162.

BUNN, H. F. (1981). Nonenzymatic glycosylation of protein: relevance to diabetes. *Amer. J. Med.*, **70**, 325–330.

MARKS, V. and ROSE F. C. (1981). *Hypoglycaemia*, 2nd edit. Oxford: Blackwell Scientific Publications.

Historical Interest

BANTING, F. G. and BEST, C. H. (1922). The internal secretion of the pancreas. *J. Lab. Clin. Med.*, **7**, 251–266; Pancreatic extracts. *ibid.* **7**, 464–472.

INVESTIGATION OF DISORDERS OF CARBOHYDRATE METABOLISM

INVESTIGATION OF SUSPECTED DIABETES MELLITUS

INITIAL TESTS

If the patient presents with the symptoms of severe diabetes mellitus or with glycosuria, or if it is desirable to exclude the diagnosis because, for example, of a strong family history, take blood for plasma glucose estimation. The finding of one random glucose level (at least 2 hours after a meal) of more than 11 mmol/l is very suggestive, and of two at such a level is diagnostic of the syndrome. However, it is preferable to take blood after a fast of at least 10 hours, especially if one abnormal random level has been found.

INTERPRETATION

	Plasma glucose (mmol/l)	
	Fasting	Random
Normal	6 or less	
Diabetic	8 or more	11 or more

(a) The diagnosis of diabetes mellitus is confirmed if:

 (i) the fasting plasma glucose level is 8 mmol/l or more *on two occasions*;

 or (ii) the random plasma glucose level is 11 mmol/l or more *on two occasions*;

 or (iii) *both* a fasting level of more than 8 mmol/l and a random level of more than 11 mmol/l are found.

(b) Random values are less reliable for excluding than for confirming the diagnosis. *Fasting* plasma glucose levels *below 6 mmol/l on at least two occasions* exclude it.

(c) If, *and only if*, any of the fasting values lie between 6 and 8 mmol/l or the random values between 8 and 11 mmol/l, proceed to an oral glucose tolerance test.

ORAL GLUCOSE TOLERANCE TEST (See p. 227)

If the plasma glucose concentration at 2 hours after an oral dose of 75 g glucose (or glucose equivalent p. 227) is:

 8 mmol/l or less the result is normal;
 11 mmol/l or more the patient is diabetic;
 between 8 and 11 mmol/l the patient has impaired glucose tolerance.

A patient *without symptoms* may be suspected of having diabetes mellitus because of the chance finding of glycosuria or hyperglycaemia. If the plasma value in such a patient 2 hours after a standard glucose load is more than

11 mmol/1, the finding should be confirmed before a definitive diagnosis is made.

INVESTIGATION OF A DIABETIC PRESENTING IN COMA

A diabetic patient may present in coma due to hyperglycaemia, hypoglycaemia, or any of the causes shown in Table XIX.

TABLE XIX

Diagnosis	Clinical features	Plasma glucose	[HCO$_3^-$]	Urine glucose	ketones
Ketoacidosis	Dehydration Hyper-ventilating	High	Low	+++	+++
Hyperosmolal coma	Dehydration May be hyper-ventilating	Very high	N or slightly reduced	+++	Neg.
Hypoglycaemia	Non-specific	Low	N	Neg.	Neg.
Cerebrovascular accident	Neurological May be hyper-ventilating	May be raised	May be low	May be +	Usually neg.

After a thorough clinical assessment:

1. Notify the laboratory that specimens are being taken, and ensure that they are delivered promptly. In this way delays in assay are minimised.

2. Take blood for:

plasma   glucose
           potassium   the most urgent;
           sodium
plasma TCO$_2$ and/or arterial blood pH and PCO$_2$.

Repeated arterial puncture is undesirable and unnecessary. If arterial pH is measured initially, plasma TCO$_2$ may be estimated if it is felt to be necessary to monitor acid-base balance.

3. Test a urine sample for glucose and ketones. *It is unwise to rely solely on urine testing to diagnose hyperglycaemia.* The urine may have been in the bladder for some time and reflect earlier and very different plasma levels.

4. If it is really necessary to obtain a rapid assessment of blood glucose using a reagent strip (for instance, because the laboratory is at such a distance that there may be delays in receiving precise results) *it is essential to remember that improperly stored reagent strips, or failure to follow the instructions for use, may give completely and dangerously wrong results.*

5. If hypoglycaemia is suspected on clinical grounds, or on the results of reagent strips, glucose should be given while awaiting the laboratory results.

It is less dangerous to give glucose to a diabetic than to give insulin to a patient falsely diagnosed as hypoglycaemic, perhaps because of results obtained with a faulty reagent strip.

In summary, institution of therapy in diabetic coma should be based on clinical diagnosis. Blood samples should be sent to the laboratory immediately, but treatment should never be delayed until the results are available. *Sideroom tests must be interpreted with caution.*

INVESTIGATION OF HYPOGLYCAEMIA

The most important test in a patient with *proven* hypoglycaemia is measurement of the plasma insulin level *when glucose levels are low.* This should differentiate exogenous administration or endogenous insulin production (the most important being *insulinoma*) from other causes of hypoglycaemia. If plasma insulin levels are inappropriately high, and doubt remains, C-peptide assay may help. A high plasma value suggests endogenous insulin secretion (including that due to pancreatic stimulation by sulphonylureas): an undetectable one suggests exogenous insulin administration.

1. *If a patient is seen during an episode of hypoglycaemia, take blood for glucose, insulin and C-peptide assay before giving glucose.* Plasma should be stored for the last two assays until hypoglycaemia has been proven (see 4). This step may considerably shorten further investigation.

2. More commonly the patient has been referred for investigation of previously documented hypoglycaemia or with a history strongly suggestive of hypoglycaemic attacks.

A full assessment should be made with special attention to:

(a) time of attacks in relation to meals (reactive hypoglycaemia);
(b) drug (especially antidiabetic) or alcohol ingestion. This information may not be freely given and a negative history does not exclude it;
(c) possible adrenocortical hypofunction or hypopituitarism. If considered likely, investigate as outlined on pp. 166 and 167;
(d) a possible non-pancreatic tumour (p. 476).

3. If no cause is identified an attempt should be made to induce hypoglycaemia by fasting, if necessary for up to 72 hours, and even accompanied by exercise under close supervision. Blood should be taken for glucose and insulin estimations after fasting overnight, and at hourly intervals thereafter, and assayed for glucose immediately. The test should be stopped if hypoglycaemia is demonstrated.

If hypoglycaemia is not induced by these means endogenous hyperinsulinism is unlikely to be a cause of the symptoms.

4. Insulin should be measured in a sample taken at the time of *proven* hypoglycaemia. If insulin administration is suspected, C-peptide should also be assayed.

The interpretation of results is summarised in Table XX.

5. If results of fasting glucose and insulin assays are equivocal, an *insulin suppression test* can be performed (p. 228). Hypoglycaemia induced by intravenous injection of insulin should suppress endogenous insulin (and C-

peptide secretion). Failure of C-peptide levels to fall confirms autonomous insulin secretion (usually due to an insulinoma).

TABLE XX

RESULTS OF PLASMA INSULIN AND C-PEPTIDE ESTIMATIONS *during hypoglycaemia*
(SPONTANEOUS OR AFTER A PROLONGED FAST).

Hypoglycaemia due to	Plasma insulin	Plasma C-peptide
Insulin administration		Low
Insulinoma or ectopic insulin secretion	Inappropriately high	High
Sulphonylurea administration		
Alcohol	Appropriately low	
Non-pancreatic non-insulin-secreting tumour		
Pituitary or adrenal failure		

ESTIMATION OF PLASMA OR BLOOD GLUCOSE

Glucose levels may be assessed by enzymatic methods, which are specific for this sugar, or by measurement of total reducing substances. Because of the presence of non-glucose reducing substances, the latter methods may over-estimate glucose by 0·3–1·1 mmol/l or more. The percentage error is greatest at low levels of glucose and *may mask hypoglycaemia*. It is important to know which method your laboratory uses.

The supply of glucose to cells depends on extracellular concentrations and these are reflected in *plasma* levels. Since intracellular levels are kept low by metabolism, inclusion of erythrocytes in the sample "dilutes" the plasma, giving results 10 per cent to 15 per cent lower than those of plasma or serum (the actual figure depending on the haematocrit). Plasma assay is therefore desirable, but blood is sometimes used.

Either whole blood or plasma must be assayed immediately, or mixed with an inhibitor of glycolysis. *In vitro* metabolism of glucose by the cellular elements of the blood will produce falsely low levels; tubes containing fluoride or iodoacetate (both of which inhibit glycolysis) with an anticoagulant are used.

For home glucose monitoring capillary blood from a finger prick is used. Results on such samples are usually between those of whole blood and venous plasma.

Unless otherwise indicated the glucose values given in this book are those using glucose-specific methods measured on venous plasma.

GLYCOSURIA

Glycosuria is best detected by enzyme reagent strips such as Clinistix (Ames). Directions for use are supplied with the strips. Clinistix detects about 5·6 mmol/l (100 mg/dl) of glucose in urine, but is not suitable for the relatively accurate quantitation required for control of diabetic therapy. The concentration of glucose in normal urine varies from 0·1 to 0·8 mmol/l (1–15 mg/dl).

Glycosuria may be due to:
    diabetes mellitus;
    glucose infusion;
    renal glycosuria.

Glucose may be present in the urine when plasma levels are normal. This is due to a low renal threshold and may be found in *pregnancy* or as an *inherited* (autosomal dominant) characteristic. The condition is usually benign. It may also be found in generalised proximal tubular defects (Fanconi syndrome). Differentiation from diabetes mellitus may be made by serial blood and urine glucose estimations.

False negative results may occur if the urine contains large amounts of ascorbic acid, as may occur after therapeutic doses, or after injection of tetracyclines which contain ascorbic acid as a preservative.

False positive results may occur if the urine container is contaminated with detergent.

REDUCING SUBSTANCES

Clinistix contains the enzyme glucose oxidase, and is specific for glucose; it is used for detection of glycosuria. **Clinitest** (Ames), gives a positive result with any reducing substance. It is semiquantitative and is valuable in the control of diabetic therapy, whereas Clinistix is unsatisfactory for this purpose. Urine screening in the newborn and infants should be done by Clinitest because of the diagnostic importance of non-glucose reducing substances, such as galactose, in this age group.

Clinitest tablets must be used as directed. They are caustic and require careful handling. The test is less sensitive than Clinistix but more accurate for quantitation.

Causes of a Positive Clinitest

The most important causes of a positive Clinitest are:

glucose ⎫
glucuronate ⎬ common
lactose   common in pregnancy
fructose ⎫
galactose ⎬
pentoses   rare
homogentisic acid ⎭

Slight reduction may occur if urate or creatinine is present in high concentration.

Most causes of a positive Clinitest in the adult are due to glucose. Clinistix will confirm the presence or absence of glucose. Non-glucose reducing substances are further identified by chromatography and specific tests.

The significance varies with the substance.

Glucose (see p. 224)

Glucuronates.—A large number of drugs, such as salicylates, and their metabolites are excreted in the urine after conjugation with glucuronic acid in the liver (p. 314). These glucuronates are relatively common reducing substances in the urine.

Galactose.—Galactose is found in the urine in galactosaemia (p. 216).

Fructose.—Fructose may appear in the urine after very large oral doses of the sugar, or excessive fruit ingestion, but in general fructosuria occurs in two rare inborn errors of metabolism, both transmitted by autosomal recessive genes.

Essential fructosuria is seen almost exclusively in Jews and is a harmless condition.

Hereditary fructose intolerance (p. 216) is a serious disease characterised by hypoglycaemia that may lead to death in infancy.

Lactose.—Lactosuria may occur in:

late pregnancy and during lactation;
congenital lactase deficiency (p. 301).

Pentoses.—Pentosuria is very rare. It may occur in:

alimentary pentosuria after excessive ingestion of fruits such as cherries and grapes. The pentoses excreted are arabinose and xylose;

essential pentosuria, a rare recessive disorder characterised by the excretion of xylulose due to a block in the metabolism of glucuronic acid. It is seen usually in Jews and is harmless.

Homogentisic acid.—Homogentisic acid appears in the urine in the rare inborn error alkaptonuria (p. 395). It is usually recognisable by the blackish precipitate formed.

KETONURIA

Most simple urine tests for ketones detect acetoacetate, because of the greater sensitivity of reagents to this than to acetone, and because acetoacetic acid is present in greater concentration. β-Hydroxybutyrate is not measured.

Rothera's test.—5 ml of urine is saturated with a mixture of ammonium sulphate and sodium nitroprusside (as a powder). 1–2 ml concentrated ammonia is added and mixed. A purple-red colour indicates the presence of acetoacetate (or acetone).

Ketostix and *Acetest* (Ames) are strips and tablets respectively, impregnated with the reagents of Rothera's test. Instructions are supplied with the reagents. They are more convenient to use.

Rothera's test detects about 0·5 mmol/l (5 mg/dl) acetoacetate while the tablet tests are slightly less sensitive (10–20 mg/dl). After a fast of several hours ketonuria may be detected by these methods. Occasional colour reactions resembling, but not identical with, that of acetoacetate, may be given by phthalein compounds such as phenolphthalein or bromsulphthalein (BSP).

TEST PROTOCOLS

Your laboratory should be contacted before starting these tests, because local details may vary.

TESTS FOR GLUCOSE INTOLERANCE

Oral Glucose Tolerance Test

The indications for and interpretation of the glucose tolerance test are discussed on p. 220; only the procedure is given here.

1. The patient fasts overnight (at least 10, but not more than 16 hours). Water, but no other beverage, is permitted.
2. A fasting blood sample for glucose is withdrawn.
3. The equivalent of 75g of anhydrous glucose (children 1·75 g/kg body weight to a maximum of 75g) is given. 75g glucose dissolved in 300 ml is hyperosmolar, and may not only cause nausea and occasionally vomiting and diarrhoea, but, because of delayed absorption, may affect the results of the test. It is therefore more usual to give a solution of a mixture of glucose and its oligosaccharides, which although all converted to glucose at the brush border, have less osmotic effect when present as the larger molecules. The patient drinks this within a period of 5 minutes.
4. Further samples are taken for glucose assay 1 and 2 hours later. During the test the patient should be at rest, and may not smoke. Glucose is measured on venous plasma by a glucose-specific method.

Factors Influencing the Result of the GTT

Previous diet.—No special restrictions are necessary if the patient has been on a normal diet for 3–4 days. If, however, the test is performed after a period of carbohydrate restriction, such as a reducing diet, abnormal glucose tolerance may be shown. The probable reason for this is that metabolism is set in the "fasted" state favouring gluconeogenesis. Mildly abnormal tests should be repeated after appropriate dietary preparation.

Time of day.—Most glucose tolerance tests are performed in the morning and the values quoted are for this time. There is evidence that tests performed in the afternoon yield higher plasma glucose values and that the accepted normals may not be applicable. This phenomenon may be due to a circadian variation in islet-cell responsiveness.

State of health.—The glucose tolerance test should not be carried out if the patient has been ill or been confined to bed during the preceding two weeks.

Drugs.—Drugs such as steroids, oral contraceptives and some diuretics may impair glucose tolerance.

TESTS FOR HYPOGLYCAEMIA

Insulin Suppression Test

1. The patient must fast for 14–16 hours overnight to ensure an adequate response to insulin, and is allowed only small amounts of water.
2. Insert an indwelling intravenous cannula and keep patent with sodium citrate solution or heparinised saline (p. 171).
3. Take fasting samples for glucose and C-peptide estimation.
4. Administer soluble insulin (0·15 U/kg body weight) intravenously.
5. Take further samples for glucose and C-peptide assay at 30, 60, 90, 120 and 150 minutes after the insulin injection.

Note:

1. The precautions outlined on p. 171 must be observed.
2. If clinically indicated, cortisol may also be measured to assess hypothalamic-pituitary-adrenal function (p. 172).

Interpretation

In normal subjects plasma C-peptide levels fall by more than 50 per cent. In most cases of insulinoma this does not occur.

PRINCIPLES OF INTRAVENOUS FEEDING

The principles discussed at the beginning of this chapter have an important application in the management of parenteral feeding. More details will be found in the reference quoted at the end of this section.

Degradation of carbohydrate, fat and, to a lesser extent, the carbon chains of some amino acids, supplies the energy needs of the body by coupling the breaking of their energy-rich bonds with ATP synthesis. Some of the constituent elements, carbon, hydrogen and oxygen, leave the body as carbon dioxide and water: a normal male loses energy at the rate of about 8400 kJ (2000 kcal) a day as heat. These losses are usually balanced by dietary intake of equivalent amounts of carbohydrate, fat and protein. If output exceeds intake, energy stores are drawn upon: once these are depleted structural compounds, such as the protein of cells and muscle, may be used to supply energy.

The energy needs of the ill patient may be much higher than normal, and yet he may not be able to take any or all of his nutritional requirements by mouth.

Short-term Intravenous Therapy (a period of days)

A well-nourished individual has enough energy stored as hepatic glycogen to last for at least a day. Stores will be replenished when he can eat again. During such a short fast, energy is derived from carbohydrate as well as fat, and ketosis is usually mild. Maintenance of electrolyte and fluid balance is all that is necessary, for instance, in the immediate post-operative period, or during an acute gastro-intestinal disturbance. Although a litre of 5 per cent glucose (dextrose) contains 850 kJ (205 kcal) it is probably unnecessary to give even this, and certainly not all the administered fluid should be in this form (p. 48).

If this so-called *catabolic phase* is likely to be followed by rapid recovery, amino acid infusion is unnecessary. Much protein breakdown is due to stress-induced cortisol secretion, sometimes contributed to by tissue damage, and nitrogen is lost in the urine, mostly as urea; the body is unable to use administered nitrogen efficiently until the factors causing negative balance are removed. Moreover the short-term nitrogen loss will be replenished when the patient starts eating. For similar reasons supplementation of other cellular constituents is ineffective in the catabolic phase.

Medium-term Intravenous Feeding (a period of weeks)

If prolongation of fasting is anticipated intravenous feeding should be used from the outset.

Hepatic glycogen stores are relatively small, and once they are used up energy is derived almost entirely from triglyceride. At this stage ketoacidosis may become a problem unless carbohydrate or substrate for gluconeogenesis is provided: in addition, at least enough energy substrate should be given to maintain balance and to replete hepatic glycogen. If it is not possible to do this by nasogastric tube, "food" must be given intravenously (*total parenteral nutrition*). 10 litres of 5 per cent glucose would be needed merely to provide normal daily energy requirements, and an even larger volume would be necessary to keep an ill patient in balance. It is impossible to infuse such volumes and a more concentrated energy source is therefore needed.

40 per cent glucose contains 6800 kJ/1, but is very hyperosmolar, and is therefore irritant to veins; this leads to local thrombosis. Insertion of a central venous catheter so that the fluid is delivered to a site of more rapid flow may overcome this problem, but is technically more difficult and carries a greater risk of infection. Severe illness may cause some insulin resistance and utilisation of such large amounts of glucose may be inefficient unless exogenous insulin is given; the consequent high extracellular levels result in urinary wastage, and may cause dangerous hyperosmolality (p. 35). Plasma glucose levels must be monitored frequently to ensure that control is adequate.

Fat may therefore be used to supplement glucose infusions. 10 per cent "Intralipid" (soybean oil emulsion) contains 4620 kJ/1, and even a 20 per cent solution is so little hyperosmolar that it can be infused into a peripheral vein. The fat particles are similar in size to, and are metabolised in the same way as chylomicrons. Provided that some carbohydrate is given at a constant rate throughout the day there is no risk of dangerous ketoacidosis. The grossly *lipaemic plasma may interfere with laboratory tests* (p. 37), but this is easily avoided if *non-lipid-containing solutions are infused for the few hours before sampling.* If lipaemia persists after these precautions the patient cannot metabolise all the fat supplements, and the rate of administration should be slowed.

Administration of carbohydrate and fat, by minimising the utilisation of amino acids for gluconeogenesis, reduces nitrogen loss in the urine. However, there is always some daily nitrogen loss and if this is not replaced healing may be delayed, there may be increased liability to infection and, eventually, there may even be profound muscle wasting and osteoporosis. Mixtures containing *essential amino acids* should be infused into any patient unable to eat for long. Because most nitrogen is lost as urinary urea, estimation of 24-hour urea nitrogen output can, if glomerular function is normal, be used as a guide to replacement. However, once tissue repair predominates, amino acid utilisation increases (the *anabolic phase*): urinary nitrogen loss may fall abruptly and much more than that retained may be needed for maximum regeneration. The onset of this phase is heralded by a sense of relative well-being and by increasing weight.

The composition of some of the available solutions is given on p. 65.

Long-term Intravenous Feeding (a period of months)

In addition to the above, vitamin supplementation is advisable in long-term parenteral feeding. Calcium rarely needs to be given if vitamin D intake is adequate: low plasma calcium levels are usually due to reduction of the protein-bound rather than of the ionised fraction. Magnesium supplementation may be necessary and is included in some amino acid solutions.

During a long illness tissue destruction leads to loss of intracellular constituents, such as phosphate and some trace metals (for example zinc and copper). Phosphate may need to be added to the infusion.

Urinary trace metal loss is high, with quantitative excretion of supplements during the catabolic phase, but falls as anabolism becomes predominant: if parenteral feeding has been very prolonged deficiencies may become apparent at this time and weight gain may be impaired if supplements are not given. A more detailed discussion of this controversial topic will be found in the reference.

Fluid and electrolyte balance should be controlled throughout as described in Chapters II and III. Potassium supplementation is almost always necessary especially during the anabolic phase.

FURTHER READING

BLACKBURN, G. L. and WOLFE, R. R. (1981). Clinical biochemistry and intravenous hyperalimentation. In: *Recent Advances in Clinical Biochemistry, No. 2*, pp. 197–228. K. G. M. M. Alberti and C. P. Price, Eds. Edinburgh: Churchill Livingstone.

Chapter X

PLASMA LIPIDS AND LIPOPROTEINS

THE association of abnormalities of plasma lipids with atherosclerosis and ischaemic heart disease has stimulated much research on the subject. Current understanding of the physiology and pathology of the plasma lipids is based on the concept of lipoproteins, the form in which lipids circulate in the blood. The first part of this chapter will describe the classifications in current usage and show how terminologies are related.

TERMINOLOGY AND CLASSIFICATION

PLASMA LIPIDS

The chemical structures of the four forms of lipid present in plasma are illustrated in Fig. 22.

Fatty acids are straight-chain compounds of varying lengths. They may be *saturated* (containing no double bonds), or *unsaturated* (with one or more double bonds). The main saturated fatty acids in plasma are *palmitic* (16 carbon atoms) and *stearic* (18 carbon atoms). Fatty acids may be *esterified* with glycerol to form glycerides, or they may be free, when they are called *free fatty acids* (FFA) or *non-esterified fatty acids* (NEFA). In the blood FFA are carried mainly bound to albumin. Free fatty acids are an immediately available energy source and provide a significant proportion of the energy requirements of the body. This aspect of fat metabolism is considered further on p. 197.

Triglycerides consist of glycerol, each molecule of which is esterified with three fatty acids.

Phospholipids are complex lipids, resembling triglycerides, but containing phosphate and a nitrogenous base. The major phospholipids in plasma are *lecithin* and *sphingomyelin*. The phosphate and nitrogenous bases are water soluble, a fact that is important in lipid transport.

Cholesterol has a steroid structure, and other steroids are derived from it. About two-thirds of the plasma cholesterol is esterified with fatty acids to form *cholesterol esters*. Assays in routine use measure the total cholesterol level, but do not distinguish between the unesterified and esterified forms.

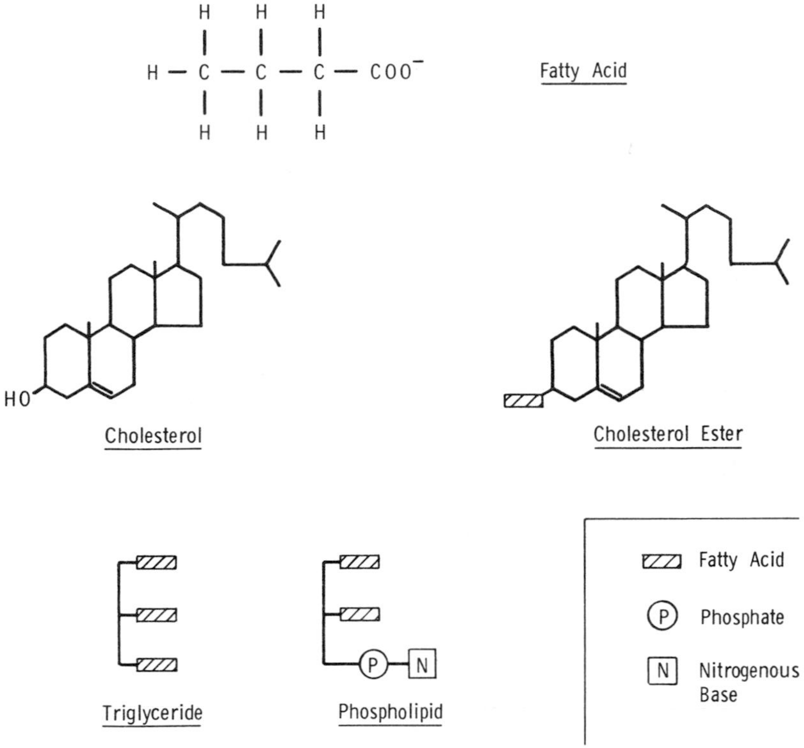

FIG. 22.—Plasma lipids.

LIPOPROTEINS

Lipids are relatively insoluble in water but are carried in the body fluids as soluble protein complexes, known as *lipoproteins*: the water-soluble (polar) groups of protein, phospholipid and free cholesterol, enclose a core of insoluble (non-polar) cholesterol esters and triglycerides. The proteins are divided into 5 groups known as *apoproteins A, B, C, D* and *E*. Subgroups, such as A-1 or C-1, with known specific functions, will be mentioned when relevant.

The lipoproteins differ in size and composition. There is extensive *in vivo* interconversion and interchange of both the lipid and protein components. They may be classified according to their densities, using ultracentrifugation, into four main classes (Fig. 23).

Two classes are most important for *cholesterol* transport:

HDL (high density lipoproteins) transport cholesterol from cells;
LDL (low density lipoproteins) transport cholesterol to cells.

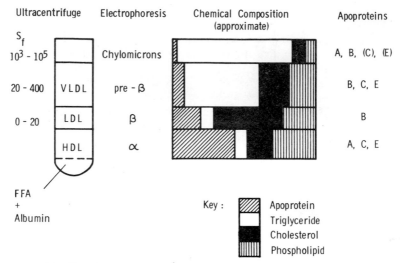

FIG. 23.—Correlation of plasma lipid and lipoprotein nomenclature.

Two classes are most important for *triglyceride* transport:

VLDL (very low density lipoproteins) transport endogenous triglycerides from the liver to the cells;

chylomicrons transport exogenous (dietary) triglycerides from the gut.

A fifth class, IDL (intermediate density lipoprotein), is usually a transient intermediate in the metabolism of VLDL to LDL and contains both cholesterol and endogenous triglycerides; it is undetectable in normal plasma.

Plasma taken from a normal fasting subject contains only HDL, LDL and VLDL. The measured plasma cholesterol level in normal subjects parallels that incorporated in LDL and the plasma triglyceride level that in VLDL. This is also true of most cases of hyperlipidaemia (but see p. 248). It may rarely be necessary to characterise the lipoproteins using one or more of the following techniques:

Ultracentrifugation is the definitive method and separates lipoproteins according to their densities. HDL sediments with other plasma proteins during ultracentrifugation. Low density lipoproteins tend to float; the rate of flotation is expressed in S_f (Svedberg flotation) units. The higher the lipid:protein ratio, the lower the density of the lipoprotein and the higher its S_f number.

Electrophoresis separates the lipoproteins according to the electrical charge of the apoprotein. This technique is more widely available than ultracentrifugation and, while the electrophoretic nomenclature is not used in this chapter, it occurs in the names of some of the disorders to be discussed. Lipoproteins separate into alpha (HDL), beta (LDL), pre-beta (VLDL), and chylomicron fractions. IDL excess may produce a broad beta band.

Simple precipitation techniques allow separation of HDL from the other lipoproteins. HDL cholesterol and LDL cholesterol may then be distinguished.

The nomenclature and composition of the main lipoprotein classes are shown in Fig. 23.

METABOLISM OF LIPOPROTEINS

Plasma lipids are derived from food (*exogenous*) or are synthesised in the body (*endogenous*).

The Fate of Exogenous (Dietary) Lipids (Fig. 24)

Fatty acids and cholesterol released by digestion of dietary fat are absorbed into the intestinal mucosal cells, where they are re-esterified to form triglycerides and cholesterol esters (p. 288). These, together with phospholipids and apoproteins B (essential for secretion from the cell) and A, are secreted as *chylomicrons* into lymphatics and pass through the thoracic duct into the systemic circulation. In the lymphatics and the blood, apoproteins C and E, originating from HDL, are added to the chylomicrons. Because of their large size (30–600 nm), chylomicrons scatter light and this accounts for the plasma turbidity which may be found after a fatty meal (postprandial lipaemia).

Most chylomicrons are metabolised by adipose and muscle tissue. The enzyme *lipoprotein lipase*, located on capillary walls, is activated by apoprotein C-2, and triglycerides in the chylomicrons are hydrolysed to glycerol and fatty acids. The fatty acids are taken up by fat or muscle cells or are bound to albumin in the plasma, and the glycerol enters the hepatic glycolytic pathway. As the chylomicron shrinks surface apoprotein A, and some apoprotein C and phospholipid, are reincorporated into HDL. The smaller *remnant particles* no longer scatter light and the plasma clears. These short-lived particles are composed mainly of cholesterol and apoproteins B, C and E. They bind to *specific receptors* in the liver and are taken up by hepatocytes where the protein is catabolised and the cholesterol released into the cells.

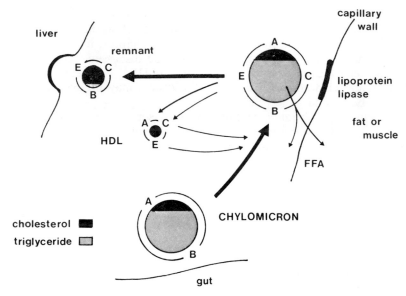

Fig. 24.—Metabolism of exogenous lipids. Lipoprotein core containing cholesterol and triglycerides, surrounded by phospholipids and apoproteins (clear area).

At the end of this process dietary triglycerides have been delivered to adipose tissue and muscle, and cholesterol to the liver.

The Fate of Endogenous Lipids (Fig. 25)

Triglycerides are synthesised in the liver from FFA derived from adipose tissue (p. 197); if carbohydrate intake is high FFA may be synthesised directly from excess glucose. These triglycerides, together with cholesterol synthesised in the liver or derived from chylomicron remnants, combine with apoproteins B and C to form *VLDL*. After secretion into the blood the VLDL takes up more apoprotein C from HDL. This activates lipoprotein lipase on the capillary wall and the triglycerides, like the exogenous triglycerides in chylomicrons, are hydrolysed and removed from the plasma, leaving *IDL*. Some IDL is taken up by receptors in the liver, but the rest is converted to LDL, which consists almost entirely of cholesterol and apoprotein B. The mechanism and site of this conversion are not known.

The further metabolism of LDL is only partly understood but is of great importance in the development of atheromatosis (p. 241). LDL is probably removed from the plasma by two pathways. In the first, after binding to *specific receptors* found on most cell membranes, LDL is taken up by the cells, and releases cholesterol which can be

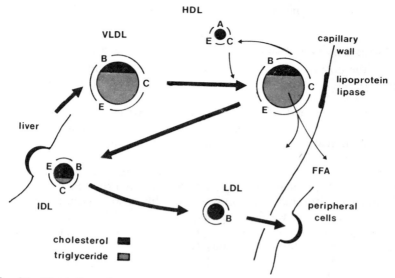

Fig. 25.—Metabolism of endogenous lipids. Lipoprotein core containing cholesterol and triglycerides, surrounded by phospholipids and apoproteins (clear area).

incorporated in the membranes. This cholesterol, by feedback inhibition of early steps in cellular cholesterol synthesis and by reduction of the synthesis of LDL receptors on the surface, regulates the cell content of cholesterol. Probably most of the cholesterol needed by peripheral cells is derived from the liver. Some LDL, especially if plasma levels are high, may also enter some cells by an unregulated, passive route. Some LDL may be taken up by hepatocytes but the mechanism and the extent of this route are not known.

Thus, at the end of this process endogenous triglycerides have been delivered to peripheral cells for local energy requirements and endogenous cholesterol for membrane synthesis.

The Control of Plasma and Cellular Cholesterol Levels and the Role of LDL and HDL

Cholesterol is both absorbed from the diet and synthesised in the body. The amount absorbed is almost proportional to intake if the diet is normal. In affluent societies dietary cholesterol is derived mainly from egg yolk (the richest source), dairy products and meat, and the daily intake is about 1·5 to 2·0 mmol (600 to 800 mg).

Most endogenous cholesterol is synthesised in the liver, and the rate of this synthesis is reduced by exogenous cholesterol reaching it.

Cholesterol synthesis in peripheral tissues is controlled by LDL uptake.

Plasma Cholesterol and LDL

In the normal subject the amount of cholesterol added to the body pool is balanced by biliary excretion, partly after hepatic degradation to bile salts and acids, and partly by excretion as free cholesterol: both can be reabsorbed in the ileum (enterohepatic circulation). Drugs which, because they bind bile salts in the intestinal lumen, interfere with their reabsorption, can be used to lower plasma cholesterol and LDL levels (p. 245).

If the dietary cholesterol intake is very high the suppression of hepatic synthesis may be inadequate to prevent a rise in plasma levels. Moreover, saturated fatty acids, which, like cholesterol, are derived mainly from dietary animal fat seem to increase LDL, and therefore plasma cholesterol levels, whilst polyunsaturated fatty acids (derived mainly from vegetable fats) may tend to lower them: the reason for this is not known.

Cell Cholesterol, HDL and LCAT

LDL uptake, local synthesis, or release by cell membrane breakdown may cause cholesterol accumulation in peripheral cells. This excess cholesterol can only be transported to the liver for excretion in the bile if incorporated into HDL.

HDL is synthesised in the hepatic and intestinal cells, from which it is secreted as small complexes of phospholipid surrounded by apoproteins A and E. Free cholesterol derived from peripheral cell membranes and from other lipoproteins is taken up by HDL and esterified: esterification is catalysed by the enzyme *lecithin-cholesterol acyltransferase (LCAT)*, and requires apoprotein A-1. Most of the esterified cholesterol is transferred to LDL, VLDL and remnant particles, and so reaches the liver: a small amount is stored in the core of HDL.

We have mentioned that HDL carries most of the plasma apoprotein C during fasting. As levels of VLDL or chylomicrons rise these particles take up apoprotein C-2 which activates lipoprotein lipase. After the triglyceride has been hydrolysed this apoprotein returns to HDL.

APOPROTEINS

As we have seen, the route of metabolism of lipoproteins is determined by the apoproteins they carry. These proteins not only make the lipids water soluble, but are necessary for lipoprotein secretion by hepatic and intestinal cells and for recognition by cell-surface receptors. They also activate the enzymes involved in lipoprotein metabolism. The functions of the main groups of apoproteins are summarised in Table XXI. Further details may be found in the references at the end of the chapter.

TABLE XXI

THE MAIN APOPROTEINS

Apoprotein	Occurrence	Known Functions
A	Chylomicrons, HDL	Cofactor for LCAT (A-1)
B	Chylomicrons, VLDL, IDL, LDL	Secretion of chylomicrons Secretion of VLDL Binding of LDL to receptors
C	HDL, VLDL, IDL Chylomicrons (from HDL)	Cofactor for lipoprotein lipase (C-2)
D	HDL	Cholesterol ester transfer
E	HDL, VLDL, IDL Chylomicrons (from HDL)	Binding of IDL and remnant particles to receptors

REFERENCE RANGES FOR PLASMA CHOLESTEROL

Plasma cholesterol at birth (cord blood) is usually below 2·6 mmol/l (100 mg/dl). Levels increase slowly throughout childhood, mostly during the first year, but do not usually exceed 4·1 mmol/l (160 mg/dl). In most affluent populations a further progressive increase occurs after the second decade. This increase is greater in men than in women during the reproductive years. The "reference range" (95 per cent) upper limit in many societies is as high as 8·4 mmol/l (330 mg/dl) in the fifth and sixth decades. This rise does not occur in less affluent communities in which the incidence of ischaemic heart disease is much lower.

DISORDERS OF LIPID METABOLISM

Most common disorders of lipid metabolism are associated with hyperlipidaemia. Very rare congenital disorders may be associated with accumulation of lipid in tissues and not blood.

Hyperlipidaemia is most commonly due to excessive calorie intake, especially in the form of alcohol, or is secondary to another disease which affects lipid metabolism. More rarely it is due to a primary (inherited) disorder, the effects of which may be exaggerated by dietary excess.

CLINICAL MANIFESTATIONS OF LIPID DISORDERS

Accumulation of lipid in tissues is usually, *but not always*, the result of severe and prolonged hyperlipidaemia, and causes cell damage. Lipid accumulations, for example under skin or mucous membranes, may be visible.

Lipid may accumulate in:

arterial walls. (*This is by far the commonest and most important manifestation of lipid disorders.*) Cholesterol accumulation and associated cellular proliferation and fibrous-tissue formation produce the *atheromatous plaque*. *Atherosclerosis* is due to the distortion and obstruction of the artery which may result from calcification and ulceration of the plaque.

subcutaneous tissue, causing *xanthomatosis*. The nature of the lipid fraction most affected usually seems to determine the clinical appearance:

eruptive xanthomata are crops of small, itchy yellow nodules. They are associated with very high *VLDL* or *chylomicron* (*triglyceride*) levels, and soon resolve if plasma lipid concentrations fall to normal.

tuberous xanthomata are yellow plaques found mainly over the elbows and knees, and can be large and disfiguring. These, and raised linear lipid deposits in the palmar creases, are associated with high levels of *IDL* (which contains both triglyceride and cholesterol).

xanthelasma is the name given to lipid deposits under the periorbital skin, and may be associated with high *LDL cholesterol* levels.

tendons. Tendinous xanthomata.

cornea. A *corneal arcus* is due to corneal deposits.

Both tendinous xanthomata and corneal arcus in relatively young subjects (less than 40 years of age) are, like xanthelasma, especially likely to be associated with high *LDL cholesterol* levels.

Hypertriglyceridaemia, whether due to chylomicrons, VLDL, or both, causes *turbid plasma*. Sustained and very high concentrations of *chylomicrons* are associated with *abdominal pain* and even *pancreatitis*, as well as eruptive xanthomata. Many cases of hypertriglyceridaemia are symptom free. These large lipoproteins are probably unlikely to cause atheroma. However, many patients with increased levels of VLDL triglyceride have reduced HDL (necessary for transport of cholesterol from tissues) and some also have increased LDL or IDL (which contain cholesterol); the effects on cholesterol metabolism may explain the slightly increased risk of atheroma which has been attributed to hypertriglyceridaemia.

FACTORS ASSOCIATED WITH CARDIOVASCULAR DISEASE

While there is considerable evidence of a positive correlation between LDL and cardiovascular disease, there is an even stronger negative correlation with HDL. The higher the plasma HDL level, the lower the risk. HDL seems to have a protective function which, because of the physiological role of HDL as a carrier of cholesterol from peripheral tissues for excretion, is not surprising. Several factors known to diminish the risk of cardiovascular disease are associated with high HDL levels. Such factors may be hormonal (levels are higher in women during the reproductive years than in men); physical exercise tends to increase HDL levels and carbohydrate-rich diets and smoking to lower them.

SECONDARY HYPERLIPIDAEMIA

Most cases of hyperlipidaemia are secondary to an abnormal dietary intake, or to underlying diseases in which lipid metabolism is affected; these include *obesity, excessive alcohol consumption* (p. 239), *diabetes mellitus* (p. 206), *hypothyroidism* (p. 187), and the *nephrotic syndrome* (p. 358). The main causes are listed in Table XXII.

TABLE XXII

THE CAUSES OF SECONDARY RAISED PLASMA CHOLESTEROL AND TRIGLYCERIDE LEVELS
(Where both are indicated it does not imply that they always occur together)

	Cholesterol	Triglycerides
Obesity		+
Excessive alcohol intake		+
Diabetes mellitus	+	+
Nephrotic syndrome	+	+
Hypothyroidism	+	+
Pancreatitis		+
Renal failure		+
Pregnancy	+	
Oral contraceptives		+
Biliary obstruction	+	
Glycogenosis Type 1		+
Acute porphyria	+	

distinction between unaffected, homozygous and heterozygous subjects; this differs from the commoner polygenic disease. In most countries monogenic hypercholesterolaemia accounts for *less than 5 per cent of all cases of primary hypercholesterolaemia.*

Apoprotein Abnormalities

Deficiency of apoprotein C-2 will be mentioned in the discussion of lipoprotein lipase deficiency.

Low apoprotein levels are usually associated with accumulation of lipid in tissues, and *not hyperlipidaemia.*

Apoprotein A is an important constituent of *HDL* and low levels cause HDL deficiency. Because HDL is important for cholesterol transport, cholesterol esters accumulate in tissues, especially the reticuloendothelial system. The syndrome, known as *Tangier disease*, consists of:

large yellow tonsils;
hepatomegaly;
lymphadenopathy.

Apoprotein B deficiency (abetalipoproteinaemia; LDL deficiency) results in impaired synthesis of chylomicrons and VLDL (and therefore of LDL). Lipids cannot, therefore, be transported from the intestine or liver. The clinical syndrome consists of:

steatorrhoea;
progressive ataxia;
retinitis pigmentosa;
acanthocytosis ("thorny" red cells).

Apoprotein B, when produced in excess by the liver, is associated with *familial combined hyperlipidaemia.* Apoprotein B is essential for secretion of chylomicrons and VLDL, and for binding of LDL to receptors. About a third of *affected* family members have increased plasma VLDL levels due to a primary or secondary increase in triglyceride synthesis. Another third have no increase in triglyceride synthesis, and have increased LDL levels only. A final third have increases of both fractions. The lipid abnormalities only become apparent after the third decade, and the risk of cardiovascular disease in all cases is higher than normal. High triglyceride levels may cause eruptive xanthomata.

A functionally abnormal apoprotein E can be demonstrated in 1 per cent of the population, but the disease *familial dysbetalipoproteinaemia ("broad-beta disease")* occurs only if there is another cause of primary or secondary hyperlipidaemia. IDL levels rise and the patients may then have:

a high incidence of vascular disease;
tuberous xanthomata;
lipid deposition in the palmar creases.

Enzyme Abnormalities

Lipoprotein lipase deficiency.—Lipoprotein lipase activity may be reduced due to:

enzyme deficiency;
failure to activate the enzyme because of *apoprotein C-2 deficiency* (p. 235).

The accumulation of *chylomicrons* results in a very turbid plasma. True lipoprotein lipase deficiency usually presents during childhood, with signs and symptoms due to excess of fat at various sites, such as:

skin (eruptive xanthomata);
liver (hepatomegaly);
retinal vessels (lipaemia retinalis);
abdominal pain (a symptom associated with hyperchylomicronaemia).

Chylomicronaemia due to apoprotein C-2 deficiency is more likely to present in adults.

LCAT deficiency.—LCAT is necessary for esterification of free cholesterol (p. 238). Deficiency results in accumulation of free, mostly unesterified, cholesterol in tissues, causing:

premature atherosclerosis;
corneal opacities;
renal damage;
anaemia, which may be due to a cell-membrane defect.

Overproduction of Triglyceride

Familial endogenous hypertriglyceridaemia is due to hepatic triglyceride overproduction. VLDL secretion from the liver into the plasma is increased. The condition seems to be transmitted as an autosomal dominant trait, and usually becomes apparent only after the fourth decade of life. It may be associated with:

obesity;
glucose intolerance;
hyperuricaemia.

Superimposed secondary factors, such as diabetes mellitus or alcoholism, may cause very high VLDL levels and often, in addition, chylomicronaemia.

PRINCIPLES OF TREATMENT

The decision as to whether to treat a patient with hyperlipidaemia must be based on clinical considerations as well as plasma lipid levels. Some forms of treatment carry risks which must be weighed against possible benefits.

General

Secondary causes of hyperlipidaemia, including obesity and excessive alcohol consumption, should be sought and treated.

The diet should be controlled whatever the cause of hyperlipidaemia. The type of diet depends on the nature of the abnormality.

Hypercholesterolaemia

Hypercholesterolaemia increases the risk of cardiovascular disease. Because the lipid deposited in the cells of the arterial wall is derived from the plasma (especially from LDL and IDL), it has been assumed that abnormally high levels will increase the rate of deposition. The further assumption that reduction of levels will slow the rate of development of atherosclerosis seems likely (see References at end of chapter). The risks of vigorous treatment must be weighed against possible benefits. It is, in any case, essential to understand the nature of the lipid disorder to ensure rational therapy.

Restriction of dietary animal fats, eggs and dairy products reduces the intake of both cholesterol and saturated fatty acids. The intake of polyunsaturated fats should be increased (p. 238). These dietary measures are not always completely successful.

Bile-salt sequestrants such as cholestyramine and colestipol are resins which bind bile salts derived from cholesterol, and so prevent their reabsorption and reutilisation. There is a compensatory increase in hepatic synthesis, but plasma cholesterol levels usually fall.

Nicotinic acid may reduce VLDL secretion and therefore the formation of LDL, but may cause uncomfortable side-effects such as flushing. It may be given together with bile-salt sequestrants. It may also be effective in those cases of familial combined hyperlipidaemia known to have increased VLDL secretion.

Hypertriglyceridaemia

Dietary restriction may be the only treatment needed for hypertriglyceridaemia.

Dietary fat restriction may be effective in lowering plasma chylomicron levels.

Dietary carbohydrate restriction reduces endogenous triglyceride synthesis and can be used to treat high VLDL levels.

Clofibrate is a drug which may activate lipoprotein lipase, and so increase the rate of clearance of VLDL and chylomicrons from plasma. It is used to treat familial dysbetalipoproteinaemia when dietary measures are ineffective, and may be used in cases of endogenous hypertriglyceridaemia (increased VLDL levels), if clinically indicated. There is an increased incidence of gall stones in patients taking clofibrate: other side-effects include muscle cramps and, rarely, impotence. As always, the risks of treatment must be weighed against its possible benefits.

SUMMARY

1. Lipids in plasma are carried in the form of lipoproteins.

2. Plasma lipids may be classified according to their chemical structure, or as lipoproteins by electrophoresis or ultracentrifugation. Lipoprotein analysis offers the best understanding of abnormalities.

3. The chemical lipid fractions are cholesterol, triglycerides, phospholipids and free fatty acids.

4. The main lipoproteins are high-density lipoprotein (HDL), low-density lipoprotein (LDL), very low-density lipoprotein (VLDL) and chylomicrons.

5. Cholesterol occurs mainly in HDL and LDL and triglycerides in VLDL and chylomicrons.

6. Hyperlipidaemia may be primary or secondary to other disease. Only in primary hyperlipidaemia may it be necessary to define the lipoprotein abnormality more fully for the purpose of treatment.

7. Different genetic defects may produce similar lipoprotein abnormalities. Extensive family studies are required to differentiate them. This is rarely practicable, because of the difficulty of tracing family members.

8. The treatment of primary hyperlipidaemia is primarily regulation of the diet. If necessary drugs may be added. The treatment is largely empirical.

FURTHER READING

HAVEL, R. J. (Ed.) (1982). Lipid disorders. *Med. Clin. N. Amer.*, **66**, 317–553.

EISENBERG, S. (Ed.) (1979). Lipoprotein metabolism. *Prog. Biochem. Pharmacol.*, **15**, 1–258.

EDITORIAL (1981). High-density lipoprotein. *Lancet*, **1**, 478–480.

OLIVER, M. F. (1981). Diet and coronary heart disease. *Brit. Med. Bull.*, **37**, 49–58.

LIPID RESEARCH CLINICS PROGRAM. The Lipid Research Clinics Coronary Primary Prevention Trial Results. (1984). I Reduction in incidence of coronary heart disease; II The relationship in incidence of coronary heart disease to cholesterol lowering. *J. Amer. Med. Ass.*, **251**, 351–364; 365–374.

INVESTIGATION OF SUSPECTED HYPERLIPIDAEMIA

There should be a logical progression in the evaluation of a patient suspected of having an abnormality of plasma lipids.

1. Is There True Hyperlipidaemia?

Hyperlipidaemia may be diagnosed because of the lipaemic appearance of the plasma, or by measuring cholesterol and triglyceride levels in plasma. While cholesterol levels do not vary greatly after fat intake, triglyceride levels do, so that specimens for analysis of both should be taken after the patient has fasted for 14 to 16 hours.

Lipid infusion is a common cause of grossly lipaemic plasma taken from hospital patients, and no lipid should be infused for several hours before taking blood for *any* such investigation.

2. Is The Cause Primary or Secondary?

Most cases of hyperlipidaemia are *secondary* to other factors and may be corrected by modification of the diet or treatment of the underlying disease.

The main secondary causes of abnormal plasma lipids are shown in Table XXII (p. 241). *Diabetes mellitus, hypothyroidism* and the *nephrotic syndrome* should be considered in all cases of hyperlipidaemia. Abuse of *alcohol* is a common cause of secondary hypertriglyceridaemia.

3. What is the Nature of the Abnormality?

If no secondary cause is found, the hyperlipidaemia must be considered a primary abnormality. Selection of appropriate therapy sometimes depends on an accurate definition of the nature of the basic lipoprotein disturbance. Valuable information may be obtained by visual inspection of the plasma.

On the basis of the results of these tests we can distinguish three main groups:

A predominant increase in LDL cholesterol is associated with *clear plasma*. It may be due to:

familial hypercholesterolaemia;
polygenic hypercholesterolaemia;
familial combined hyperlipidaemia.

A predominant increase in triglycerides is associated with *turbid* or even *milky plasma* due to large lipoproteins, which scatter light. It may be due to:

familial combined hyperlipidaemia (VLDL);
familial endogenous hypertriglyceridaemia (VLDL);
familial hyperchylomicronaemia.

The treatment of these conditions differs and it is important to determine whether VLDL or chylomicrons are causing the turbidity.

The turbid plasma sample is left standing for 18 hours at 4° C. During this time the large, low-density chylomicrons form a creamy layer on the surface. The smaller, denser VLDL particles do not rise and the sample remains diffusely turbid. If hyperlipidaemia is gross the distribution may be less easy to distinguish.

An increase in both cholesterol and triglycerides in the same proportions may be due to:

familial combined hyperlipidaemia (LDL and VLDL);

familial dysbetalipoproteinaemia (IDL).

In such cases lipoprotein electrophoresis or ultracentrifugation are often necessary to distinguish the relative contribution from LDL, IDL and VLDL.

4. Family Studies

These are necessary both to identify the disorder and to detect other affected, possibly asymptomatic individuals.

SAMPLING BLOOD FOR PLASMA LIPID INVESTIGATION

Plasma lipid levels and lipoprotein patterns are labile and affected by eating, smoking, alcohol intake, changes in diet, posture and stress. For the initial evaluation and follow-up studies it is *essential* that the sample be taken under *standard conditions*. It may, in some cases, be necessary to repeat the analysis to establish the most consistent pattern. The following points are important.

1. The patient must have fasted for 14 to 16 hours. Obviously he should not be receiving a lipid infusion.

2. The patient should have been on his "normal" diet and his weight should have remained constant for two weeks before the test.

3. The patient must not be on any treatment designed to lower plasma lipid levels, unless therapy is being monitored.

4. Like all large particles, lipoprotein concentrations are affected by venous stasis and posture (pp. 498 and 512). Plasma cholesterol concentration may be up to 10 per cent higher in the upright than in the recumbent position; triglyceride concentration may change slightly more than this. A standardised collection procedure is important in serial estimations to assess the effect of treatment. For example, the patient should remain seated for 30 minutes before blood is taken.

5. Investigation for and typing of hyperlipidaemia is preferably deferred for three months after a myocardial infarction, a major operation, or any serious illness because stress may alter plasma lipid levels. However, samples taken within 12 hours of myocardial infarction seem to reflect "true" values.

6. The blood sample should *not* be heparinised and plasma or serum must be separated from cells as soon as possible.

THE FREDRICKSON (WHO) CLASSIFICATION

Fredrickson introduced a classification of hyperlipidaemia based on electrophoretic patterns. However, even the same individuals with primary hyperlipidaemia ·may have different patterns at different times and the same pattern may be found in different disorders. As the terms still appear in the medical literature the student may use Table XXIII for comparison.

TABLE XXIII
(See also Fig. 23 p. 234)

Fredrickson Type	Electrophoretic picture	Lipoprotein increased
I	Increased chylomicrons	Chylomicrons
II	Increased β lipoprotein	LDL
IIb	Increased pre-β and β lipoprotein	VLDL and LDL
III	"Broad-β" band	IDL
IV	Increased pre-β lipoprotein	VLDL
V	Increased chylomicrons and pre-β lipoprotein	chylomicrons and VLDL

Chapter XI

CALCIUM, PHOSPHATE AND MAGNESIUM METABOLISM

MOST of the body calcium is in bone, and prolonged deficiency causes bone disease. However, the extra-osseous fraction, although amounting to only 1 per cent of the whole, is of great importance because of its effect on neuromuscular excitability and on cardiac muscle.

Phosphate is the most important anion associated with calcium *in vivo* and a knowledge of plasma levels is needed to interpret disturbances of calcium metabolism. Net renal tubular secretion of hydrogen ion and formation of bicarbonate depend on the presence of phosphate and sodium in the glomerular filtrate (p. 100). Cellular "high energy" phosphate compounds are of great biological importance.

More than half the body magnesium is in bone: most of the remainder is intracellular but, as in the case of calcium, the small extracellular fraction (about 1 per cent) is important because of its effect on neuromuscular activity. Because of this similarity in distribution and physiological action calcium and magnesium metabolism will be discussed in one chapter.

CALCIUM METABOLISM

TOTAL BODY CALCIUM

The total amount of calcium in the body depends on the balance between intake and loss.

Factors Affecting Intake

The amount of calcium entering the body depends on the amount in the *diet*, which is normally about 25 mmol (1 g) a day. Between 25 and 50 per cent of this is absorbed. *Vitamin D*, in the form of 1,25-dihydroxycholecalciferol (p. 254), is needed for adequate calcium absorption.

Factors Affecting Loss

Calcium is lost in urine and faeces. *Faecal calcium* consists of that which has not been absorbed from the diet, with a much larger amount which has entered the intestinal lumen from the extracellular fluid.

Calcium in the intestine, whether endogenous or exogenous in origin, may be rendered insoluble by large amounts of *phosphate* or *fatty acids*. Oral phosphate may be used therapeutically to control calcium absorption and reabsorption (p. 272): an excess of fatty acids in the intestinal lumen in steatorrhoea contributes to calcium malabsorption.

Urinary calcium excretion depends on the *amount of calcium circulating* through the glomerular vessels, *renal function* (p. 269), *parathyroid hormone* and *1,25-dihydroxycholecalciferol levels* and to a lesser extent on *urinary phosphate excretion*. The amount of calcium circulating through the glomerulus is increased after a calcium load or during decalcification of bone which is not due to calcium deficiency (osteoporosis or acidosis). Such decalcification rarely causes hypercalcaemia because of the rapid renal clearance of calcium. Conversely, hypercalcaemia, whatever the aetiology, causes hypercalciuria if renal function is normal. Glomerular insufficiency reduces calcium loss in the urine, and urinary calcium excretion is then low, even if there is hypercalcaemia.

PLASMA CALCIUM

Estimation of plasma calcium should be accurate to about 0·05 mmol/l (0·2 mg/dl) and changes of less than this are probably insignificant. Calcium circulates in the plasma in two main forms. The albumin-bound fraction accounts for a little less than half the total calcium as measured by routine analytical methods: it is physiologically inactive and is a transport form comparable to iron bound to transferrin (p. 417). Most of the remaining plasma calcium is free ionised (Ca^{++}), and is the physiologically important fraction (compare with thyroxine, p. 178, and cortisol, p. 141). Ideally, free ionised rather than total calcium should be measured, but techniques for this, although now becoming more widely available, are still not of proven value for routine use. Whenever a total plasma calcium value is assessed an effort should be made to decide whether the free ionised fraction is normal by comparing the value for total calcium with that for albumin on the same specimen. The *total* (but not free ionised) calcium concentration is lower in the supine than in the erect position, because of the effect of posture on plasma protein concentration.

The liberation of free ionised calcium from its salts in aqueous solution falls as the pH rises: in plasma an increase in pH also enhances protein binding. In alkalosis, therefore, the proportion of free ionised to total calcium is reduced for two reasons, and tetany may occur despite a normal total calcium level. In clinical states associated with prolonged acidosis (for instance after transplantation of the ureters into the colon or in renal tubular acidosis) bone salts become more soluble and more calcium than normal is released from bone into the extracellular fluid, and so into the urine; osteomalacia can therefore occur even with an adequate supply of the ion to bone and with normal parathyroid function.

Control of Plasma Calcium

The most important factor controlling the level of circulating free ionised calcium within narrow limits is parathyroid hormone (PTH), secreted by the parathyroid glands. Calcitonin, produced in the thyroid, has an opposite action to that of PTH on calcium levels: its importance in calcium homeostasis is less certain than that of PTH. Both these hormones use bone as a reservoir from which plasma free ionised calcium levels can be controlled. 1,25-dihydroxycholecalciferol increases calcium absorption from the intestine, and in physiological amounts seems to be necessary for the action of PTH.

Action and control of parathyroid hormone.—Parathyroid hormone raises the plasma free ionised calcium concentration. It has two direct actions on plasma calcium and phosphate levels. By:

acting directly on osteoclasts it releases bone salts into the extracellular fluid. This action tends to increase both calcium *and* phosphate in the plasma;

acting on the renal tubular cells it decreases the reabsorption of phosphate from the glomerular filtrate, causing phosphaturia. This action tends to *decrease* the plasma phosphate level.

PTH may also increase renal tubular reabsorption of calcium, perhaps by stimulating conversion of 25-hydroxycholecalciferol to the biologically active 1,25-dihydroxycholecalciferol (1,25-DHCC).

The secretion of PTH, like that of insulin from the pancreas, is not controlled by any other endocrine gland. It is influenced mainly by the concentration of free calcium ions circulating through the parathyroid glands. Reduction of the concentration of these ions increases the rate of release of hormone, and hormone secretion is maintained until the free ionised calcium level returns to normal, when it stops. 1,25-DHCC deficiency may increase the sensitivity of this feedback mechanism.

Bearing in mind the actions of PTH on bone and on the renal tubules, and the fact that its secretion is controlled by the level of free ionised calcium, the consequences of most disturbances of calcium metabolism on plasma calcium and phosphate levels can be predicted. A low free ionised calcium concentration, unless due to hypoparathyroidism, causes phosphaturia by stimulating the parathyroids: this loss of phosphate overrides the tendency to hyperphosphataemia due to the direct action of PTH on bone, and plasma phosphate levels are low normal or low. Conversely, a high calcium concentration, unless due to an excess of PTH, cuts off production of the hormone and causes a high phosphate level. As a general rule, therefore, calcium and phosphate levels tend to vary in the same direction unless there is an inappropriate excess or deficiency of PTH (as in primary hyperparathyroidism or hypoparathyroidism), or in the presence of renal failure when the phosphaturic effect of PTH is reduced.

Metabolism and action of vitamin D (Fig. 26).—Most cholecalciferol (vitamin D_3) is formed in the skin from 7-dehydrocholesterol by the action of ultraviolet light and in this form is found in animal tissues; the liver is particularly rich in it. Ergocalciferol (D_2) can be obtained from plants after artificial ultraviolet irradiation. These D vitamins are transported in the plasma bound to specific carrier proteins: they are inactive until metabolised.

In normal adults much more cholecalciferol is derived from the skin than from food: dietary sources may be important during growth or pregnancy, or in subjects such as the elderly who are confined indoors and not exposed to ultraviolet light.

In the *liver* cholecalciferol is hydroxylated to *25-hydroxycholecalciferol (25-HCC)*, the main circulating form of the vitamin. The activity of the 25-hydroxylase does not seem to be subject to feedback control, and the rate of formation of 25-HCC depends mainly on the supply of cholecalciferol derived from the skin or intestine. Some is excreted in the bile after inactivation, but most circulates bound to protein; this protein-bound 25-HCC is the largest store of "vitamin D".

Further hydroxylation of circulating 25-HCC by the *kidney* to *1,25-dihydroxycholecalciferol (1,25-DHCC)* is necessary to confer biological activity on the vitamin. 1,25-DHCC acts on the intestine, and probably the kidney, increasing calcium absorption and reabsorption; in conjunction with PTH it releases calcium from bone. The kidney is thus an endocrine organ, manufacturing and releasing the hormone 1,25-DHCC; failure of the final hydroxylation probably explains the hypocalcaemia of renal disease.

The lag between administration of therapeutic vitamin D and its action (p. 282) may be due to the time taken for the two-stage hydroxylation.

24,25- and 25,26-DHCC are also formed in the kidney. Their functions have not yet been elucidated, and they will not be discussed further.

Control of renal vitamin D metabolism.—The enzyme *lα-hydroxylase* catalyses the hydroxylation of 25-HCC in the renal tubular cells.

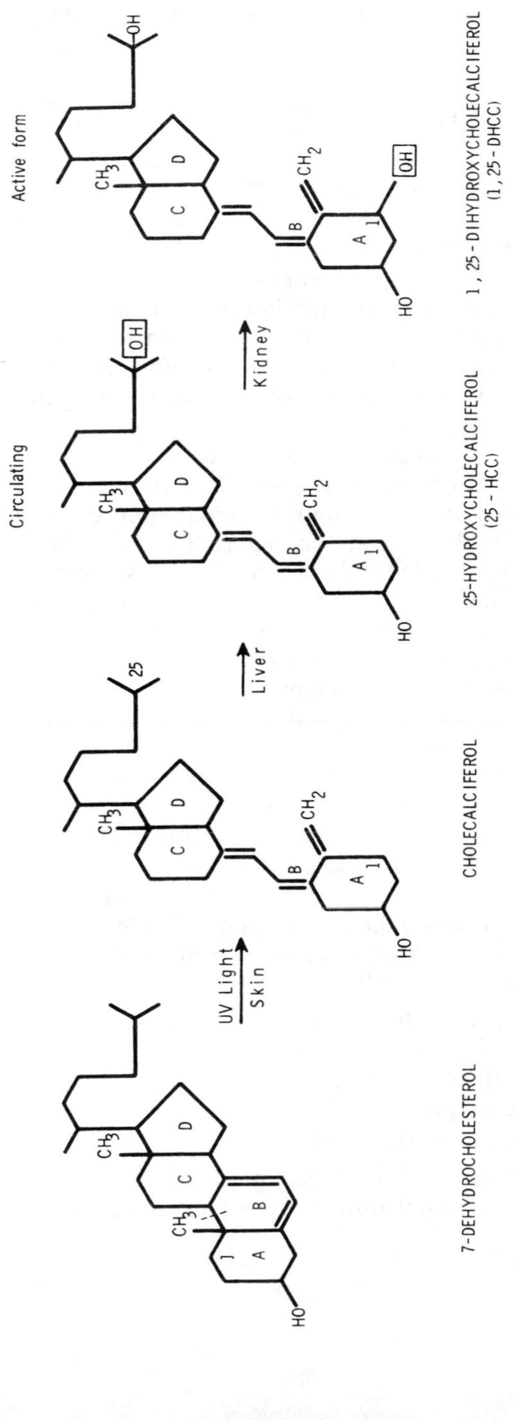

Fig. 26.—Formation of active "vitamin D" in the body from 7-dehydrocholesterol.

Its activity, and hence the production of 1,25-DHCC, may be *stimulated* by:

cholecalciferol deficiency;
an increase in circulating PTH.

Interrelationship of parathyroid hormone and vitamin D.—The action of PTH on bone is impaired in the absence of 1,25-DHCC.

A reduction of extracellular free ionised calcium levels stimulates PTH production by the parathyroid glands. PTH stimulates 1,25-DHCC synthesis, and the two hormones act synergistically on the bone "reservoir" releasing calcium into the circulation; 1,25-DHCC also increases calcium absorption from the intestinal and renal tubular lumina. In short-term homeostasis the bone effect is more important: after prolonged hypocalcaemia more efficient absorption becomes important. As soon as the free ionised calcium concentration is corrected PTH secretion ceases, and that of 1,25-DHCC is reduced.

Action and control of calcitonin.—Calcitonin reduces plasma calcium by decreasing osteoclastic activity, and therefore bone resorption. It is produced in the C cells of the thyroid and its secretion is stimulated by high free ionised calcium levels. Its clinical importance is doubtful. Although circulating concentrations may be very high in patients with medullary carcinoma of the thyroid, associated hypocalcaemia has very rarely been reported.

Action of thyroid hormone on bone.—Thyroid hormone excess may be associated with the histological appearance of osteoporosis, and with increased faecal and urinary excretion of calcium, probably due to release from bone. Unless there is very gross excess of thyroid hormone, effects on plasma calcium levels are overridden by the homeostatic reduction of PTH secretion and by urinary loss, and hypercalcaemia is very rare in this condition.

Calcium homeostasis follows the general rule that extracellular levels, rather than total body content, is controlled. The effectiveness of this control depends mainly on:

normal parathyroid function;
normal renal function;
normal intestinal function;
an adequate supply of calcium;
an adequate supply of vitamin D.

If any of these factors is impaired, plasma levels may be controlled at the expense of the total amount of calcium in the body.

CLINICAL EFFECTS OF DISORDERS OF CALCIUM METABOLISM

Clinical Effects of Increased Free Ionised Calcium Levels

Effects on the kidney.—The most dangerous consequence of prolonged mild hypercalcaemia is *renal damage*. Because of the high plasma free ionised calcium, the solubility product of calcium phosphate may be exceeded and the salt precipitated in extra-osseous sites, of which the kidney is the most important. Renal tubular calcification may impair the ability to form a concentrated urine and thus contribute to the *polyuria* so characteristic of hypercalcaemia: in acute hypercalcaemia impairment may be due to reversible inhibition of the tubular response to ADH rather than to tubular damage.

Because of this danger of renal damage every attempt should be made to diagnose the cause of even mild hypercalcaemia and treat it at an early stage (but see p. 272).

The high free ionised calcium concentration in the glomerular filtrate may, if circumstances favour it, result in precipitation of calcium salts in the urine: the patient may present with *renal calculi* without significant renal parenchymal damage. All patients with renal stones should have plasma calcium estimated on more than one occasion.

Effects on neuromuscular excitability.—High free ionised calcium levels depress neuromuscular excitability in both voluntary and involuntary muscle. The patient may complain of *constipation* and *abdominal pain*. Muscular hypotonia may also be present.

Effects on the central nervous system.—*Depression* is a common complaint even in patients with only mild hypercalcaemia. This, and the *anorexia, nausea* and *vomiting* associated with higher plasma calcium levels are probably due to an effect on the central nervous system.

Effects on the stomach.—Calcium stimulates gastrin, and therefore gastric acid secretion (p. 302), and there is an association between peptic ulceration and chronic hypercalcaemia.

Effects on the blood pressure.—Some patients with hypercalcaemia have hypertension, which may sometimes respond to lowering of the plasma calcium level.

Effects on the heart.—Severe hypercalcaemia causes changes in the electrocardiogram. If levels exceed about 3·75 mmol/l (15 mg/dl) there is a risk of sudden *cardiac arrest*, and for this reason such marked hypercalcaemia must be treated as a matter of urgency.

Clinical Effects of Reduced Free Ionised Calcium Levels

Low free ionised calcium levels (including those associated with a normal total calcium in alkalosis, p. 252) cause increased neuromuscular activity leading to *tetany*.

Prolonged hypocalcaemia, even when mild, interferes with the metabolism of the lens and causes *cataracts*: it may also cause depression and other *psychiatric symptoms*. Because of the danger of cataracts, asymptomatic hypocalcaemia should be sought when there has been a known risk of damage to the parathyroid glands (for example, after partial or total thyroidectomy).

Effects of High Parathyroid Hormone Levels on Bone

High PTH levels with a normal supply of vitamin D and calcium (for example, *primary hyperparathyroidism*).—Parathyroid hormone acts on the osteoclasts of bone, increasing their activity. In cases in which the supply of vitamin D and calcium is normal and in which PTH has been circulating in excess *over a long period of time* the effects are as follows:

> *clinical*:
> bone pains;
> bony swellings.
> *radiological*:
> generalised decalcification;
> subperiosteal erosions;
> cysts in the bone.
> *histological*:
> increased number of osteoclasts.

A secondary osteoblastic reaction may occur: osteoblasts manufacture the enzyme alkaline phosphatase which, when these cells are increased in number, is released into the extracellular fluid in abnormal amounts and leads to a *rise in plasma alkaline phosphatase activity*.

The effects of excess parathyroid hormone on bone are only evident in some long-standing cases. In disease of short duration (for instance, when the excess of PTH is due to malignant disease, or to early primary hyperparathyroidism) they are absent. An osteoblastic reaction is a late manifestation and plasma *alkaline phosphatase levels are usually normal or only moderately raised* in primary hyperparathyroidism.

High PTH levels in vitamin D and calcium deficiency (causing *rickets and osteomalacia*).—1,25-dihydroxycholecalciferol must be present in adequate amounts for the complete action of PTH on bone. In cases of vitamin D deficiency PTH levels may be very high, but the effects on bone differ from those described above. Despite marked osteoblastic proliferation the supply of calcium is inadequate, and uncalcified osteoid is a characteristic histological finding: the plasma

alkaline phosphatase activity is increased because of osteoblastic proliferation.
The effects are therefore as follows:

clinical:
bone pains;
rarely bony swelling.

radiological:
generalised decalcification;
pseudofractures (cortical fractures);
rarely subperiosteal erosions or cysts in bone.

histological:
uncalcified osteoid: wide osteoid seams;
an increased number of osteoblasts;
occasionally an increased number of osteoclasts.

in the plasma:
raised alkaline phosphatase activity.

These changes in the bone may sometimes be found in cases with impaired renal tubular reabsorption of phosphate (p. 267).

DISORDERS OF CALCIUM METABOLISM

In this section disorders of calcium metabolism will be considered in relation to circulating parathyroid hormone levels or PTH-like effects, to explain the plasma findings; *this does not usually mean that PTH assay is indicated.* The disorders may be summarised as follows:

diseases associated with high PTH levels:

inappropriate secretion of PTH (*causing a raised plasma calcium concentration*);
appropriate secretion of PTH (*with a low normal or low plasma calcium concentration*).

diseases associated with low PTH levels:

disease of the parathyroid glands (*causing a low plasma calcium concentration*);
appropriate suppression of PTH secretion by a *high plasma calcium concentration* due to causes other than inappropriate PTH levels.

Diseases Associated with High Circulating Parathyroid Hormone Levels

Diseases due to inappropriate secretion of parathyroid hormone.— The term "inappropriate secretion of parathyroid hormone" is used here to mean the production of the hormone under circumstances in which it should normally be absent (that is, in the presence of a high free ionised calcium concentration).

Such inappropriate secretion occurs in three circumstances:

> primary hyperparathyroidism;
> tertiary hyperparathyroidism;
> (In these two cases the hormone is produced by the parathyroid gland.)
> ectopic production of PTH (other terms which have been used are "pseudohyperparathyroidism" and "quaternary" hyperparathyroidism).

The findings, if renal function is normal, are those due to high circulating PTH levels—a raised plasma calcium concentration, with a normal or low plasma phosphate level. If renal glomerular damage develops, usually due to hypercalcaemia, both of these tend to return to normal, the phosphate because of the inability of the kidney to respond to the phosphaturic effect of PTH and the calcium because of the lowering of plasma calcium in renal disease (p. 263). Diagnosis at this stage may be very difficult.

The *clinical features* of these cases are due to:

> excess of circulating free ionised calcium (p. 257);
> the effect of PTH on bone (p. 258).

The differences between these syndromes depend mainly on the duration of the disease.

Primary hyperparathyroidism.—Primary hypersecretion of PTH from the parathyroid glands is usually due to an adenoma, occasionally in an ectopic gland. Parathyroid tumours are almost always benign, although very rarely primary hyperparathyroidism is due to parathyroid carcinoma. Occasionally it results from diffuse hyperplasia of the glands.

The incidence of parathyroid adenoma increases with age, but it may be found in young subjects and occasionally in children. The apparent preponderance in elderly women may merely be due to the longer life-expectancy of females; in lower age groups the sex incidence seems to be equal.

Primary hyperparathyroidism presents most commonly with signs and symptoms associated with high free ionised calcium levels, such as depression, nausea, anorexia and abdominal pain, or with renal changes (calculi, polyuria or even renal failure). Other cases may be diagnosed after the chance finding of a high plasma calcium and low phosphate level. *In many of these cases bone changes are absent and the plasma alkaline phosphatase concentration is normal or only slightly elevated.* On the other hand, patients presenting with overt bony lesions may have no demonstrable renal impairment. It may be that when conditions in the urine favour calcium precipitation (for instance, if it is alkaline), or when a patient complains of symptoms early, the disease is diagnosed before bony changes occur: if these warning signs are absent the patient presents with the overt picture of bone disease. However, some patients with primary hyperparathyroidism do not develop bone disease even after many years without treatment.

Occasionally the patient is admitted as an emergency with abdominal pain, vomiting and constipation due to severe hypercalcaemia.

Tertiary hyperparathyroidism.—Tertiary hyperparathyroidism is the name given to the condition in which the parathyroid has been under prolonged feedback stimulation by low free ionised calcium levels, and the resulting hypersecretion has become partially autonomous. Except for the history of previous hypocalcaemia, the disease is identical with primary hyperparathyroidism. It is usually diagnosed when the cause of the original hypocalcaemia is removed (for example, by renal transplantation or correction of calcium or vitamin D deficiency). The typical biochemical findings due to high circulating levels of PTH are superimposed on those of osteomalacia, and because the condition is of long duration *the plasma alkaline phosphatase level is usually very high*. In some cases calcium levels later return to normal.

Ectopic production of parathyroid hormone or of a parathyroid-hormone-like substance.—Sometimes malignant tumours of non-endocrine tissues manufacture peptides foreign to them: PTH can be one of these. The production of hormones at these sites is not subject to normal feedback control, and the initial findings are those of inappropriate PTH excess. Because of the nature of the underlying disease the condition is rarely of long standing, and the bony lesions due to excess circulating PTH are not evident: the alkaline phosphatase, however, may be raised because of bony or hepatic metastases, or both. Plasma calcium levels may rise from normal to dangerously high levels within a day in patients with hypercalcaemia due to malignancy.

The subject of ectopic hormone production is discussed more fully in Chapter XXII.

Familial hypocalciuric hypercalcaemia (familial benign hypercalcaemia).—Hypercalcaemia has been reported in blood relations, in none of whom was a parathyroid adenoma found at operation. They were said to have inappropriately high plasma PTH levels in the presence of hypercalcaemia, and to have hypocalciuria. The condition seems to be inherited as an autosomal dominant trait. The combination of reported findings is very difficult to explain, and it is extremely rare.

Secondary hyperparathyroidism ("appropriate" secretion of para-

thyroid hormone).—In secondary hyperparathyroidism, the parathyroids are stimulated to produce hormone in response to low plasma free ionised calcium levels. This secretion is therefore "appropriate" in that it is necessary to restore free ionised calcium levels to normal. If it is effective in doing so the stimulus to production is removed, and the glands are "switched off". In secondary hyperparathyroidism, therefore, plasma calcium concentration is low (if the increased production of hormone is inadequate to correct the hypocalcaemia) or normal: it is *never* high. Hypercalcaemia, or a plasma calcium level in the upper normal range, suggests either a diagnosis of primary hyperparathyroidism (and this may be a *cause* of renal failure), or that prolonged calcium deficiency has led to tertiary hyperparathyroidism.

If renal function is normal the high PTH levels cause phosphaturia with hypophosphataemia. In all cases of secondary hyperparathyroidism, except those due to renal failure, both the plasma calcium and phosphate levels tend to be low.

If the PTH feedback mechanism is normal any factor which tends to lower the free ionised calcium concentration causes secondary hyperparathyroidism. In long-standing cases, if the bone cells can respond to PTH, decalcification of bone leads to *osteomalacia* in adults, or, before fusion of the epiphyses, to *rickets*. The findings are described on p. 259. In very prolonged or severe depletion the high concentration of PTH may fail to maintain plasma calcium levels and the reduced free ionised calcium concentration may produce the effects described on p. 258.

Secondary hyperparathyroidism with osteomalacia or rickets occurs when there is both chronic calcium and vitamin D deficiency. Predisposing factors include:

   reduced dietary intake of vitamin D and calcium in *malnutrition*;
   reduced absorption of vitamin D in *steatorrhoea* and after-
      gastrectomy;
   reduced metabolism of vitamin D to 1,25-DHCC due to *renal*, or,
      more rarely, *liver disease*;
   increased hepatic inactivation of vitamin D due to *anticonvulsant
      therapy*.

Secondary hyperparathyroidism without osteomalacia or rickets occurs in:

   early calcium and vitamin D deficiency;
   most cases of pseudohypoparathyroidism (very rare).

Osteomalacia and rickets without secondary hyperparathyroidism may rarely occur in phosphate depletion due to:

   renal tubular disorders of phosphate reabsorption (p. 267).

Calcium and vitamin D deficiency.—1,25-DHCC is essential for normal absorption of calcium from the intestinal tract and calcium depletion is more commonly the result of deficiency of intake of vitamin D than of calcium itself. The commonest precipitating cause of calcium and vitamin D deficiency in relatively affluent countries is the malabsorption syndrome: dietary deficiency is more important in the world as a whole. It is most likely to occur in children and pregnant women, in whom the supply from the skin may be insufficient for the increased needs, and in subjects confined indoors and not exposed to sunlight.

In the presence of steatorrhoea vitamin D, which is fat soluble, is poorly absorbed. In addition, calcium combines with the excess of fatty acids in the intestine to form insoluble soaps.

In Britain there is a relatively high incidence of osteomalacia among Asian immigrants, especially during puberty and pregnancy, but also in younger children. The reasons for this are still not clear. The condition is commonest in those Asian Muslims not integrated into the Western community; dietary habits, lack of sunlight and even genetic factors may all play a part.

Malnutrition is rarely selective, and protein deficiency may reduce the level of the protein-bound fraction of the calcium. The free ionised calcium level may not, therefore, be as low as that of total calcium would suggest.

Patients on prolonged *anticonvulsant therapy* (barbiturates and phenytoin) sometimes develop hypocalcaemia, and even osteomalacia; it has been shown that, in rats, these drugs induce hepatic synthesis of enzymes catalysing inactivation of vitamin D.

Cholecalciferol is inactive until finally hydroxylated, as needed, in the kidney. Patients with *renal disease* are therefore relatively resistant to calciferol and hypocalcaemia may develop within a few days of the onset of renal failure. Low plasma protein levels sometimes contribute to the reduction of total calcium. Tetany rarely occurs, possibly because the accompanying acidosis keeps the free ionised calcium above tetanic levels. Bicarbonate should not be given to correct the acidosis of renal failure, since if the pH rises calcium may be precipitated in non-osseous sites, including the kidney, and cause further deterioration of renal function.

Osteomalacia, usually mild, is common in *chronic liver disease*, especially when there is cholestasis. Although 25-HCC levels are usually low, they can be corrected by vitamin D supplements. It is therefore unlikely that there is sufficient impairment of 25-hydroxylation to account for the osteomalacia. Factors such as reduced vitamin D intake and impaired absorption due to bile-salt deficiency are probably more important.

In all these conditions the low plasma free ionised calcium concentration stimulates PTH production, usually with the consequences described above. However, in renal glomerular failure phosphate retention causes hyperphosphataemia. In chronic cases a rising alkaline phosphatase level heralds the onset of the bone changes of osteomalacia or rickets.

Pseudohypoparathyroidism.—In this very rare inborn error the response of both the kidney and bone to PTH is impaired. Plasma calcium levels therefore fall and stimulate PTH secretion. The high PTH levels are ineffective in raising the plasma calcium and lowering the plasma phosphate concentrations and the biochemical changes of hypoparathyroidism, rather than of hyperparathyroidism, are present. The stimulatory effect of PTH on the renal 1α-cholecalciferol hydroxylase is also impaired.

Diseases Associated with Low Circulating Parathyroid Hormone Levels

A low level, or absence, of circulating PTH leads to the changes in plasma phosphate levels described on p. 254.

Primary parathyroid hormone deficiency (hypoparathyroidism).— Hypoparathyroidism may be due to autoimmune destruction of the parathyroid glands, or more commonly to surgical damage, either directly to the glands or to their blood supply during partial thyroidectomy (during total thyroidectomy or laryngectomy removal of the glands is almost inevitable). Post-thyroidectomy hypocalcaemia is not always due to parathyroid damage, but may be due either to the rapid entry of calcium into depleted bone as thyroid hormone levels fall to normal or, more commonly, to postoperative hypoalbuminaemia.

Removal of a parathyroid adenoma to treat primary hyperparathyroidism also carries a slight risk of damage to the remaining parathyroid tissue. However, apparent hypocalcaemia after parathyroidectomy (and after thyroidectomy) is usually transient, and due to the temporary hypoalbuminaemia that is so common after any surgery. In the rare cases which present with severe bone disease, rapid entry of calcium into bone as PTH levels fall may contribute to true but temporary postoperative hypocalcaemia. If the antecedent hypercalcaemia has been severe and prolonged the activity of the remaining parathyroid tissue may have been suppressed for so long by the feedback mechanism that there may be true hypoparathyroidism with a rising plasma phosphate level, which recovers within a few days. Evidence for asymptomatic hypocalcaemia should always be sought after partial thyroidectomy, or after removal of a parathyroid adenoma, because of the danger of cataracts (p. 258): early postoperative parathyroid insufficiency may recover, but a low calcium level persisting for more than a few weeks should be treated. Even with highly skilled surgery there is a danger of parathyroid damage. The clinical symptoms of hypoparathyroidism are those due to a low free ionised calcium concentration (p. 258).

Secondary suppression of parathyroid hormone secretion.—Just as hypocalcaemia from any cause (except hypoparathyroidism) leads to secondary hyperparathyroidism, so hypercalcaemia (unless due to inappropriate PTH secretion) suppresses PTH secretion, with the consequences described on p. 254.

The causes of hypercalcaemia with suppression of PTH secretion are:

    vitamin D excess;
    idiopathic hypercalcaemia of infancy;
    sarcoidosis;
    very severe thyrotoxicosis (very rare);
    possibly some cases of bony secondary deposits and myeloma.

The *clinical picture* in these cases is due to the high free ionised calcium level (p. 257). Generalised decalcification of bone only occurs in this group if the hypercalcaemia is due to thyrotoxicosis or sarcoidosis, or sometimes to myeloma or bony deposits. The alkaline phosphatase level is usually normal in myeloma and in conditions in which bone is not involved.

Hypercalcaemia due to vitamin D excess.—Overdosage with vitamin D increases calcium absorption and may cause dangerous hypercalcaemia. PTH secretion is suppressed, and phosphate levels are usually normal or high.

Such overdosage may be caused by over-vigorous treatment of hypocalcaemia, and this therapy should always be controlled by frequent plasma calcium and, if there is osteomalacia, alkaline phosphatase estimations.

Occasionally patients overdose themselves with vitamin D. In any obscure case of hypercalcaemia a drug history should be taken.

The hypercalcaemia associated with two diseases is said to be due to hypersensitivity to the action of vitamin D. These diseases are:

    idiopathic hypercalcaemia of infancy;
    hypercalcaemia of sarcoidosis.

Idiopathic hypercalcaemia of infancy.—In these days of milk and vitamin supplementation rickets is a rare disease in Great Britain. By contrast, the number of infants presenting with hypercalcaemia of obscure origin increased during the 1950s. The incidence of this syndrome has declined since the recommended intake of vitamin D for infants was reduced in 1957, although the correlation between these two variables is uncertain. An even rarer and more severe form—an inborn error—is associated with, amongst other stigmata, mental deficiency and "elfin" facies.

Hypercalcaemia of sarcoidosis.—Hypercalcaemia as a complication of sarcoidosis may be due to vitamin D sensitivity. *Chronic beryllium poisoning* produces a granulomatous reaction very similar to that of sarcoidosis, and may also be associated with hypercalcaemia: beryllium is used in the manufacture of fluorescent lamps and in several other industrial processes.

Other diseases sometimes associated with hypercalcaemia are:

Hypercalcaemia of thyrotoxicosis.—Thyroid hormones probably have a direct action on bone. Rarely very severe thyrotoxicosis may cause hypercalcaemia. If the hypercalcaemia fails to respond when hyperthyroidism is controlled the possibility of the co-existent hyperparathyroidism, which may be part of the syndrome of multiple endocrine adenopathy (p. 470), should be remembered.

Malignant disease of bone.—The hypercalcaemia due to ectopic production of PTH, or a PTH-like substance, by malignant tumours has been described on p. 261. A different syndrome has been reported in cases with multiple bony metastases, or with myeloma: in these the parallel rise of phosphate is said to indicate that the hypercalcaemia is caused by direct breakdown of bone by the local action of malignant deposits. Critical examination of the findings in most of these cases shows a rise of plasma urea accompanying that of phosphate, and such findings are compatible with the action of excess PTH (or PTH-like factor) in renal glomerular failure. Moreover, there is little correlation between the incidence of hypercalcaemia and the rate of extension of the bony lesions. Probably most if not all such cases are due to ectopic hormone production.

The raised alkaline phosphatase level associated with secondary deposits in bone tissue is due to the resultant stimulation of local osteoblasts (p. 259). This osteoblastic reaction does not occur with the bone erosion due to the marrow expansion of myelomatosis, in which the plasma level of alkaline phosphatase of bony origin is therefore normal. This fact may be a useful pointer to the diagnosis of myeloma in the presence of extensive bone deposits of unknown origin.

The paraproteins of myeloma very rarely bind calcium to any significant extent, and are unlikely to account for hypercalcaemia in most of these cases.

The Hypocalcaemia of "Shock"

Temporary hypocalcaemia may follow any acute episode, including *acute pancreatitis*, which causes "shock". In such conditions plasma albumin levels also fall (p. 342). Free ionised calcium levels very occasionally fall in acute pancreatitis, but they are almost always normal. Calcium should only be given if there is tetany.

Diseases of Bone Not Affecting Plasma Calcium Levels

Diseases producing biochemical abnormalities, or important in the differential diagnosis from those already mentioned, are:

Osteoporosis.—In this disease the primary lesion is a reduction in the mass of bone matrix with a secondary loss of calcium. Clinically and radiologically it may resemble osteomalacia, but normal plasma calcium, phosphate and alkaline phosphatase concentrations (especially the latter) are more in favour of osteoporosis;

Paget's disease of bone.—Plasma calcium and phosphate levels are rarely affected in Paget's disease. The alkaline phosphatase level is typically very high;

Renal tubular disorders of phosphate reabsorption.—This group of diseases comprises a number of inborn errors of renal tubular function in which *phosphate is not reabsorbed normally* from the glomerular filtrate. The cases usually present with *rickets or osteomalacia*, but, unlike the usual form, respond poorly to vitamin D therapy. The syndrome has, therefore, been called *"resistant rickets".* The defect in phosphate reabsorption may be an isolated one, as in *familial hypophosphataemia* (an X-linked dominant trait), or part of a more general reabsorption defect as seen in the *Fanconi syndrome* (p. 19). In these cases phosphate deficiency is the primary lesion, and failure to calcify bone may be due to this. Plasma levels of phosphate are usually very low and fail to rise when vitamin D alone is given: there is a relative phosphaturia. The osteomalacia is reflected in the *high plasma alkaline phosphatase* levels. Plasma calcium concentration is usually normal, and in this respect the findings differ from those of classical osteomalacia; because of the normocalcaemia there is rarely evidence in the bone of hyperparathyroidism. The condition responds to large doses of oral phosphate, perhaps together with a small dose of calciferol.

TESTS FOR DISORDERS OF CALCIUM METABOLISM

In our experience, if careful attention is paid to the clinical picture, plasma calcium, phosphate *in relation to urea*, alkaline phosphatase, and haematological and radiological findings, other tests are unnecessary and may even be misleading.

Plasma Parathyroid Hormone Levels

The specificity and sensitivity of radioimmunoassay of plasma parathyroid hormone has improved. However, results of these assays can still be misleading, and we rarely find the test to be necessary. We only use it in carefully defined circumstances:

In hypercalcaemia:

if, after considering all the factors mentioned at the beginning of this section, the cause of hypercalcaemia is still in doubt;

if the neck vessels are cannulated in an attempt to localise the adenoma by radiological means: if the PTH level is higher in blood derived from one gland than from the others this *may* help to localise the adenoma pre-operatively. The procedure is not without risk, and should not be performed for this purpose alone.

If the feedback mechanism is normally responsive, PTH levels are suppressed when there is true (free ionised) hypercalcaemia. However, raised values are found not only in parathyroid adenomas, but also in some tumours of non-parathyroid origin, which can secrete immuno-reactive parathyroid hormone inappropriate to the calcium level; by contrast, some proven parathyroid adenomas secrete biologically active hormone which does not react *in vitro* even with antibodies to more than one peptide sequence. It should be remembered that false "negative" results of PTH assay may be obtained in primary hyper-parathyroidism: moreover, we have seen that high PTH levels do not necessarily confirm the diagnosis.

In hypocalcaemia:

if the exceedingly rare syndrome of pseudohypoparathyroidism (p. 264) is suspected. Very high PTH levels despite hypocal-caemia *and hyperphosphataemia* suggest end-organ unrespon-siveness.

Urinary Calcium

This frequently requested estimation is of little diagnostic value in the differential diagnosis of hypercalcaemia. In the presence of normal renal function, hypercalcaemia from any cause increases the load on the glomerulus and causes hypercalciuria: although excess PTH, by increasing renal tubular reabsorption of calcium, might be expected to reduce urinary excretion below that appropriate to the plasma level, the "normal" range is too wide for the test to be helpful. Causes of a high urinary calcium excretion without hypercalcaemia are so-called "*idiopathic hypercalciuria*", some cases of *osteoporosis* in which calcium cannot be deposited in normal amounts in bone because of a deficient matrix, and acidosis, in which release of free ionised calcium is increased.

Calcium excretion will be normal or, more often, low if renal glomerular function is impaired, even if there is hypercalcaemia (whether due to hyperparathyroidism or to other causes).

Steroid Suppression Test

This test, although by far the most useful one in the differential diagnosis of hypercalcaemia, is rarely necessary if the factors discussed above are taken into account. It is difficult to interpret the results if the plasma calcium concentration is less than about 3 mmol/l.

In most cases it differentiates primary or tertiary hyperparathyroidism from any other cause. Large doses of hydrocortisone (or cortisone) cause high plasma calcium levels to fall to within the normal range in almost all cases *except primary* (or tertiary) *hyperparathyroidism.* Exceptions do occasionally occur, but in most of them the clinical diagnosis is clear and the test unnecessary (for instance in very advanced malignant disease of bone or very severe hyperthyroidism). Details of the test are given on p. 284.

The reason for this difference in reaction to steroids is not clear. It is particularly interesting that the calcium level even in cases of proven ectopic PTH production is suppressed by steroids, whereas it fails to be so in those in which the hormone is produced in the parathyroid glands.

The findings which may help in the differential diagnosis of hypercalcaemia and hypocalcaemia are shown in Tables XXIV and XXV respectively. *The investigation of disorders of calcium metabolism is considered further on p. 278.*

BIOCHEMICAL BASIS OF TREATMENT

Hypercalcaemia

Mild hypercalcaemia.—If plasma calcium levels are less than about 3·75 mmol/l (15 mg/dl) in a patient without serious symptoms, or electrocardiographic changes of hypercalcaemia, there is no *immediate* need for urgent therapy. However, treatment should be instituted as soon as a diagnosis is made because of the danger of renal damage.

The patient should be rehydrated and, if possible, the cause (such as a parathyroid adenoma, a primary malignant lesion or hyperthyroidism) should be treated. "Loop" diuretics such as frusemide have been claimed to lower plasma calcium levels.

TABLE XXIV

DIFFERENTIAL DIAGNOSIS OF HYPERCALCAEMIA BY SIMPLE LABORATORY TESTS

Diagnosis	Plasma			Steroid suppression	Comments
	Phosphate	Proteins	Alk. phos.		
Due to Raised Free Ionised Calcium					
Group 1. Inappropriate Parathyroid Hormone					
Primary or tertiary hyperparathyroidism	N or ↓	N	N or ↑	No	Calcium and phosphate changes masked in uraemia.
Malignancy with ectopic hormone production	N or ↓	N (↑ myeloma)	N or ↑ (N in myeloma)	Yes (usually)	
Group 2. Appropriately low PTH					
Vitamin D					
Vitamin D overdosage	Variable	N	N	Yes	
Sarcoidosis	Variable	γ glob. ↑	N or ↑	Yes	Rare
Idiopathic hypercalcaemia of infancy	Variable	N	N	Yes	
Thyrotoxicosis	N or ↑	N	N or ↑	Yes (usually)	*Severe* thyrotoxicosis. Thyroid function tests abnormal.
Due to Raised Protein Bound Calcium					
Dehydration	N or ↑	↑	N		Clinical signs of volume depletion. Urea ↑ Corrected by dehydration.
Artefactual (excess stasis)	N or ↑	↑	N		Specimen taken without stasis gives normal values.

Hypercalcaemia due to raised protein-bound calcium should *NOT* be treated.

TABLE XXV

DIFFERENTIAL DIAGNOSIS OF HYPOCALCAEMIA BY SIMPLE LABORATORY TESTS

Diagnosis	Plasma				Comments
	Phosphate	Proteins	Urea	Alk. phos.	
Due to Low Free Ionised Calcium					
Hypoparathyroidism	↑	N	N	N	
Calcium and vitamin **D** deficiency	↓	N or ↓	N	↑	May be accompanied by low protein-bound calcium.
Renal failure	↑	N or ↓	↑	N or ↑	Should be treated with caution in absence of bone disease.
Pseudohypoparathyroidism	↑	N	N	N	*Very rare.* PTH levels high.
Due to Low Protein-Bound Calcium					
Overhydration	N	↓	N or ↓	N	Responds to fluid restriction.
Hypoalbuminaemia	N	↓	N or ↓	N	

Hypocalcaemia due to lowered protein-bound calcium should *NOT* be treated.

Not everyone agrees that patients with apparently asymptomatic, mild hypercalcaemia due to primary hyperparathyroidism need parathyroidectomy. It is argued that prolonged hypercalcaemia by no means always causes obvious renal dysfunction, and that the risk of surgery is greater than that of mild hypercalcaemia. However, we find that almost all patients volunteer the information that a successful operation has improved their feeling of well-being, and has made them aware of previous vague ill-health. We find the morbidity of the operation, in the hands of an experienced surgeon, to be minimal. The decision as to whether to operate must be made on clinical grounds; of particular importance are the age, and the fitness of the patient for operation when other disease is present. Symptomatic hypercalcaemia in a patient deemed unfit for operation may be treated medically.

Medical measures used to treat mild hypercalcaemia include oral sodium phosphate, steroids and calcitonin.

Oral sodium phosphate, by forming calcium phosphate in the intestinal lumen, prevents reabsorption of calcium entering the gut; it therefore removes calcium from the body. We find this treatment effective and cheap. The tendency to osmotic diarrhoea can be minimised by dissolving the phosphate in enough water: side-effects are then rare. It may be used in conjunction with steroids to treat intractable hypercalcaemia due to extensive malignancy;

Steroids may lower the calcium level in some cases. They are more likely to have undesirable side-effects than phosphate, and are most useful if there are other indications for giving them (for example sarcoidosis, or some types of malignant disease);

Calcitonin is very expensive and should only be used if oral phosphate therapy is not possible.

Severe hypercalcaemia.—Both oral phosphate and steroids only start to be effective after about 24 hours. If plasma calcium levels exceed about 3·75 mmol/l (15 mg/dl) treatment is indicated as a matter of urgency, because of the danger of cardiac arrest. The exact level varies in different subjects. If there is any doubt about the degree of urgency the electrocardiogram should be inspected for abnormalities associated with hypercalcaemia.

The use of *intravenous phosphate* to lower calcium within a few hours depends, partially at least, on precipitation of insoluble calcium salts. Since some of this precipitation may occur in the kidney, the treatment carries a slight theoretical risk of initiating or aggravating renal failure. In practice we do not find this to be a problem; the immediate danger of cardiac arrest due to severe hypercalcaemia is greater. The disadvantages of steroids and of calcitonin have been outlined above: steroids act too slowly to be useful in the immediate treatment of severe hypercalcaemia.

Solutions for intravenous administration contain a mixture of sodium and potassium phosphates.

As with most other extracellular constituents, rapid changes in calcium level may be dangerous because time is not allowed for equilibration across cell membranes. The aim of emergency treatment should be to lower plasma calcium temporarily to safe levels, while initiating treatment for mild hypercalcaemia. Too rapid a reduction of calcium concentration may induce tetany, or, more seriously, hypotension, even though the calcium is within or above normal levels. (Tetany in the presence of normocalcaemia may also occur during the rapid fall of serum calcium concentration after removal of a parathyroid adenoma.) A slow lowering of calcium level also reduces the risk of renal calcification.

Hypocalcaemia

Postoperative hypocalcaemia.—Whatever its cause (p. 264), hypocalcaemia during the first week after thyroidectomy or parathyroidectomy should be treated only if there is overt tetany, and then only with calcium supplements (with a rapid effect and short half-life), rather than with calciferol: a slightly low free ionised calcium level is of no immediate danger, and helps to stimulate recovery of suppressed parathyroid cells. Persistent hypocalcaemia may indicate that the glands are permanently damaged and that long-standing or life-long calciferol therapy is necessary.

Asymptomatic hypocalcaemia.—Hypocalcaemia, whatever the cause, if asymptomatic or accompanied by only mild clinical symptoms, can usually be effectively treated with large doses of vitamin D by mouth. It is difficult to give enough oral calcium by itself to make a significant lasting difference to plasma calcium levels, and vitamin D, by increasing absorption of calcium, is usually adequate without calcium supplementation.

The hypocalcaemia of renal failure should be treated with caution in the absence of such signs of osteomalacia as a raised plasma alkaline phosphatase activity, because of the danger of ectopic calcification in the presence of hyperphosphataemia. The plasma phosphate may be lowered first by giving aluminium hydroxide orally. 1,25-DHCC, or 1α-hydroxycholecalciferol (which is converted to 1,25-DHCC by the liver) are now available and have been used, especially to treat the hypocalcaemia of renal disease. Such cases usually respond to very high doses of the much cheaper calciferol: the risk of ectopic calcification is the same whichever compound is used.

Apparent hypocalcaemia due to low albumin levels should not be treated.

Hypocalcaemia with severe tetany.—In the presence of severe tetany hypocalcaemia should be treated, as an emergency, with intravenous calcium (usually as the gluconate).

ABNORMALITIES OF PLASMA PHOSPHATE CONCENTRATION

Hypophosphataemia

Hypophosphataemia associated with disturbances of calcium metabolism is usually due to *excess of circulating PTH*. In these conditions, and in *renal disorders of phosphate reabsorption*, phosphate is lost from the body.

Phosphate, like potassium, enters cells from the extracellular fluid under the influence of *insulin* (this occurs during the treatment of diabetic coma). This redistribution of phosphate is a common cause of hypophosphataemia in patients on parenteral nutrition with insulin and glucose. *Long-term artificial feeding* without adequate phosphate supplementation may cause true phosphate depletion. Hypophosphataemia during artificial feeding, whether due to true deficiency or to redistribution, has been said to cause convulsions. These patients are inevitably very ill; if the hypophosphataemia by itself is the cause of the symptoms it is difficult to explain why patients with an excess of circulating PTH, or with hypophosphataemic rickets, and with equally low plasma phosphate levels, do not have the same symptoms and signs.

Hyperphosphataemia

The commonest cause of hyperphosphataemia is *renal glomerular failure*; it is important not to correct hypocalcaemia until measures have been taken to lower the plasma phosphate level (p. 273).

Less common causes of hyperphosphataemia are *hypoparathyroidism* and *acromegaly*.

Normal levels are higher in children than in adults.

MAGNESIUM METABOLISM

Magnesium is present with calcium in bone salts, and tends to move in and out of bone with calcium. It is also present in all cells at much higher concentrations than in the extracellular fluid, and therefore tends to enter and leave cells under the same conditions as do potassium and phosphate.

Magnesium can be lost in large quantities in the faeces in diarrhoea or in fluid lost through intestinal fistulae.

Plasma Magnesium and Its Control

About 35 per cent of the plasma magnesium, like calcium, is protein bound. However, less is known about the importance of this than in the case of calcium.

The mechanism of control of magnesium levels is poorly understood, but aldosterone is known to increase its renal excretion; PTH may also affect its absorption and excretion.

Clinical Effect of Abnormal Plasma Magnesium Levels

Hypomagnesaemia

Hypomagnesaemia causes symptoms very similar to those of hypocalcaemia. If a patient with tetany has normal plasma calcium, protein and bicarbonate (or, more accurately, blood pH) levels, blood should be taken for magnesium estimation. If there is good reason to suspect magnesium deficiency (for example, in presence of severe diarrhoea, or loss through intestinal fistulae), and if the estimation cannot be performed reasonably quickly, intravenous magnesium should be administered as a therapeutic test. Less severe magnesium deficiency should be treated orally (p. 283).

Hypermagnesaemia

Hypermagnesaemia causes muscular hypotonia, but as it is rarely found without other abnormalities, such as hypercalcaemia or renal failure, it is not always easy to distinguish the findings due to the abnormal magnesium level alone from those of the coexistent abnormalities.

Causes of Abnormal Plasma Magnesium Levels

Hypomagnesaemia

Excessive loss of magnesium.—Excessive loss of magnesium occurs in *severe*, prolonged diarrhoea, or loss through intestinal fistulae, and these are by far the most important causes of a clinical disturbance of magnesium metabolism severe enough to warrant treatment.

Hypomagnesaemia accompanied by hypocalcaemia.—Magnesium tends to move in and out of bones in association with calcium, and hypocalcaemia is often accompanied by hypomagnesaemia. This may occur in hypoparathyroidism, or during the fall of calcium concentration after the removal of a parathyroid adenoma associated with severe bone disease.

Hypomagnesaemia accompanied by hypokalaemia.—Since magnesium moves in and out of cells with potassium, hypomagnesaemia tends to occur in association with hypokalaemia. These conditions include diuretic therapy and primary aldosteronism. Such hypomagnesaemia is rarely of clinical importance.

Hypermagnesaemia

The commonest cause of hypermagnesaemia is probably renal glomerular failure (when plasma potassium is also high). It rarely, if ever, needs specific treatment, and responds to measures to treat the underlying condition. Magnesium salts should never be administered in renal glomerular failure.

SUMMARY

Calcium Metabolism

1. About half the calcium in plasma is bound to protein, and half is in the free ionised form.

2. The free ionised calcium is the physiologically important fraction and calcium levels should be interpreted together with albumin levels.

3. Plasma calcium levels are controlled by parathyroid hormone. Parathyroid hormone secretion is increased if free ionised calcium concentrations are reduced.

4. Parathyroid hormone acts on:

> bone, releasing calcium and phosphate into the plasma: if prolonged, this increases osteoblastic activity and increases plasma alkaline phosphatase levels;
>
> kidneys, causing calcium reabsorption and phosphaturia, which lowers the plasma phosphate. It may also increase 1α-hydroxylation of 25-OHCC.

5. Symptoms and findings in diseases of calcium metabolism can be related to circulating levels of free ionised calcium, to levels of parathyroid hormone, and to renal function.

6. Excessive "inappropriate" levels of parathyroid hormone are present in primary parathyroid disease, or if the hormone is produced at ectopic sites. In these circumstances plasma calcium is high.

7. Increased "appropriate" levels of parathyroid hormone are present whenever plasma calcium levels fall.

8. Parathyroid hormone levels are low in hypoparathyroidism (associated with a low plasma calcium level), or in the presence of

hypercalcaemia other than that of inappropriate parathyroid hormone production. The causes and differential diagnosis of hypercalcaemia and hypocalcaemia are summarised in Tables XXIV and XXV (pp. 270 and 271).

Magnesium Metabolism

1. Hypomagnesaemia may cause tetany in the absence of hypocalcaemia.

2. The commonest cause of significant hypomagnesaemia is severe diarrhoea.

3. Magnesium levels tend to follow those of calcium (in and out of bone) and potassium (in and out of cells).

FURTHER READING

FRASER, D. R. (1980). Regulation of the metabolism of vitamin D. *Physiol. Rev.*, **60**, 551–613.

DE LUCA, H. F. and SCHNOES, H. K. (1983). Vitamin D. Recent advances. *Ann. Rev. Biochem.*, **52**, 411–439.

GOODWIN, F. J. (1981). Symptomless abnormalities: hypercalcaemia. *Brit. J. Hosp. Med.*, **28**, 50–58.

HEATH, D. A. (1981). Vitamin D therapy. *Prescribers' J.*, **21**, 164–171.

FRASER, D. R. (1983). The physiological economy of vitamin D. *Lancet*, **1**, 969–972.

KANIS, J. A., PATERSON, A. D and RUSSELL, R. G. G. (1983). Disorders of calcium and skeletal metabolism. In: *Biochemistry in Clinical Practice*, pp. 299–325. D. L. Williams and V. Marks, Eds. (Scientific Foundations of Clinical Biochemistry, Vol. II.) London: Heinemann Medical.

INVESTIGATION OF DISORDERS OF CALCIUM METABOLISM

We have discussed disorders of calcium metabolism according to their aetiology in an attempt to explain the biochemical and clinical findings. The problem most often presents as one of the differential diagnosis of a sometimes unexpected finding of an abnormal plasma calcium level.

Plasma calcium should never be interpreted without taking account of albumin and phosphate levels, nor plasma phosphate without that of urea.

DIFFERENTIAL DIAGNOSIS OF HYPERCALCAEMIA

In the presence of hypercalcaemia three groups of causes must be differentiated:

raised protein-bound, with normal free ionised calcium;

raised free ionised calcium due to inappropriate PTH excess;

raised free ionised calcium due to other causes, and associated with appropriately low PTH levels.

We have found the following procedure to be simple and very effective. The diagnosis is often obvious before all steps have been followed.

A. Is the rise in the total plasma calcium level due only to a high protein-bound fraction?

1. What is the plasma calcium level?

2. *Is the plasma albumin level high?* If so and if the total calcium concentration is less than about 3 mmol/l (12 mg/dl), *in vivo* or *in vitro* haemoconcentration may be the cause. If the calcium is much higher than this, true hypercalcaemia is likely although the true calcium level may be lower than it seems.

   (*a*) If indicated, rehydrate the patient: this will correct *in vivo* haemo-concentration.

   (*b*) Take a specimen without venous stasis (p. 498) to eliminate *in vitro* haemoconcentration.

   (*c*) Repeat the plasma calcium and albumin assays. If there is still hypercalcaemia a cause for a high free ionised calcium must be sought.

   (*d*) If the calcium is now in the high "normal" range repeat the assays at three-monthly intervals, to exclude developing hypercalcaemia.

3. *Is the plasma albumin level low?* If so, plasma calcium levels within the high "normal" range may indicate a high free ionised fraction.

Many formulae have been proposed in an attempt to "correct" total calcium for abnormal protein levels. Unfortunately these may yield misleading results: there is good evidence that the relationship between albumin concentration and the amount of calcium bound is not simple.

B. What is the cause of a raised free ionised calcium concentration?

Hypercalcaemia in a specimen with a normal albumin concentration, or a plasma calcium in the high "normal" range in a specimen with hypoalbuminaemia, should be assumed to be due to a rise in the physiologically active free ionised fraction. The commonest causes are primary hyperparathyroidism and malignancy; the latter is usually, but not always, obvious on other grounds.

1. Take a careful drug history, with special reference to vitamin-D-containing preparations.

2. *Is the plasma phosphate low?*

   (*a*) If it is in the *low or low normal range* there is likely to be *inappropriate PTH secretion*.

   (*b*) If it is normal or high, *is the urea level in the same specimen high?* Phosphate and urea are both retained in glomerular failure. *If there is hypercalcaemia* Fig. 27 is a rough guide to the significance of the plasma phosphate level.

3. *If the level of phosphate is low in relation to that of urea* seek a cause of a PTH-like effect. The most important differential diagnosis is between primary hyperparathyroidism and malignancy.

   (*a*) A history of *renal calculi or peptic ulceration suggests chronic hypercalcaemia* and therefore primary hyperparathyroidism. Either of these conditions may be coincidental.

   (*b*) Repeat the clinical examination, paying special attention to palpation

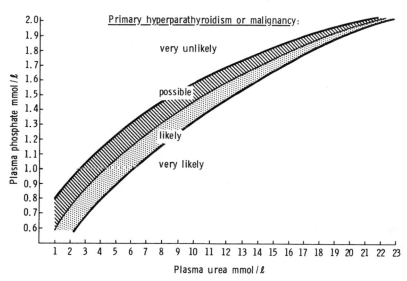

FIG. 27.—Plasma phosphate in relation to urea in the differential diagnosis of *PROVEN* hypercalcaemia. The lower the phosphate the greater, and the higher the phosphate the lower, the probability of primary hyperparathyroidism or malignancy. The differentiation between the two diagnoses can usually be made on non-chemical evidence (see text) or if doubt remains, by PTH assay. *Use this graph ONLY in cases with PROVEN hypercalcaemia.*

of the breasts and pelvic examination. Look at the chest X-ray for evidence of carcinoma of the bronchus.

(c) If myeloma is suspected, and especially if the plasma or serum total protein level is high despite a normal or low albumin, perform serum and urine electrophoresis.

(d) A very high plasma alkaline phosphatase activity is unlikely to be due to primary hyperparathyroidism; it may be slightly raised if there is radiological evidence of bone disease, but a normal level is the rule. If it is very high it suggests malignancy, or some other intercurrent disease. Many elderly subjects have a high alkaline phosphatase, perhaps because of Paget's disease of bone.

(e) If all the findings in (b), (c) and (d) are negative, if the haemoglobin level and ESR are normal, and especially if the history suggests chronic hypercalcaemia, primary hyperparathyroidism is by far the most likely cause. We rarely find other investigations necessary.

(f) In rare cases with a plasma calcium level of 3 mmol/l (12 mg/dl) or above, a steroid suppression test may help (p. 284). Failure of suppression in a case without obvious malignancy is highly suggestive of primary hyperparathyroidism. Results are difficult to interpret at lower calcium levels.

(g) PTH assay, in our experience, rarely helps in making a diagnosis (p. 268). Remember:

the results can only be interpreted if there is concomitant hypercalcaemia; a high PTH level may be appropriate in its absence;

failure to detect a high PTH level does not exclude primary hyperparathyroidism because the circulating hormone may not be immunoreactive PTH. The finding of a high level does not prove primary hyperparathyroidism.

4. *Is the plasma phosphate high in relation to the urea?*

(a) By far the most common cause is *ingestion of vitamin D*, alone or in multivitamin preparations, of which the patient may be unaware. Some people seem to be sensitive to diets high in calcium and vitamin D. A fall in both calcium and phosphate during a supervised stay in hospital suggests vitamin D ingestion as a likely cause.

(b) Look for evidence of *sarcoidosis*.

(c) Is there severe thyrotoxicosis?

(d) In the infant consider "idiopathic" hypercalcaemia (p. 265).

In all these conditions the calcium levels fall to normal during a steroid suppression test.

It is unwise to attempt parathyroidectomy on hypercalcaemic patients unless the plasma phosphate level is low in relation to that of urea.

DIFFERENTIAL DIAGNOSIS OF HYPOCALCAEMIA

As in the case of hypercalcaemia, the causes of hypocalcaemia fall into three groups:

reduced protein-bound, with normal free ionised, calcium;

reduced free ionised calcium due to primary PTH deficiency;

reduced free ionised calcium due to other causes and associated with appropriately high PTH levels.

A. Is the fall in total plasma calcium level due only to a low protein-bound fraction?

A low plasma albumin level may be due to overhydration, or to any condition associated with prolonged low protein intake, protein loss, failure to synthesise albumin (liver disease) or redistribution of albumin ("shock"). These cases should not be given calcium or vitamin D unless there is clinical, radiological or other evidence of an associated low free ionised level.

B. What is the cause of a low free ionised calcium level?

1. *Is the plasma phosphate low?* If so, calcium deficiency with normal secretion of PTH in response to feedback to the parathyroid gland is likely.

   (*a*) *Is the alkaline phosphatase level high?* This finding suggests prolonged secondary hyperparathyroidism due to calcium deficiency.

   (*b*) *Do bone X-rays show signs of osteomalacia?* This confirms very prolonged calcium deficiency.

   (*c*) Look for causes of malnutrition, especially malabsorption.

2. *Is the plasma phosphate high?* If so:

   (*a*) If *the urea level is high* glomerular failure is the likely cause.

   (*b*) If the urea level is normal hypoparathyroidism is the most likely cause. Is there a history of an operation on the neck?

   (*c*) Hypoparathyroidism may be autoimmune in origin. If there is doubt it may usually be distinguished from the even rarer "pseudo-hypoparathyroidism" by PTH assay (low levels in true hypoparathyroidism and very high levels in the end-organ unresponsiveness of pseudohypoparathyroidism).

TREATMENT

HYPERCALCAEMIA

Emergency Treatment of Hypercalcaemia

The solution for intravenous infusion contains a mixture of mono- and dihydrogen phosphate such that the pH is 7·4.

Na_2HPO_4 (anhydrous)—11·50 g
KH_2PO_4 (anhydrous)— 2·58 g } made up to 1 litre with water.

500 ml of this should be infused over 4–6 hours. This 500 ml will contain a total of 81 mmol of sodium, 9·5 mmol of potassium and 50 mmol of phosphorus.

Long-Term Oral Phosphate Treatment of Hypercalcaemia

Oral phosphate is given as the disodium or dipotassium salt. The choice depends on the serum potassium level.

1. The *solution* should contain 32 mmol of phosphorus in 100 ml.

This is   46 g/litre of Na_2HPO_4 (anhydrous)
or   56 g/litre of K_2HPO_4 (anhydrous)

The dose is 100–300 ml per day in divided doses. 100 ml contains approximately 65 mmol of sodium or potassium respectively.

2. *Phosphate Sandoz Effervescent tablets* contain:

Phosphorus	16 mmol (500 mg)	
Sodium	20 mmol	per tablet
Potassium	3 mmol	

The dose is 1–6 tablets daily. These *must* be dissolved in water, as directed, to reduce the osmotic concentration and therefore the danger of diarrhoea.

3. Alternatively the phosphate can be given as *sodium cellulose phosphate* 5 g (the contents of one sachet), three times a day, dissolved in water.

Note that hypokalaemia is a common accompaniment of hypercalcaemia. When this is present the phosphate preparation of choice is K_2HPO_4.

HYPOCALCAEMIA

Emergency Treatment of Hypocalcaemia with Severe Tetany

Calcium Gluconate Injection (B.P.)—0.22 mmol (10 mg) calcium/ml.
Dose—10 ml intravenously in the first instance.

Long-Term Treatment of Hypocalcaemia

Vitamin D Therapy
Warning.—There may be several days' lag in response to either starting or stopping therapy with vitamin D: plasma calcium levels may continue to rise for weeks after stopping therapy. Treatment should be carried out with caution, using intelligent anticipation based on laboratory assessment of

plasma calcium and, in osteomalacia, alkaline phosphatase levels. *Patients on maintenance doses should be seen at regular intervals, because requirements may change, with the danger of the development of hypercalcaemia.* Calcium levels must be assessed together with those of plasma albumin.

Osteomalacia with raised alkaline phosphatase levels.—1. *Initially*— 12·5 mg (500 000 IU) daily orally or IM. Rarely 18·75 to 25 mg (750 000 to 1 000 000 IU) may be needed.

2. When the plasma calcium level reaches about 1·75 mmol/l (7·0 mg/dl), and the alkaline phosphatase is near normal, reduce to 100 000 to 250 000 (2·50–6·25 mg) daily. At this stage bone calcification is nearing normal, and hypercalcaemia may develop rapidly if high dosage is continued.

3. *Maintenance doses* vary between 0·63 and 2·5 mg (25 000 and 100 000 IU), and are determined by trial and error. The aim should be to keep plasma calcium, phosphate, and alkaline phosphatase levels normal.

Hypocalcaemia of hypoparathyroidism with normal alkaline phosphatase levels.—The lower dosage of 100 000 IU (2·50 mg) daily should be used from the outset, monitoring being based on plasma calcium levels; again, the exact dose is found by trial and error.

Oral Calcium Tablets

Calcium Gluconate (B.P.) (600 mg) = 1·35 mmol (54 mg) of calcium per tablet.

Calcium Gluconate Effervescent (B.P.) (1000 mg) = 2·25 mmol (90 mg) of calcium per tablet.

"Sandocal" Effervescent (4·5 g calcium gluconate) = 10 mmol (400 mg) of calcium per tablet.

Calcium Lactate (B.P.) (300 mg) = 1·4 mmol (55 mg) of calcium per tablet.

Calcium Lactate (B.P.) (600 mg) = 2·8 mmol (110 mg) of calcium per tablet.

MAGNESIUM

Emergency Treatment of Magnesium Deficiency

Magnesium Chloride ($MgCl_2.6H_2O$)—20 g/100 ml.

This contains 1 mmol of magnesium per ml and may be added to other intravenous fluid. If renal function is normal, up to 40 mmol (40 ml) may be infused in 24 hours.

Oral Magnesium Therapy

Magnesium Chloride ($MgCl_2.6H_2O$)—20 g/100 ml.

The dose is 5–10 ml, q.d.s. Each 5 ml contains 5 mmol of magnesium. Note, however, that magnesium is poorly absorbed and may therefore cause osmotic diarrhoea.

TEST PROTOCOL

STEROID SUPPRESSION TEST IN DIFFERENTIAL DIAGNOSIS OF HYPERCALCAEMIA

All specimens for calcium estimations should be taken without venous stasis.

Procedure

1. At least two specimens are taken on two different days before the test starts for estimation of plasma calcium and albumin concentrations.

2. The patient takes 120 mg of oral hydrocortisone a day in divided doses (40 mg, 8-hourly), for 10 days.

3. Blood is taken for estimation of plasma calcium and albumin concentrations on the 5th, 8th and 10th day.

4. After the 10th day hydrocortisone is stopped.

Interpretation

Failure of plasma calcium concentration to fall to within the normal range by the end of the test is strongly suggestive of *primary hyperparathyroidism*. For exceptions see p. 269.

During administration of steroids there may be significant fluid retention. This will dilute the protein-bound calcium and may lead to an apparent fall of calcium. This effect must be allowed for when the calcium results are interpreted.

Chapter XII

INTESTINAL ABSORPTION: PANCREATIC AND GASTRIC FUNCTION

THE most important function of the gastro-intestinal tract is the digestion and absorption of nutrients. Some nutrients enter the intestinal tract from the body and net absorption is less than true absorption: for instance, dietary fat is almost completely absorbed under physiological circumstances, and almost all that found in normal faeces is derived from intestinal cells—a fact of importance in interpretation of faecal fat values (p. 310).

Digestion of larger molecules involves the action of intestinal enzymes: if these are to act effectively the nutrient molecules must be dispersed as much as possible, firstly by mechanical action such as chewing, and secondly by mixture of the food with large amounts of fluid of optimal electrolyte and hydrogen ion concentrations. We saw in Chapters II, III and IV that most fluid entering the gastro-intestinal tract is derived from plasma by passive ultrafiltration; active processes produce different bicarbonate, hydrogen ion, and chloride concentrations at different levels of the tract, so facilitating the action of digestive enzymes and the active transport of products of digestion from the lumen into the extracellular fluid.

Maintenance of normal fluid, solute and hydrogen ion balance depends on active reabsorption of about 99 per cent of this large volume of filtrate, and on distal bicarbonate : chloride exchange. Disturbances of water, electrolyte and hydrogen ion homeostasis are common in diarrhoea due to extensive small intestinal or colonic disease, or when there is loss of a large amount of fluid and electrolyte from the upper intestinal tract through vomiting or fistulae. Such disturbances also occur if absorption from the upper intestinal cells is so grossly impaired that the amounts of fluid and electrolyte entering distal parts exceed the reabsorptive capacity.

In this chapter we are concerned with disturbances of the absorptive mechanisms, either due to abnormality of the absorptive cells of the small intestine, or to failure of normal digestion of food. Unless gross malabsorption causes severe intestinal hurry, water, electrolyte and hydrogen ion disturbances are relatively unimportant in malabsorption syndromes. The effects are largely those of disturbed nutrition, since nutrient cannot pass through the intestinal cells.

NORMAL DIGESTION AND CONVERSION OF NUTRIENT TO AN ABSORBABLE FORM

Complex molecules such as those of protein, polysaccharide and fat are usually broken down by digestive enzymes. This process starts in the mouth, where food is mechanically broken down by chewing and is mixed with saliva containing α-amylase. In the stomach further fluid is added and the low pH initiates protein digestion by *pepsin*. The stomach also secretes intrinsic factor necessary for absorption of vitamin B_{12}. However, quantitatively by far the most important digestion takes place in the duodenum and upper jejunum, where alkaline fluid is added to the already liquid food. Pancreatic enzymes in this fluid convert protein to amino acids and small peptides, polysaccharides to mono- and disaccharides, and oligosaccharides (consisting of a small number of monosaccharide units), and fat to monoglycerides and fatty acids.

Severe generalised malabsorption due to failure of *digestion* is rare, but if present is most commonly due to pancreatic disease.

NORMAL ABSORPTION

Normal absorption depends on:

the integrity and normal surface area of absorptive cells;
the presence of nutrient in an absorbable form (and therefore on normal digestion);
a normal ratio of speed of absorption to speed of passage of contents through the intestinal tract.

The *absorptive area* of the small intestine is normally very large. Macroscopically the mucosa forms *folds*, increasing the area considerably. Microscopically, these folds are covered with *villi* lined with absorptive cells: this further increases the area about eight-fold.

If these villi are flattened, as they are in, amongst other conditions, gluten-sensitive enteropathy, the absorptive area is much reduced.

Each intestinal absorptive cell has on its surface a large number of minute projections (*microvilli*), detectable with the electron microscope, further increasing absorptive area about 20-fold.

Minute spaces exist between the microvilli (*microvillous spaces*).

To be absorbable, the substances must be in the form of relatively *small molecules*, such as result from normal digestion. Vitamin B_{12} can only be absorbed after forming a complex with intrinsic factor, which, by binding to intestinal cells, brings the vitamin molecule into a spatially advantageous position for absorption. The method of absorption depends on whether a molecule is *water soluble* or *lipid soluble*. Lipid-soluble nutrients share the mechanisms for fat absorp-

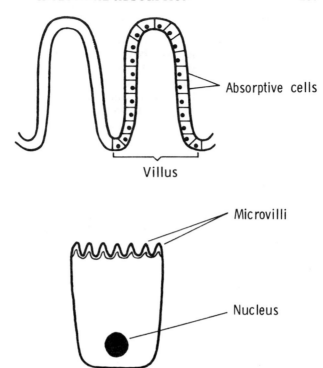

Absorptive cells

Villus

Microvilli

Nucleus

tion. Absorption may be *active*, in which case it can occur against a physicochemical gradient, or *passive* when it passes along physico-chemical gradients.

LIPID ABSORPTION

Digestion of Neutral Fats

Triglycerides are esters of glycerol with three, usually different, fatty acids.

Bile salts emulsify fat entering the duodenum. They inhibit pancreatic *lipase*, but this inhibition is counteracted by a peptide coenzyme, *colipase*. Colipase also reduces the pH optimum of lipase from about 8·5 to about 6·5. A lipase/colipase/bile-salt complex forms at the fat/water interface, and the lipase catalyses the hydrolysis of the fatty-acid/glycerol bonds, especially in positions 1 and 3 of the triglyceride molecule. The end result is production of some diglycerides, but mainly the 2-monoglycerides together with free fatty acids.

Micelle Formation

The monoglycerides and fatty acids aggregate with bile salts to form a *micelle*: the micelle also contains free *cholesterol* (liberated by hydrolysis of cholesterol esters in the lumen) and *phospholipids*, as well as *fat-soluble vitamins* (A, D and K). The diameter of the negatively charged micelle is between 100 and 1000 times smaller than that of the emulsion particle. Both its small size and its charge allow it to pass through the microvillous spaces.

Lipids in the Intestinal Cell

In the intestinal cell triglycerides are resynthesised from monoglycerides and fatty acids. Cholesterol is re-esterified. The triglycerides, cholesterol esters and phospholipids, together with the fat-soluble vitamins, are coated with a layer of apoprotein (manufactured in the cell) to form *chylomicrons* (p. 235). These are readily suspended in water and pass into the lymphatic circulation.

Absorption of Free Fatty Acids

Some short- and medium-chain free fatty acids pass through the intestinal cell into the portal blood stream.

From this account it will be seen that absorption of *lipids and fat-soluble vitamins* depends on normal:

> emulsification of fats, and therefore on the presence of *bile salts*;
> digestion of triglyceride by *lipase* and therefore on normal pancreatic function;
> *absorptive area of the intestinal mucosa* for chylomicron formation.

Primary bile acids (p. 327) are synthesised in the liver from cholesterol. They enter the intestinal lumen in the bile conjugated with glycine or taurine. In the alkaline duodenal fluid they form sodium salts, which, after bacterial conversion to secondary bile salts, are actively reabsorbed in the *distal ileum*. They are recirculated to the liver and resecreted into the bile (the "enterohepatic circulation"). Some bacteria contain enzymes which catalyse luminal deconjugation of bile salts: unconjugated bile salts emulsify fat less effectively than conjugated ones.

CARBOHYDRATE ABSORPTION

Polysaccharides such as starch and glycogen are hydrolysed by salivary and pancreatic α-amylase, the latter being of the greater importance, mainly to 1:4 disaccharides such as maltose (glucose +glucose): a few larger branch-chain saccharides remain.

Disaccharides (maltose, sucrose (glucose+fructose) and lactose (glucose+galactose)) are hydrolysed to their constituent monosaccharides by the appropriate disaccharidase (maltase, sucrase or lactase). Sucrase also hydrolyses the 1:4, and isomaltase the 1:6 linkages of limit dextrins. These enzymes are not present in significant amount in intestinal secretions, but are located on the surface ("brush border") of the intestinal cell where the hydrolysis takes place: concentrations are highest in the proximal jejunum.

Monosaccharides are absorbed in the duodenum. Glucose and galactose are probably absorbed by a common active process, while fructose is absorbed by a different mechanism.

Thus carbohydrate absorption depends on:

the presence of *amylase*, and therefore on normal pancreatic function (polysaccharides only);

the presence of *disaccharidases* on the luminal membrane of the intestinal cell (disaccharides);

normal *intestinal mucosa* with normal active transport mechanisms (monosaccharides).

Note that absorption of polysaccharides requires normal function of all three mechanisms.

PROTEIN ABSORPTION

The diet is not the only source of protein in the intestinal lumen: a significant proportion of that absorbed originates from intestinal secretions and desquamated mucosal cells.

Ingested protein is broken down by pepsin, followed by trypsin and the other proteolytic enzymes of pancreatic juice: pancreatic trypsinogen is converted to the active trypsin by enterokinase on the luminal cell surface. The products of digestion are amino acids and small peptides. Many peptides are further hydrolysed by peptidases on the brush border.

Amino acids are actively absorbed in the small intestine.

Small peptides are actively absorbed into the cell intact: their absorption is independent of that of amino acids and, with a few exceptions, they are hydrolysed intracellularly.

Protein absorption therefore depends on:

the presence of *pancreatic proteolytic enzymes* and therefore on normal pancreatic function;

normal *intestinal mucosa* with normal active transport mechanisms.

VITAMIN B_{12} ABSORPTION

Vitamin B_{12} can be absorbed only when it has formed a complex with *intrinsic factor*, a glycoprotein secreted by the stomach: in this form it can bind to the intestinal cells in the distal ileum, where it is absorbed. Its absorption therefore depends on normal:

> *gastric secretion of intrinsic factor;*
> *intestinal mucosa in the distal ileum.*

Some intestinal bacteria require vitamin B_{12} for growth and probably prevent its absorption by competing for it with intestinal cells. Normal absorption therefore also depends on normal *intestinal flora*.

ABSORPTION OF OTHER WATER-SOLUBLE VITAMINS

Most of the water-soluble vitamins (C and all those of the B group except B_{12}) are absorbed, probably by specific transport mechanisms, mainly in the upper small intestine. Clinical deficiencies of all but folate are relatively uncommon in malabsorption syndromes, probably because absorption of these vitamins is independent of that of fat.

CALCIUM AND MAGNESIUM ABSORPTION

Calcium is actively absorbed in the upper small intestine, particularly the duodenum and upper jejunum. Normal calcium absorption depends on its presence in a free ionised form (it is inhibited by the formation of insoluble salts with fatty acids and phosphate), and on the presence of vitamin D in the form of 1,25-dihydroxycholecalciferol (p. 254). Much of the faecal calcium is endogenous and is derived from calcium in intestinal secretions.

Magnesium is also absorbed by an active process which may be shared with calcium.

Normal calcium and magnesium absorption depends on:

> a *low concentration of fatty acids* and phosphate in the intestine;
> the absorption and metabolism of *vitamin D* and therefore on normal fat absorption;
> normal *intestinal mucosa*.

IRON ABSORPTION

Iron is absorbed by an active process in the duodenum and upper jejunum, and absorption is stimulated by anaemia (p. 417). Some of the iron absorbed into the intestinal cell enters the blood but some is lost into the lumen with desquamated cells (p. 415).

Intestinal handling of *electrolytes and water* is discussed in the first four chapters of this book.

We are now in a position to understand the consequences of disorders of digestion and absorption. From the above account it should be noted that:

> *pancreatic failure* affects absorption of large molecules only (fats and substances sharing fat absorption, polysaccharides and protein);
>
> *disease of the intestinal mucosa* affects absorption of small molecules and products of digestion of large molecules;
>
> absence of *bile salts* causes malabsorption of fats and of those substances sharing mechanisms for fat absorption.

MALABSORPTION SYNDROMES

Clinically the malabsorption syndromes can be divided into those associated with generalised malabsorption of fat (*steatorrhoea*), protein, carbohydrate and other nutrients, and those associated with failure to absorb one or more specific substances.

GENERALISED MALABSORPTION (ASSOCIATED WITH STEATORRHOEA)

Intestinal Disease

Reduction of absorptive surface or *general impairment of transport mechanisms*:

due to villous atrophy:

gluten sensitivity causing coeliac disease;
tropical sprue;
idiopathic steatorrhoea;

extensive surgical resection of small intestine.

Extensive infiltration or inflammation of the small intestinal wall (for example, Crohn's disease, Whipple's disease or small intestinal lymphoma).

Increased rate of passage through the small intestine:

post-gastrectomy;
carcinoid syndrome.

Pancreatic Dysfunction (Failure of Digestion)

Chronic pancreatitis.

Fibrocystic disease of the pancreas.

Inactivation of lipase by low pH in the Zollinger-Ellison syndrome (p. 469).

FAILURE OF ABSORPTION OF SPECIFIC SUBSTANCES

Altered bacterial flora (mainly fat, p. 288; and vitamin B_{12}, p. 290):

"contaminated bowel" ("blind-loop") syndrome due to:
diverticula;
surgery;
broad-spectrum antibiotic therapy.

Biliary obstruction (fat).

Local disease or surgery (for instance, disease of the distal ileum affects vitamin B_{12} absorption and bile-salt reabsorption).

Gastric atrophy causing malabsorption of vitamin B_{12} (pernicious anaemia).

Disaccharidase deficiency (carbohydrate):

congenital;
acquired.

Protein-losing enteropathy.

Isolated transport defects resulting from inborn metabolic errors (p. 400).

GENERALISED MALABSORPTION

Due to villous atrophy.—Malabsorption may be due to reduction of the absorptive area due to flattening of the intestinal villi; histological demonstration of flat villi on biopsy specimens is the definitive evidence for this type of disease.

Gluten-sensitive enteropathy (coeliac disease) occurs at any age. Sensitivity to gluten (in wheat germ) causes the villous atrophy. The flattening of the villi, and therefore the symptoms, are reversible after treatment with a gluten-free diet: the response may be delayed for some months.

In *tropical sprue* there is also atrophy of the villi, but there is no response to a gluten-free diet. The condition does respond to broad-spectrum antibiotics and folate: this suggests that a bacterial factor is important in the aetiology of the disease.

The term *idiopathic steatorrhoea* should be reserved for those cases of steatorrhoea with flattened villi which do not respond either to a gluten-free diet or to broad-spectrum antibiotics.

Extensive surgical resection of the small intestine may so reduce the absorptive area as to cause malabsorption.

Extensive infiltration and inflammation of the small intestinal mucosa may impair its ability to carry out absorptive processes. This may be aggravated by the presence of altered bacterial flora in these conditions.

After gastrectomy normal mixing of food with fluid, acid and pepsin does not occur in the stomach, so that the activity of enzymes in the small intestine is less efficient than usual. Passage of intestinal contents through the duodenum is also more rapid than usual. In spite of this, post-gastrectomy malabsorption is rarely severe, although iron deficiency is relatively common.

The carcinoid syndrome (p. 466) is associated with excessive production of 5-hydroxytryptamine (5-HT) by tumours of argentaffin cells usually arising in the small intestine, and by their metastases, usually in the liver. This is a rare disease which even more rarely presents a problem of differential diagnosis of malabsorption. The malabsorption is probably the result of increased intestinal motility induced by 5-HT.

In pancreatic disease (due most commonly to *chronic pancreatitis*) failure of absorption is due to failure of digestion and predominantly affects large molecules. *Fibrocystic disease of the pancreas (cystic fibrosis; mucoviscidosis)* is an inherited disease, and usually presents in early childhood but, more rarely is first recognised in the adult patient. Pancreatic and bronchial secretions are viscid and, by blocking pancreatic ducts and bronchi, cause obstructive disease of these organs. Sweat glands are also affected and the diagnosis depends on the demonstration of an *increased sweat sodium concentration,* often to about twice normal. The patient may present with pulmonary disease, or with malabsorption.

Detection of Steatorrhoea

In intestinal malabsorption fat digestion is normal, but absorption of the products of digestion is impaired. In the rarer pancreatic steatorrhoea the intestinal wall is normal but fat cannot be digested. In either case absorption of fat is impaired. However, unequivocal steatorrhoea (a daily fat excretion consistently more than 18 mmol, or 5 g of "fat", measured as fatty acids) can only be demonstrated when disease is extensive and has usually already been elucidated by radiological or histopathological means, and the test then contributes no clinically useful information. If there is gross steatorrhoea the stools are bulky, pale and greasy, and there may be diarrhoea. Faecal fat assay is not only an insensitive test for small intestinal or pancreatic disease, but also, because of the difficulty of ensuring an accurately timed faecal collection, imprecise (p. 309).

Absorbed fat is ultimately metabolised to carbon dioxide and water. If ^{14}C-labelled triglyceride is given by mouth the ratio of $^{14}CO_2$ to unlabelled CO_2 can be measured in expired air. A low ratio should indicate impaired fat absorption. This test eliminates the imprecision due to collection of timed specimens. However, various assumptions have to be made when interpreting results, and the assay needs expensive isotopes and special expertise. Its value as a routine test remains to be proved.

Results of Steatorrhoea

Severely impaired fat absorption is accompanied by malabsorption of the substances mentioned on p. 288. *Vitamin D deficiency* impairs calcium absorption, and this is one of the causes of a low level of circulating free ionised calcium and of osteomalacia in malabsorption syndromes; calciferol is derived from the skin, and overt vitamin D, and therefore calcium deficiency is most likely if, in addition to malabsorption, calcium requirements are increased during growth or pregnancy, or if the patient is house-bound. If ionised calcium levels have been low long enough to cause osteomalacia the *plasma alkaline phosphatase* activity will rise. Secondary hyperparathyroidism due to the low free ionised calcium level causes phosphaturia with a low plasma phosphate (p. 262). Rarely, the free ionised calcium levels fall low enough to cause *tetany*.

Vitamin K is required for hepatic synthesis of prothrombin and other clotting factors, and patients with severe malabsorption may develop a bleeding tendency associated with a prolonged prothrombin time: this prothrombin deficiency, unlike that of liver disease, can be reversed by parenteral administration of vitamin K. *Vitamin A* deficiency is rarely clinically evident, although malabsorption of an oral dose of vitamin A can sometimes be demonstrated.

Amino acid and peptide malabsorption occurs in intestinal disease because intestinal transport mechanisms are impaired: in pancreatic disease the digestion of protein is severely impaired. In either case prolonged disease may cause generalised *muscle and tissue wasting* and *osteoporosis* (p. 267), and a reduced level of all protein fractions in the blood. The low albumin may cause *oedema* (p. 343) and result in a reduction of *protein-bound calcium*. The total calcium level may therefore give a false idea of the severity of the hypocalcaemia.

Reduced *antibody formation* (immunoglobulins) may predispose to infection.

Thus in *severe or long-standing* generalised malabsorption, whether pancreatic or intestinal, the *clinical picture* may be:

bulky, fatty stools, with or without diarrhoea;
generalised wasting and malnutrition (protein deficiency);
osteoporosis and osteomalacia (protein and calcium deficiency);
oedema (hypoalbuminaemia due to protein deficiency);
haemorrhages (vitamin K deficiency);
tetany (hypocalcaemia due to vitamin D and calcium deficiency);
recurrent infections (immunoglobulin deficiency).

The *laboratory findings* may be:

increased excretion of fat in the stools;
anaemia (iron, and/or folate, and/or vitamin B_{12} deficiency);
hypocalcaemia (free ionised and protein-bound)—with hypophosphataemia;
raised plasma alkaline phosphatase levels in cases with osteomalacia;
generalised lowering of the concentration of all serum protein fractions:
prolonged prothrombin time.

The complete picture is only seen in advanced cases of the syndrome. Milder cases present with vague abdominal symptoms and perhaps anaemia, but with normal chemical findings.

Differential Diagnosis of Generalised Intestinal and Pancreatic Malabsorption

The differential diagnosis of steatorrhoea is usually made by clinical, haematological, histological and radiological means, which may be supplemented by chemical tests.

When considering the significance of the results of laboratory tests, remember that pancreatic malabsorption affects mainly those large molecules which must be digested before absorption.

Malabsorption of fat occurs in both types of disease.

Anaemia is more common in intestinal than in pancreatic malabsorption, since iron, vitamin B_{12} and folate are absorbed without digestion by pancreatic enzymes. In intestinal malabsorption the blood and bone-marrow films typically show a mixed iron deficiency and megaloblastic picture. Malabsorption of iron, vitamin B_{12} and folate may be demonstrable and the absorption of vitamin B_{12} is not increased by the simultaneous administration of intrinsic factor, as it is in pernicious anaemia (see p. 300). Anaemia may be aggravated by protein deficiency.

Differences in carbohydrate metabolism.—Polysaccharide absorption is impaired in both conditions. However, some of the carbohydrate in the diet is in the form of mono- and disaccharides and these can be absorbed in pancreatic disease when neither intestinal disaccharidase activity nor active monosaccharide absorption is affected. Hypoglycaemia is rare in either condition. Significant loss of pancreatic function is part of a disease process affecting the gland as a whole, including the insulin-secreting β-cells: by contrast, diabetes mellitus due to β-cell dysfunction, but with apparently normal exocrine secretion is very common. Although hyperglycaemia is suggestive, *it is by no means diagnostic of a pancreatic origin* of malabsorption.

Xylose absorption test.—Xylose is a pentose which, like glucose, can be absorbed without digestion, but the metabolism of which is not controlled by hormones: for this reason, once xylose has been absorbed the levels are, unlike those of glucose, not affected by such endocrine diseases as diabetes mellitus.

An oral dose of xylose will be absorbed by normally functioning upper small intestinal cells. Metabolism of the absorbed pentose is very slow, and, because it is freely filtered by the glomeruli, most of it will appear in the urine. In the xylose absorption test either the amount excreted, or the plasma level at a defined time after the dose, is measured. In theory, pancreatic dysfunction should not affect xylose absorption and excretion, whilst intestinal disease should impair it. Unfortunately, whether the test is based on plasma or on urine assays, there are several sources of imprecision (p. 310):

Xylose is mostly absorbed in the upper small intestine and absorption may therefore be normal despite significant ileal dysfunction (for example in Crohn's disease).

The urine test depends on accurate collection over two and five hours. Potential errors are even greater than those in tests depending on 24-hour collections.

Even mildly impaired renal function decreases glomerular filtration of xylose and may give a falsely low result of the urine test or, perhaps, a falsely high plasma level. This is a common cause of misleading results in the elderly. The xylose absorption test should not be performed if the plasma urea and/or creatinine levels are even marginally raised.

Both plasma levels and urine excretion depend partly on the volume through which the absorbed xylose is distributed. In oedematous or very obese patients results may be falsely low.

Tests of Exocrine Pancreatic Function

Plasma enzymes.—*Plasma trypsin* can now be measured by

immunoassay and the finding of low levels may help in the diagnosis of fibrocystic disease of the pancreas. *Plasma amylase* in normal subjects consists mostly of the salivary, not the pancreatic, isoenzyme: the total enzyme activity is therefore not significantly lowered when cell mass is reduced by chronic pancreatic disease and the assay is of no value in the diagnosis of such conditions. However, in acute pancreatitis, release of enzyme from damaged cells usually increases *total* plasma activity (see below).

Faecal trypsin levels are extremely variable, probably because of bacterial action. The estimation is of no value in diagnosis of pancreatic hypofunction in adults. In infants with diarrhoea its absence is suggestive of fibrocystic disease of the pancreas, although plasma levels are more reliable.

Duodenal enzymes.—Measurement of pancreatic enzymes and bicarbonate in duodenal aspirate before and after stimulation of the pancreas with secretin or pancreozymin (hormones stimulating the pancreas which are normally secreted during digestion) is not very suitable for routine use because of difficulty in positioning the duodenal tube correctly and in quantitative sampling of the secretions. In the *Lundh test* a mixture of corn oil, milk powder and glucose is given, and samples of duodenal secretion analysed for trypsin: this test, too, is only useful if performed by an expert.

Disorders of the Pancreas Rarely Associated with Malabsorption

It is convenient to digress to discuss three relatively common pancreatic diseases which are only rarely associated with malabsorption.

Carcinoma of the pancreas is very difficult to diagnose by laboratory tests unless the lesion is in the head of the organ and causes obstructive jaundice: extensive gland destruction may cause late-onset diabetes.

In *acute pancreatitis* necrosis of pancreatic cells results in release of their enzymes into the peritoneal cavity and blood stream. The presence of pancreatic juice in the peritoneal cavity causes *severe abdominal pain* and *shock*: this picture is common to many acute abdominal emergencies. A vicious cycle is set up as more pancreatic cells are digested by the released enzymes.

Acute pancreatitis is most commonly the result of obstruction of the pancreatic duct, or of regurgitation of bile along this duct. The most important predisposing factors are *alcoholism* and *biliary-tract disease*. *Trauma* to the pancreas, by damaging the cells, may also initiate the vicious cycle. *Hypercalcaemia* and *hypertriglyceridaemia* may precipitate acute pancreatitis.

Typically plasma amylase values increase five-fold or more. However, it is important to realise that concentrations of up to, and even above, this value may be reached in any acute abdominal emergency, but especially after gastric perforation into the lesser sac: very high levels may occur in renal glomerular failure, or in macroamylasaemia (p. 374), when the enzyme cannot be excreted normally, and these cases are usually asymptomatic. Conversely, levels in acute pancreatitis may not be very high, and usually fall very rapidly as the enzyme is lost in the urine. High plasma amylase levels are therefore only a rough guide to the presence of acute pancreatitis and normal or only slightly raised values do not exclude the diagnosis.

Chronic pancreatic failure, with steatorrhoea, may rarely follow a severe attack: it is more likely to follow repeated acute or subacute attacks.

The Zollinger-Ellison syndrome is discussed on p. 469.

The Post-Gastrectomy Syndrome

Malabsorption after gastrectomy is usually mild. However, rapid passage of the contents of the small gastric remnant into the duodenum may have two clinical consequences:

the **"dumping syndrome"**.—Soon after a meal the patient may experience abdominal discomfort and feel faint and sick. The syndrome is thought to be due to the sudden passage of fluid of high osmotic content into the duodenum. Before this abnormally large load can be absorbed, water passes along the osmotic gradient from the extracellular fluid into the lumen of the intestine. The reduction in plasma volume causes faintness and the large volume of duodenal fluid causes abdominal discomfort;

post-gastrectomy hypoglycaemia.—If a meal containing much glucose passes more rapidly than normal into the duodenum glucose' absorption is very rapid. The plasma glucose level rises suddenly and causes an outpouring of insulin. The resultant "over-swing" of plasma glucose concentration may cause hypoglycaemic symptoms which typically occur at about two hours after a meal.

Both these disabilities can be mitigated if meals low in carbohydrate content are taken "little and often".

FAILURE OF ABSORPTION OF SPECIFIC SUBSTANCES

Altered Bacterial Flora (Malabsorption of Vitamin B_{12} and Fat)

The **"contaminated bowel"** (**"blind-loop"**) **syndrome** is associated with stagnation of intestinal contents with a consequent alteration of bacterial flora, and occurs when the loops are the result of surgery, or in the presence of small intestinal *diverticula*.

Treatment with broad-spectrum antibiotics may, by altering the balance of small intestinal bacteria, cause a similar syndrome.

Many intestinal bacteria need vitamin B_{12} for growth, and *megaloblastic anaemia* is common in these syndromes of bacterial colonisation. Some organisms contain enzymes which catalyse deconjugation of bile salts, and if the flora is altered, either quantitatively or qualitatively, so increasing the number of such bacteria, there may be *steatorrhoea* due to impaired emulsification of fat. Protein and carbohydrate are usually normally absorbed.

Biliary Obstruction (Malabsorption of Fat)

Bile salts are synthesised and secreted by the liver. In biliary obstruction these cannot reach the intestinal lumen in normal amounts and *steatorrhoea* results. Because of extreme jaundice there is rarely any difficulty in the differential diagnosis.

Differential Diagnosis of Steatorrhoea

All the malabsorptive conditions described so far may be associated with steatorrhoea. As has been mentioned, biliary obstruction is rarely a problem of diagnosis. The important points in laboratory findings in differential diagnosis of these conditions are summarised in Table XXVI.

No chemical test is sensitive or precise in the detection or differential diagnosis of generalised malabsorption or steatorrhoea. Recent developments such as endoscopy and ultrasound have reduced the need for laboratory tests which, however, continue to be important to detect

TABLE XXVI

DIFFERENTIAL DIAGNOSIS OF STEATORRHOEA

	Upper Small Intestinal Disease	*Pancreatic Disease*	*"Contaminated Bowel" Syndrome*
Anaemia	Mixed megaloblastic and iron deficiency common ("dimorphic anaemia")	Rare	Megaloblastic common
Intestinal Biopsy	May show flattened villi or other cause	Normal	Normal
Xylose absorption	Reduced	Normal	Usually normal

and to monitor treatment of those metabolic effects of prolonged and severe malabsorption which have been described on p. 294. *A proposed scheme for the investigation of suspected steatorrhoea is outlined on p. 307.*

Local Disease or Surgery

Local disease or resection may cause selective malabsorption of substances whose absorption occurs predominantly at these sites. The lower ileum is concerned with vitamin B_{12} absorption and with reabsorption of bile salts, and resection or disease (for example, Crohn's or tuberculosis) of this area may cause megaloblastic anaemia.

Pernicious Anaemia

Vitamin B_{12} cannot be absorbed in the absence of intrinsic factor. In pernicious anaemia antibodies to parietal cells and intrinsic factor may both be present, and the stomach cannot secrete this substance. Significant deficiency may also occur after total gastrectomy or if there is extensive malignant infiltration of the stomach.

The Schilling test.—Malabsorption of vitamin B_{12} can be demonstrated if a small dose of the radioactively labelled vitamin is given and its excretion in the urine measured. If the malabsorption is due to pernicious anaemia, administration of the labelled vitamin together with intrinsic factor results in normal absorption: if it is due to intestinal disease malabsorption persists. A "flushing" dose of non-radioactive vitamin B_{12} is given parenterally at the same time as, or just after, the labelled dose, to ensure quantitative urinary excretion. Haematological tests such as examination of blood and marrow films should have been completed before the vitamin B_{12} is given.

Disaccharidase Deficiency

Because disaccharidases are localised on the brush border, generalised disease of the intestinal wall usually causes a non-selective disaccharidase deficiency. This is relatively unimportant compared with general malabsorption and tests for this syndrome are therefore useful only in the absence of steatorrhoea, when a selective rather than a generalised malabsorption of carbohydrate may be present.

The *symptoms* of disaccharidase deficiency are those of the effects of unabsorbed, osmotically active sugars in the intestinal tract, and include faintness, abdominal discomfort and severe diarrhoea after ingestion of the offending disaccharide. (Compare the "dumping syndrome", p. 298.) Diarrhoea may be severe enough to cause volume depletion in infants. Stools are typically acid because bacteria metabolise sugars to acids; unabsorbed sugars may be detectable in the faeces, or disaccharides may be absorbed intact and appear in the urine.

Lactase deficiency.—*Acquired lactase deficiency* is much more common than the congenital form and is probably the commonest type of disaccharidase deficiency: it may not present until childhood, or even adult life, possibly in genetically predisposed subjects. It is common in non-Caucasians, especially the Chinese; this may be due to genetic factors, or may be secondary to a low dietary intake of dairy products, with consequent failure to induce synthesis of the enzyme after weaning.

Lactase deficiency associated with prematurity.—In premature infants lactase may not be present in normal amounts in intestinal cells. Initially these cases resemble the congenital ones. However, the sensitivity to milk usually disappears within a few days of birth.

Congenital lactase deficiency.—This is very rare. Infants present with severe diarrhoea soon after the introduction of milk feeds. The syndrome is cured by removal of milk and milk products from the diet.

Sucrase and isomaltase deficiency usually coexist.

Congenital sucrase-isomaltase deficiency is more common than congenital lactase deficiency.

Acquired sucrase-isomaltase deficiency and *maltase deficiency* of any kind are very rare.

The *diagnosis* of disaccharidase deficiency is made most reliably by estimation of the relevant enzymes in intestinal biopsy tissue.

The relevant disaccharide may be given orally and plasma glucose estimated as in the glucose tolerance test. If the disaccharide cannot be hydrolysed the constituent monosaccharides (glucose, or those converted to glucose in the intestinal cell) cannot be absorbed, and the curve is flat. The result should usually be compared with a glucose tolerance curve. If the patient experiences typical symptoms, or if a child excretes disaccharides in the urine, when the offending sugar is given, the comparison need not be made.

Radiological examination may assist in the diagnosis of disaccharidase deficiency. Barium is administered first without, and then with, the disaccharide. When the sugar is not absorbed the flocculation pattern of barium is altered by the osmotic effect.

Protein-losing Enteropathy

This is a very rare syndrome in which absorption is normal, but the intestinal wall is abnormally permeable to large molecules (like the glomerulus in the nephrotic syndrome). In many cases of malabsorption, whatever the aetiology, protein exudes into the lumen: isolated protein-losing enteropathy is not really a malabsorption syndrome, but is due to excessive loss of protein from the body into the gut. It occurs in a variety of conditions in which there is ulceration of the bowel, lymphatic obstruction and intestinal lymphangiectasis, or hypertrophic lesions of the bowel. The clinical picture, like that of the nephrotic syndrome (p. 358), is due to hypoalbuminaemia.

The diagnosis can be made by measuring the loss into the bowel after intravenous injection of a substance of a molecular weight approximating that of albumin: substances that have been used are radioactive polyvinylpyrrolidone (PVP), dextran or albumin. Typically the electrophoretic pattern is similar to that found in the nephrotic syndrome: as in that syndrome the relatively high molecular weight components of the α_2 fraction are retained in the blood stream while all other fractions are lost from the body.

GASTRIC FUNCTION

The important components of gastric secretion are *hydrochloric acid, pepsin* and *intrinsic factor*. All these factors have already been mentioned as being of importance in digestion and absorption, and loss of hydrochloric acid in pyloric stenosis has been discussed as a cause of metabolic alkalosis (p. 113).

Stimulation of gastric secretion occurs by two main pathways:

through the *vagus nerve*, which in turn responds to stimuli from the cerebral cortex, normally resulting from the sight, smell and taste of food. Hypoglycaemia can stimulate this pathway and this fact can be used to assess the completeness of vagotomy;

by *gastrin*, a hormone normally produced by G cells in the gastric antrum in response to the presence of food; it is carried by the blood stream to the parietal area of the stomach where it stimulates secretion, perhaps through the mediation of histamine. Acid in the pylorus in turn inhibits gastrin secretion, providing feedback control. Calcium stimulates gastrin secretion; this may explain the relatively high incidence of peptic ulceration in patients with chronic hypercalcaemia.

Histamine acts after binding to receptors on the surface of cells. There are probably two types of such receptors:

those on which antihistamines compete for the sites with histamine (H_1 receptors). These are found on smooth muscle cells;

those on which antihistamines have no effect. These *H_2 receptors* are found on *gastric parietal* cells.

HYPERSECRETION

Hypersecretion of gastric juice may be associated with *duodenal ulceration*, when it may be neurogenic in origin. However, there is overlap between acid secretion in normal subjects and in those with duodenal ulceration, and its estimation is of very limited diagnostic value in this condition.

In the rare *Zollinger-Ellison syndrome* (p. 469) acid secretion by the stomach is very high. The consequent ulceration of the stomach and upper small intestine may cause severe diarrhoea: the low pH, by inhibiting lipase activity, may cause steatorrhoea.

HYPOSECRETION

Hyposecretion of gastric juice occurs in *pernicious anaemia* (p. 300) and is probably the result of antibodies to the parietal cells of the gastric mucosa: this is by far the commonest cause, and the achlorhydria is usually "pentagastrin fast" (p. 311). In extensive *carcinoma of the stomach* and in *chronic gastritis* there may also be gastric hyposecretion. However, in none of these conditions is estimation of gastric acidity of help in diagnosis: the diagnosis of pernicious anaemia should be made on haematological grounds, and on the result of the Schilling test, and results in the other two conditions are too variable to be useful.

TESTS OF GASTRIC FUNCTION

Routine tests of gastric function involve measurement of the acid secretion by the stomach, either at rest or in response to stimuli. Passage of a tube into the stomach is unpleasant for the patient: moreover, valuable results will only be obtained when the operator is skilled and able to recognise when the specimens obtained are incomplete, either because of blockage or malpositioning of the tube. For these reasons the tests should only be carried out in those very rare cases when it is really necessary for diagnosis and management, and only by a skilled operator.

Resting Juice or Overnight Secretion

Gastric contents may be aspirated overnight and the total night secretion measured, or a timed specimen may be obtained without a stimulus and the hourly secretion determined. Both these secretions will usually be high in volume and acid content in duodenal ulceration. The diagnosis of the Zollinger-Ellison syndrome depends on finding a high fasting plasma gastrin level despite a very high rate of acid secretion in a one-hour basal collection (more than 15 mmol/hour).

Stimulation of Gastric Secretion

Direct stimulation of parietal cells is used to demonstrate achlorhydria, which is typically "pentagastrin fast" in pernicious anaemia: as already pointed out, the test is of limited use in the diagnosis of this condition, and is unpleasant for the patient. *Pentagastrin* is a pentapeptide consisting of the physiologically active part of the gastrin molecule. It acts directly on the parietal cells of the stomach and can be used to stimulate acid secretion.

Vagal stimulation of gastric secretion may be used in patients who are still symptomatic after vagotomy to test completeness of the section of the nerve. The stimulus is stress due to insulin-induced hypoglycaemia (compare the use of insulin to stimulate cortisol secretion). If vagotomy is complete there should be no acid secretion even when the plasma glucose level falls below 2·5 mmol/l (45 mg/dl), and when there is clinical evidence of hypoglycaemia, and therefore of stress.

Treatment of Hypersecretion

Until a few years ago only palliative therapy with antacids and carbenoxolone, or surgery, was available for peptic ulceration due to hypersecretion of acid. More recently the drugs *cimetidine* and *ranitidine* (Tagamet and Zantac) have been used. They compete with histamine for H_2 receptors, so directly reducing acid and pepsin secretion. The initial clinical response is usually good, but relapse is common when treatment is stopped.

SUMMARY

Intestinal Absorption

1. Normal intestinal absorption depends on adequate digestion of food (and therefore on normal pancreatic function), and on a normal area of functioning intestinal cells.

2. Normal digestion and absorption of fat depends on the presence of bile salts as well as of lipase.

3. Absorption of cholesterol, phospholipids and fat-soluble vitamins depends on normal triglyceride absorption.

4. In *intestinal* malabsorption there is malabsorption of small molecules, usually due to a reduced absorptive area.

5. In *pancreatic* malabsorption there is malabsorption of fats,

proteins and polysaccharides, but small molecules are usually absorbed normally.

6. An *abnormal bacterial flora* may cause steatorrhoea (because of deconjugation of bile salts), and megaloblastic anaemia (because of competition for vitamin B_{12}).

7. There may be steatorrhoea in biliary obstruction (because of lack of bile salts).

8. Selective malabsorption of vitamin B_{12} occurs in pernicious anaemia (deficiency of intrinsic factor).

9. Selective disaccharidase deficiencies cause malabsorption of disaccharide. These deficiencies are more commonly acquired than congenital in origin.

Pancreatic Function

1. Chemical tests for pancreatic hypofunction are unsatisfactory unless performed in special centres.

2. Acute pancreatitis is associated with a transient rise in plasma amylase levels. This enzyme can also reach a high concentration in many acute abdominal emergencies and in renal failure. The assay of total plasma amylase is useless for detecting chronic pancreatic disease.

Gastric Function

1. Hypersecretion of acid occurs in:

duodenal ulceration;
the Zollinger-Ellison syndrome.

2. Hyposecretion of acid ("pentagastrin fast") occurs in:

pernicious anaemia;
extensive gastric infiltration.

3. The parietal cells can be directly tested by stimulation with pentagastrin.

4. The vagal stimulation of gastric secretion can be tested by producing insulin-induced hypoglycaemia.

FURTHER READING

READ, N. W. and CORBETT, C. L. (1983). The function of the human gastro-intestinal tract and its laboratory assessment. In: *Biochemistry in Clinical Practice*, pp. 67–95. D. L. Williams and V. Marks, Eds. (Scientific Foundations of Clinical Biochemistry, Vol. II.) London: Heinemann Medical.

MARSH, M. N. (1981). The small intestine: mechanisms of local immunity and gluten sensitivity. *Clin. Sci.*, **61**, 497–503.

THEODOSSI, A. and GAZZARD, B. G. (1984). Have chemical tests a role in diagnosing malabsorption? *Ann. Clin. Biochem.*, **21**, 153–165.

HADORN, B., GREEN, J. R., STERCHI, E. E. and HAURI, H. P. (1981). Biochemical mechanisms in congenital enzyme deficiencies of the small intestine. *Clin. Gastroenterol.*, **10**, 671–690.

HARRIES, J. T. (1982). Disorders of carbohydrate absorption. *Clin. Gastroenterol.*, **11**, 17–30.

SILK, D. B. A. (1982). Disorders of nitrogen absorption. *Clin. Gastroenterol.*, **11**, 47–72.

MULLER, D. P. R. (1982). Disorders of lipid absorption. *Clin. Gastroenterol.*, **11**, 119–140.

BARON, J. H. (1977). Are gastric secretion tests worthwhile? *Proc. Roy. Soc. Med.*, **70**, 223–224.

INVESTIGATION OF SUSPECTED MALABSORPTION

This outline of a proposed scheme of investigation will discuss the use of chemical tests in the detection and differential diagnosis of malabsorption. However, it must be stressed that such tests are often unsatisfactory.

Malabsorption may be suspected if the patient presents with one or both of:

a history of chronic diarrhoea of unknown origin;

clinical, radiological, haematological and/or biochemical findings suggestive of malnutrition, with no obvious cause.

In either case, the diagnostic procedure is similar.

1. Has the patient a *history* which may suggest bacterial overgrowth (for example, has he had intestinal surgery which may have resulted in "blind loops", or long-term treatment with wide-spectrum antibiotics)? Has he been in the tropics? If so, consider the possibility of tropical sprue.

2. What is the appearance of the *stools*?

   (a) *Watery stools*, especially if blood-stained, suggest colonic disease such as ulcerative colitis; the possibility of purgative abuse should be remembered;

   (b) *Bulky, pale, greasy stools* suggest steatorrhoea;

   (c) *Constipation* with hard, dry stools makes the diagnosis of malabsorption syndrome highly improbable.

3. *Plasma electrolyte* abnormalities, especially severe hypokalaemia and signs of extracellular volume depletion without obvious evidence of malnutrition, favour colonic disease as a cause of diarrhoea, rather than a small intestinal malabsorption syndrome.

4. *Anaemia* is more likely to be due to intestinal than to pancreatic disease:

   (a) A hypochromic, microcytic picture may be the result of blood loss at any level of the gastro-intestinal tract;

   (b) A normochromic, normocytic picture is a non-specific finding in any chronic disease;

   (c) *A dimorphic (mixed iron deficiency and macrocytic) picture is very suggestive of intestinal malabsorption*;

   (d) Low red cell folate and/or plasma B_{12} levels suggest intestinal malabsorption.

5. If any or all the above tests suggest steatorrhoea, a cause should be sought. The most definitive tests are endoscopy, with histological examination of an intestinal biopsy specimen (for example, flattened villi suggest gluten-sensitive enteropathy, "idiopathic" steatorrhoea, or tropical sprue; infiltration of the mucosa may be detectable) and radiological examination.

6. If doubt remains, faecal fat estimation *may* help. However, because of the difficulty of obtaining accurately timed faecal collections, only very high results are of unequivocal significance: in such cases steatorrhoea will probably be obvious visually.

7. If steatorrhoea is obvious, and if the intestinal histology is normal, the three most important possibilities are:

   (a) localised intestinal disease (for instance Crohn's disease);

(b) the "contaminated small bowel" ("blind-loop") syndrome;

(c) pancreatic disease. This is a rare cause of malabsorption, unless there is a history of recurrent attacks of pancreatitis.

Tests such as CT scanning may help, especially if pancreatic carcinoma is suspected.

8. The xylose absorption test is rarely helpful. If it is *unequivocally* normal an upper intestinal lesion is unlikely. Such a result does not, however, exclude ileal or pancreatic disease, or the "contaminated small bowel" or "blind-loop" syndrome, or even some cases of gluten-sensitive enteropathy.

9. If malabsorption is proven, and if there is laboratory evidence of malnutrition, blood haemoglobin and plasma calcium, phosphate, albumin and alkaline phosphatase should be monitored to assess the efficacy of treatment.

TEST PROTOCOLS

COLLECTION OF SPECIMENS FOR FAECAL FAT ESTIMATION

We have discussed the disadvantages of faecal fat estimation on p. 293. In theory the result represents the difference between the fat absorbed and that entering the intestinal tract from the diet and from the body (p. 285). Absorption of fat and addition of fat to the intestinal contents occurs throughout the small intestine. Unless there is gross intestinal hurry (in which case the test is not usually indicated) a single 24-hour collection of faeces gives very inaccurate results for two reasons:

the transit time from the duodenum to the rectum is variable;
rectal emptying is variable and may not be complete.

It has been shown that consecutive 24-hour collections yield answers which may vary by several hundred per cent, and are therefore useless as an estimate of daily loss from the body into the gut. The longer the period of collection, the nearer does the calculated daily mean value approach the "true" one.

For obvious reasons patients cannot be kept in hospital indefinitely. A usual compromise is to collect *at least a 3-day* and preferably a 5-day specimen of stools. Precision may be increased by collecting between "markers"—usually dyes which can be taken orally and which colour the stool.

Procedure

Day 0.—The first "marker" (usually two capsules of carmine) is given.

As soon as the "marker" appears in the stool the collection is started. This "marker" will gradually disappear.

Day 5.—The second "marker" is given.

As soon as the second "marker" appears in the stool the collection is stopped.

The estimation is made on all the specimens passed between the appearance of the two "markers" and including one of the marked stools.

Important.—1. It is very important that *all* stools should be collected during this period. To ensure that none is missing it is best to label each specimen with the following information:

Name of Patient
Ward
Date of Specimen ⎤
Time of Specimen ⎬ A record of these should also be kept on the
Number in the series ⎦ ward.

The patient must use bedpans during the period of the test, and the importance of a complete collection should be explained.

2. The time for collection of specimens will be about a week (including the time for appearance of the marker). Before starting the test try to make sure that during this time the patient is *not to be discharged*, is *not going to be operated on* (except in emergency) and does *not receive enemas or aperients*: any patient requiring aperients to keep his bowels open is most unlikely to

PRIMARY ABNORMALITIES OF LIPID METABOLISM

Polygenic Hypercholesterolaemia

In most families with an abnormally high incidence of hypercholesterolaemia a graph of the plasma cholesterol values of all the individual members shows a continuous distribution: this contrasts with the clear trimodal distribution found with the monogenic pattern of true familial hypercholesterolaemia, in which values for homozygotes for the normal gene, heterozygotes, and homozygotes for the abnormal gene form three distinct peaks. The continuous distribution, with a high *mean* value within a family, is thought to be due to several gene abnormalities affecting LDL or cholesterol synthesis and disposal. It is therefore called polygenic. Environmental and dietary factors may determine the expression of the defect. Xanthomata are relatively rare, but there is an increased risk of cardiovascular disease.

Because the abnormalities found in the other primary disorders can be used to illustrate the steps in the metabolic pathways described at the beginning of the chapter, they will be discussed in more detail than some would otherwise merit, and the findings in each disorder will be related to the underlying physiology. The student should not attempt to memorise the details of the rarer disorders.

Although these disorders are usually inherited, there may be no family history. Some of the rarer ones are associated with hypo- rather than hyperlipaemia. Secondary factors may influence the expression of a genetic abnormality.

Receptor Abnormalities

LDL receptor deficiency (familial hypercholesterolaemia) is inherited as an autosomal dominant trait. Because entry of *LDL cholesterol* into cells is reduced, plasma levels rise. Triglyceride levels are normal, or only slightly raised. It is the most lethal of the inherited disorders. Raised cholesterol levels may be found in cord blood.

In homozygotes LDL receptors are virtually absent and plasma LDL cholesterol levels are three or four times higher than those in normal subjects; patients rarely survive beyond the age of 20, because of cardiovascular disease. In heterozygotes the number of LDL receptors is reduced by about 50 per cent and the plasma levels are about twice those in normal subjects, and there is a 10- to 50-fold higher risk of cardiovascular disease. Tendinous xanthomata and xanthelasma develop in early childhood in homozygotes, but only after the second decade in heterozygotes.

In families with this monogenic mode of inheritance there is a clear have significant steatorrhoea at that time. *Neither a barium enema nor barium meal* should be performed during this time because barium interferes with the estimation.

The collection and estimation of faecal fat excretion is time-consuming and unpleasant for all concerned (including the patient). It is important that specimen collection is carefully controlled so that the answer may be meaningful.

Interpretation

A *mean* daily fat excretion of clearly more than 18 mmol (5 g) indicates steatorrhoea. Because in the normal subject almost all the faecal fat is of endogenous origin, it is not affected by diet within very wide limits.

XYLOSE ABSORPTION TEST

An oral dose of 25 g of xylose has been given for this test. However, the absorption of xylose is relatively inefficient and the presence of this large dose in the intestine may, by its osmotic effect, cause abdominal discomfort and diarrhoea, and further interfere with absorption. For this reason a 5 g dose is preferable.

Warning. In the presence of poor renal function the test is invalid. Oedema, because the volume through which the xylose is distributed is increased, also invalidates the test.

The test is imprecise because of the problems of urine collection (p. 23).

Procedure

The patient is fasted overnight.

08.00h.—The bladder is emptied and the *specimen discarded*. 5 g of xylose dissolved in a glass of water is given orally;

All specimens passed between 08.00h and 10.00h are put into Bottle 1;

10.00h.—The bladder is emptied and *the specimen put into Bottle 1* which is now complete;

All specimens passed between 10.00h and 13.00h are put into Bottle 2;

13.00h.—The bladder is emptied and *the specimen is put into Bottle 2*, which is now complete.

Both bottles are sent to the laboratory for analysis.

Interpretation

In the normal subject more than 23 per cent of the dose (1·15 g) should be excreted during the five hours of the test. Fifty per cent or more of the total excretion should occur during the first two hours. In mild intestinal malabsorption the total 5-hour excretion may be normal, but delayed absorption is reflected in a 2- to 5-hour excretion ratio of less than 40 per cent.

In pancreatic malabsorption the result should be normal (p. 296).

TESTS OF GASTRIC FUNCTION

PENTAGASTRIN STIMULATION

Procedure (applies to all types of gastric secretion test)

The patient fasts from 22.00h on the evening before the test. On the morning of the test a radio-opaque Levin tube is passed into the stomach (preferably under x-ray control) until it lies in the gastric antrum: the tube is attached to the side of the face with adhesive tape.

08.00h.—All the gastric juice is aspirated from the stomach and is put in a bottle marked *"Resting Juice"*;

08.00h to 09.00h.—Gastric juice is aspirated continuously for the next hour and put in a bottle marked *"Basal Secretion"*;

09.00h.—Pentagastrin 6 µg/kg body weight is injected intramuscularly;

09.00h to 09.15h
09.15h to 09.30h
09.30h to 09.45h
09.45h to 10.00h
} Four 15-minute samples are aspirated and put into bottles marked with the appropriate times.

All specimens are sent to the laboratory for analysis.

Interpretation

1. In pentagastrin-fast *achlorhydria* no specimen has a pH as low as 3·5.

2. A *basal* acid secretion of greater than 15 mmol of hydrogen ion in the hour, with no further response to stimulation, is suggestive of the *Zollinger-Ellison syndrome*. Fasting plasma gastrin levels should be measured, and if high *in the presence of hyperchlorhydria*, confirm the diagnosis.

3. If any specimen is of pH 3·5 or less the hydrogen ion secretion is checked by titration. In *hyperchlorhydria* the highest mean acid secretion (calculated by adding the results of the two consecutive specimens of highest acidity and by multiplying by two, to give an hourly secretion), is greater than 40 mmol/hour.

The resting juice is inspected for *blood*. A large quantity of altered blood is suggestive of carcinoma. Small flecks may be due to trauma during aspiration.

INSULIN STIMULATION OF GASTRIC SECRETION

Insulin-induced hypoglycaemia stimulates gastric secretion via the vagus nerve.

Procedure

The preparation of the patient and positioning of the tube is described above.

08.00h.—The stomach is emptied and the specimen discarded;

08.00h to 09.00h.—A 1-hour "basal secretion" is collected;

09.00h.—Soluble insulin (0·15 units/kg body weight) is injected intravenously;

09.00h to 11.00h.—Specimens are collected for plasma glucose estimation every 30 minutes for two hours. During this time gastric juice is aspirated every 15 minutes.

The specimens are sent to the laboratory for analysis.

Warning.—1. The test is potentially dangerous and should be done only under *direct medical supervision. Glucose* for intravenous administration should be *immediately available* in case severe hypoglycaemia develops. At the conclusion of the test the patient should be given something to eat.

2. If it is necessary to administer glucose, *continue with the sampling.* The stress has certainly been adequate.

Interpretation

Analysis of gastric samples is only useful if the plasma glucose has fallen to levels below 2·5 mmol/l (45 mg/dl) and if symptomatic hypoglycaemia is present.

If more than 2 mmol of hydrogen ion is present in any 60-minute sample the response is positive, suggesting the presence of intact vagal fibres.

Chapter XIII

LIVER DISEASE AND GALL STONES

LIVER DISEASE

OUTLINE OF FUNCTIONS OF THE LIVER

The liver shares with the kidney the task of excreting or detoxifying end-products of metabolism. In addition it has important synthetic and metabolic functions, a few of which are summarised below. Failure of these mechanisms accounts for many of the features of liver disease described in this chapter.

General Metabolic Functions

The liver receives blood from the portal vein, and all nutrient from the gut except fat passes through the hepatic sinusoidal vascular spaces before entering the systemic circulation. *Postprandially*, when levels of glucose are high in portal plasma, *glycogen* is synthesised and stored, or glucose may be converted into fatty acids (Fig. 18, p. 195). During *fasting*, systemic plasma glucose concentrations are maintained by breakdown of stored glycogen and by conversion of substrates such as glycerol, lactate and amino acids into *glucose* (Figs. 19 and 20, pp. 198 and 199). *Fatty acid* reaching the liver from fat stores may be metabolised in the tricarboxylic acid cycle, incorporated into *triglycerides*, or converted to *ketones* (Fig. 19).

Synthetic Functions

The hepatocytes synthesise:

plasma proteins, except immunoglobulins;
most coagulation factors; of these prothrombin and factors VII, IX and X cannot be synthesised without vitamin K;
the lipoproteins VLDL and HDL (Chapter X).

The fall in plasma albumin which may occur in advanced liver disease can be difficult to interpret because of the many extrahepatic factors which have the same effect; however, unless there is evidence of gross protein loss from the kidneys, gut or skin, very low levels do suggest some failure of synthetic function. Prothrombin levels (as assessed by the prothrombin time) may be reduced either due to hepatocellular impairment, or to failure to absorb vitamin K: if liver-cell function is adequate, parenteral administration of the vitamin will reverse the abnormality (see "Cholestatic jaundice" p. 322).

Excretion and Detoxification

The excretion of *bilirubin* is considered below. Some other substances inactivated and excreted by the liver include:

> *amino acids*, which are deaminated in the liver; the amino groups, and any ammonia produced by intestinal bacterial action and absorbed into the portal vein, are converted to *urea*;
>
> *cholesterol*, which is excreted in the bile either unchanged or after conversion to *bile acids* (p. 327);
>
> *steroid hormones*, which are inactivated by conjugation with glucuronate and sulphate and excreted in the urine in this water-soluble form;
>
> many *drugs*, which are metabolised and inactivated by the liver; some are excreted in the bile.

Efficient excretion of bilirubin and other end-products of metabolism depends on:

> normally functioning liver cells;
> normal blood flow to and through the liver;
> patent biliary passages.

Because of the very large hepatic reserve, tests for impairment of metabolic, including synthetic and secretory, function are relatively insensitive indicators of liver disease.

Filtering Function

Because portal blood leaving the intestine supplies the liver, the reticulo-endothelial Kupffer cells in the hepatic sinusoids are well placed to extract absorbed toxic substances. Disruption of the normal architecture, due, for example, to cirrhosis or to surgical portocaval shunting, allows such substances to enter the systemic circulation.

BILIRUBIN METABOLISM AND JAUNDICE

Jaundice is the clinical manifestation of a raised plasma bilirubin level. It occurs when bilirubin production exceeds the hepatic capacity to excrete it. This may be because:

> an increased rate of bilirubin production exceeds normal excretory capacity;
>
> the normal load of bilirubin cannot be conjugated and/or excreted by damaged liver cells;

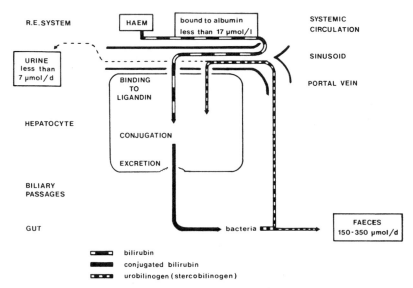

FIG. 28.—Bile pigment metabolism.

the biliary flow is so reduced by disease of the tract that conjugated bilirubin cannot flow into the intestine.

These causes will be discussed in the context of the normal sequence of events outlined in Fig. 28.

Formation of Bilirubin

At the end of their life circulating red cells are broken down in the reticulo-endothelial system, mainly in the spleen. The released haemoglobin is split into globin, which enters the general protein pool, and haem, which is converted to bilirubin after removal of iron. The iron is re-utilised.

About 80 per cent of all bilirubin is derived from such reticulo-endothelial destruction. Other sources include the breakdown of immature red cells in the bone marrow, and of compounds chemically related to haemoglobin such as myoglobin and cytochromes.

This *unconjugated* bilirubin (that is, bilirubin which has not yet been made water-soluble by hepatic conjugation with glucuronate) is carried to the liver bound to albumin, and, in adults, accounts for the bilirubin found in normal plasma: because it is protein-bound, it cannot pass the renal glomerulus into the urine; nor can it pass brain-cell membranes. *Unbound (free)* unconjugated bilirubin, though still water-insoluble, is, however, lipid-soluble and can enter and damage brain cells; levels exceeding the protein-binding capacity are most likely to occur in the newborn, in whom hepatic conjugating mechanisms are immature (p. 325). Low plasma albumin levels, or displacement of unconjugated bilirubin from binding sites on albumin by high levels of fatty acids, or drugs such as salicylates and sulphonamides, increase the proportion of the unconjugated bilirubin which is free, and hence the risk of cerebral damage at any measured level.

In the liver, bilirubin is transferred from plasma albumin, through the freely permeable vascular sinusoidal membrane, into hepatocytes, where it is bound to *ligandin* (Y protein); from there it is actively transported into the smooth endoplasmic reticulum and conjugated with glucuronic acid by a process catalysed by uridyl diphosphate (UDP) glucuronyl transferase. This sequence of events, which involves active transport, and which can only continue if there is free flow of bile, maintains the gradient from the plasma through the cell. Other anions, including drugs, may compete for binding to ligandin and so impair bilirubin conjugation and hence excretion.

Normally about 500 μmol (300 mg) of bilirubin reaches the liver and is conjugated daily: in the adult healthy hepatic cells are capable of handling much greater loads than this. *Jaundice due to unconjugated hyperbilirubinaemia only occurs if there is*:

> a marked increase in the bilirubin load as a result of haemolysis, or breakdown of large amounts of blood in tissues after haemorrhage;
> deficient binding of bilirubin to ligandin, or deficient conjugation with glucuronic acid.

Because unconjugated bilirubin is not water-soluble and is normally all protein-bound it cannot enter the urine: patients do not have bilirubinuria despite a high plasma level (acholuric jaundice).

Biliary Excretion of Bilirubin

Bilirubin monoglucuronide passes to the canalicular surface of the hepatocyte where, after addition of a second glucuronate residue, it is actively secreted into the bile canaliculus. Further transport depends on a normal bile flow, which in turn is largely dependent on the active secretion of bile acids from the hepatocyte. These energy-dependent steps are the ones *most likely to be impaired by liver damage (and by hypoxia and septicaemia) and by increased pressure in the biliary tract.* A rise in plasma conjugated bilirubin is the earliest manifestation of impaired excretion. *In most cases of jaundice in the adult, both fractions of plasma bilirubin are elevated*, but conjugated bilirubin predominates. *Conjugated bilirubin* is water-soluble and is less strongly protein-bound, and can be *excreted in the urine. Bilirubinuria is always pathological* because it reflects the presence of *conjugated* bilirubin in plasma. Dark urine is an early symptom of hepatobiliary disease.

Urobilin

The conjugated bilirubin enters the gut in bile. In the distal ileum and the colon it is broken down by bacteria to a group of products known collectively as *stercobilinogen* (or *faecal urobilinogen*). A small amount is absorbed into the portal circulation and most of this is re-excreted in bile: a very small fraction appears in the urine as *urobilinogen* which in turn can be oxidised to *urobilin*.

Urobilin(ogen), by contrast with bilirubin, *is* often detectable in normal urine by routine tests: this is especially likely if concentrated urine is passed. Urinary urobilin(ogen) is increased if:

> *haemolysis* is excessive, when large amounts of bilirubin enter the bowel and are converted to stercobilinogen. An increased amount of urobilinogen is formed and absorbed and, if the hepatic capacity to resecrete it is exceeded, is passed in the urine;
>
> *liver damage* impairs re-excretion of normal amounts of urobilinogen into the bile.

Urinary urobilinogen excretion is so variable in the normal subject that only the very high urinary levels found during acute haemolytic episodes are clinically significant.

Unabsorbed stercobilinogen is oxidised to stercobilin, a pigment which contributes to the brown colour of the faeces. Pale stools may, therefore, be evidence of biliary obstruction.

Bilirubin, *urobilin* and *stercobilin* are *coloured* ("bile pigments").

Urobilinogen and stercobilinogen are *colourless*.

Conjugated and *unconjugated* bilirubin are sometimes referred to as *direct* and *indirect* bilirubin respectively.

Bile acids, end-products of cholesterol metabolism, are excreted in bile, reabsorbed in the ileum, and re-excreted by the liver (entero-hepatic circulation) (p. 327).

BIOCHEMICAL TESTS IN LIVER DISEASE

Although many agents may impair hepatobiliary function, they can only do so in a few ways. The groups of tests, sometimes misleadingly called liver "function" tests, may indicate the nature of the main underlying pathological process. The results, *in the context of the clinical findings*, may suggest further tests, biochemical or other, that could assist in making a specific diagnosis.

The pathological processes that may be present, singly or in combination, are:

liver-cell damage;
cholestasis;
reduced functioning tissue mass.

Liver-cell Damage

Damage may range from areas of focal necrosis to extensive destruction and be caused by:

infections, most commonly *viral*;
toxins or drugs, such as *paracetamol* (acetaminophen) or *alcohol*. Several industrial chemicals such as chlorinated hydrocarbons are hepatotoxic;
hypoxia and/or *congestion* such as occur in congestive cardiac failure or "shock".

Cell damage may also occur *secondarily to prolonged biliary obstruction*.

Damage to liver cells, with or without necrosis, causes the acute release of intracellular constituents into the blood stream. Sensitive indicators of such damage are the plasma levels of the *enzymes* aspartate (AST; SGOT) and alanine (ALT; SGPT) transaminases. The highest plasma levels are found when many cells are damaged at once, for instance due to tissue hypoxia ("shock") or viral hepatitis. During recovery levels of both enzymes fall rapidly; ALT, with a longer half-life, is the last to return to normal values. Two variations may be of prognostic significance:

persistently elevated values, however slight, indicate continuing destruction that may occasionally progress to chronic liver disease (p. 323);
a sudden fall from very high levels without clinical improvement may rarely indicate massive destruction; only living cells can continue to produce enzymes.

The *relative plasma activities of AST and ALT* may also indicate the nature of the damage. The liver cell contains more AST than ALT but AST is present in both mitochondria and cytoplasm. ALT is confined to the cytoplasm where its concentration is greater than that of AST. Consequently, in conditions such as viral hepatitis, in which cell-membrane damage is primary, relatively more cytoplasmic than mitochondrial enzymes leak into the ECF: the activity of plasma ALT is therefore usually relatively more raised than that of AST. In conditions in which damage, however focal, involves the whole cell (for example, due to tumour deposits, cirrhosis and hypoxia), AST levels are relatively higher.

Active hepatic cell damage may cause a rise in the levels of other enzymes such as lactate dehydrogenase, but the changes are too non-specific or too insensitive to be of diagnostic value.

Cholestasis

Cholestasis is due to impaired secretion of bile by hepatocytes with resultant accumulation in the plasma of some or all of the substances normally excreted by this route. The primary lesion is an impaired secretion of bile acids which may be due to:

back-pressure from obstructed biliary passages (intrahepatic or extrahepatic);

inhibition of the secretory mechanism by infections, drugs, hypoxia or other, often unidentified, metabolic factors.

The most generally available indicator of *cholestasis* is the plasma level of the enzyme *alkaline phosphatase* (ALP). Not only is there increased synthesis of ALP at the sinusoidal surface of the hepatocyte but, if intraductal pressure rises, that which has been secreted into the canaliculi may pass back into the sinusoids with other bile constituents, and so enter the systemic circulation. Plasma ALP levels tend to be highest with extrahepatic obstruction.

The *consequences of cholestasis* depend on its severity and duration. Raised plasma bile-acid levels may cause severe *itching* due to deposition in the skin, and a reduced concentration in the intestinal lumen leads to *malabsorption* of fat and fat-soluble vitamins. Decreased synthesis of those coagulation factors which depend on the fat-soluble vitamin K, can be detected at an early stage by demonstrating that the *prothrombin time* is prolonged. If parenteral administration of vitamin K shortens a prolonged prothrombin time, this indicates functioning liver cells. Prolonged absence of bile acids leads to steatorrhoea and its consequences (p. 295).

In cholestasis there is usually, but not always, retention of bilirubin (predominantly conjugated) with jaundice and bilirubinuria. Cholesterol excretion is also impaired and plasma levels rise; in severe cholestasis, especially in established biliary cirrhosis (p. 324), xanthomata may develop.

Reduced Functioning Tissue Mass

The liver has great functional reserve and powers of regeneration. In *acute* liver disease, most of the liver is involved and *excretory failure and cell damage predominate*. The cells usually recover rapidly before other evidence of functional failure becomes clinically significant.

Chronic liver damage may, however, lead to such extensive *reduction of the total mass of functioning tissue* that synthetic and metabolic functions are detectably impaired. Because the reserve excretory capacity is large the remaining cells may eliminate most of the bilirubin reaching them, and jaundice is usually mild or absent.

If synthetic function is much reduced the plasma albumin level falls and the prothrombin time is prolonged: the liver cannot synthesise prothrombin even after parenteral vitamin K is given (compare the prolonged prothrombin time of cholestatic jaundice in which the inherent synthetic ability is often adequate, but vitamin K is not absorbed).

If there is very extensive destruction the metabolic features of liver failure appear (p. 324).

BIOCHEMICAL CHANGES IN DIFFERENT LIVER DISORDERS

Tests other than those discussed may help to detect specific disease processes.

ACUTE HEPATITIS

Acute hepatitis is most commonly caused by viruses that may be acquired by the faecal-oral route (usually hepatitis A) or from infected blood or blood products (hepatitis B or "non-A, non-B"). Infectious mononucleosis, or drugs and toxins can produce a similar clinical picture.

The primary pathological abnormality is cell damage: this may be detectable only by plasma enzyme estimations or may cause detectable disturbances in function.

The patient gives a history of nausea, anorexia, and tenderness or discomfort over the liver. There is commonly a cholestatic element, and he may notice that he is yellow and that his *stools are pale* (impaired entry of bilirubin into the intestine) and *urine dark* (reflecting a rise in plasma conjugated bilirubin). *Transaminases* (both AST and ALT) are *very high* by the time the patient feels ill, and reach their peak at about the time jaundice is detectable: they usually remain elevated for several weeks.

Plasma *bilirubin* levels rarely exceed 350 μmol/l (about 20 mg/dl). Because jaundice is the result of cell damage *and* cholestasis *both unconjugated and conjugated bilirubin levels rise.* The *alkaline phosphatase activity is usually only moderately increased, or even normal.* If cholestasis is predominant the picture *may resemble obstructive jaundice*, with very high plasma alkaline phosphatase and bilirubin levels. In anicteric hepatitis the only abnormal laboratory finding may be markedly raised transaminases.

The *course* of hepatitis may be followed by plasma transaminase estimations (ALT is the last to return to normal). If relapse is suspected clinically, it may be confirmed by enzyme levels which remain high, or which show a second rise. The development of a chronic phase is suggested by an increase in γ-globulin levels.

The *severity and extent* of liver-cell impairment are shown best by the finding of a prolonged prothrombin time, although this is also affected by bile-salt deficiency (p. 295): occasionally, in very severe damage, albumin levels may be unequivocally low.

The *aetiology* of hepatitis may often be determined by *serological* tests which either detect part of the virus (viral antigens) or antibodies synthesised by the body to the virus. In the preclinical and early symptomatic phases of hepatitis B, an antigen from the surface of the virus, HB_s Ag, is present in the blood. Its disappearance is followed, usually after several weeks, by the development of an antibody, anti-HB_s. During the intervening weeks when neither HB_s Ag nor anti-HB_s can be detected, an antibody to the viral core, anti-HB_c, is usually present. In some patients, particularly those with an impaired immune response in whom the acute attack is often mild, HB_s Ag persists. These subjects may carry the infection: this carrier state is especially important if they are blood donors.

The A virus is the commonest cause of epidemic hepatitis. Unfortunately anti-HA can be detected in the serum of many normal subjects, and serological tests are less useful in diagnosis than they are for hepatitis B. Now that subjects with detectable HB_s Ag are excluded as blood donors, serological tests for post-transfusion hepatitis are usually negative ("non-A, non-B hepatitis").

Acute alcoholic hepatitis occurs in heavy drinkers, often after a period of greatly increased alcohol intake. The clinical features may resemble those of acute viral hepatitis superimposed on those of chronic liver disease, but plasma bilirubin and AST levels may be only moderately increased. If acute alcoholic hepatitis is suspected it may be worth measuring gamma-glutamyltransferase (GGT); if it is raised out of proportion to the transaminase and bilirubin levels, this is compatible with, but not diagnostic of, the condition. Macrocytosis, hyperuricaemia and hypertriglyceridaemia are also suggestive, but are also not diagnostic of prolonged high alcohol intake. Alcoholic hepatitis may rarely present as cholestatic jaundice.

CHOLESTATIC JAUNDICE

Cholestatic jaundice may be due to extra- or intrahepatic causes. The most important underlying conditions are:

> *extrahepatic* (obstruction of the main biliary tract):
> gall stones in the common bile duct;
> carcinoma of the head of the pancreas, or, more rarely, of the bile duct or ampulla of Vater.
> *intrahepatic*:
> some forms of viral hepatitis;
> primary biliary cirrhosis;
> some drug-induced jaundice.

The jaundice of cholestasis is often severe, and may be associated with itching due to retention of bile acids. The patient may have noticed that his stools are pale (reduced faecal bile pigment) and urine dark (urinary excretion of conjugated bilirubin). The jaundice of extrahepatic obstruction due to malignant disease is typically painless and progressive, but there may be a history of vague, persistent back pain and loss of weight. By contrast, if the jaundice is due to intraluminal obstruction by a gall stone there may be a history of severe pain, which, like the jaundice, is intermittent: sometimes gall stones cause no such symptoms. A history of acute hepatitis may suggest a cause for intrahepatic cholestasis.

Plasma bilirubin levels, due mainly to conjugated bilirubin, may reach 850 μmol/l (50 mg/dl) or more. The urine is then very dark. Plasma alkaline phosphatase levels rise, often progressively, in chronic obstruction, but, unless prolonged back pressure has caused hepatocellular damage, those of transaminases are relatively little elevated.

It is essential to distinguish between extra- and intrahepatic causes of cholestasis. Surgery is often indicated for extra-, but is usually contraindicated for intrahepatic cholestasis. The biochemical findings are similar in both types, and unless the diagnosis is clinically obvious, evidence for the dilated bile ducts of extrahepatic obstruction should be sought using ultrasound, CT scanning or cholangiography.

Cholestasis Without Jaundice

A raised plasma alkaline phosphatase activity (of hepatic origin), the biochemical marker of cholestasis, may occur with little or no increase in bilirubin and variable increases in transaminase levels. Three conditions need to be considered:

> *obstruction* to only part of the biliary system by intrahepatic lesions, such as tumour deposits, granulomata or an intrahepatic biliary calculus, leads to retention of bile constituents. *Bilirubin*

but not ALP can be re-excreted by hepatocytes in unaffected parts of the liver;

cholangitis (inflammation of the bile ducts) causes increased synthesis of ALP and leakage of bile into the circulation. The bilirubin can be re-excreted in the same way as it is in partial obstruction;

primary biliary cirrhosis, characterised by bile-duct destruction and proliferation and increased ALP synthesis, is considered on p. 324.

CHRONIC LIVER DISEASE

Chronic or recurrent symptoms suggestive of liver disease or, more commonly, the finding of persistent, often mildly raised transaminase levels may be caused by several conditions. We will discuss the commoner ones.

Chronic Hepatitis

In chronic hepatitis the level of ALT is usually increased more than that of AST, and the former may be the only abnormal finding. If the disease progresses the pattern changes.

Chronic persistent hepatitis is the term used to describe the finding of such slightly raised transaminase levels without clinical signs or symptoms, and without a significant rise over many years. Levels rarely exceed three times the upper limit of "normal". Jaundice is rare. The biochemical abnormality may be a chance finding, or may persist after acute B or "non-A, non-B" hepatitis. The condition seems to be benign.

Chronic active hepatitis is due to active and progressive hepatocellular destruction, and may finally cause cirrhosis. It may be associated with, or perhaps be caused by $HB_s Ag$, it may be part of an autoimmune process sometimes involving more than one organ, or it may have no obvious cause. At first the transaminase pattern is that described for chronic persistent hepatitis: as more cells become completely destroyed, and as cirrhosis develops, the AST level rises, sometimes to above that of the ALT, and there may be slight jaundice. A rising IgG level (perhaps detected by an increasing γ globulin), or the presence of circulating smooth muscle, mitochondrial and antinuclear antibodies are the earliest findings differentiating it from benign chronic persistent hepatitis. Albumin levels fall if there is much liver destruction.

Primary Biliary Cirrhosis

Primary biliary cirrhosis, an autoimmune disease, is characterised by destruction and proliferation of bile ducts and variable degrees of cellular destruction. Although in the early stages the clinical picture may resemble that of chronic active hepatitis, it is usually *cholestatic* (p. 319) with itching and *very high alkaline phosphatase levels*. Jaundice may be minimal at first but develops later in most patients. Circulating mitochondrial antibodies are usually present and the IgM level is usually raised. As the disease progresses the steatorrhoea and xanthomata of prolonged cholestasis may develop.

Cirrhosis

Cirrhosis is the end-stage of a number of inflammatory liver diseases or may occur as the result of prolonged toxic damage, usually by alcohol. In some cases the aetiology is unknown ("cryptogenic cirrhosis"). The normal architecture of the liver is replaced by fibrous tissue and regenerating nodules of hepatocytes. The main biochemical changes are due to *loss of functioning tissue*, causing hypoalbuminaemia and a prolonged prothrombin time. Other biochemical tests may be normal but during active phases of cellular destruction transaminase levels (especially, and sometimes only, the AST) are usually raised and there is hyperbilirubinaemia. Diffuse hypergammaglobulinaemia with β-γ fusion is a typical electrophoretic finding. This is due to the immunoglobulin response, involving IgG and IgA, and is probably a result of the distorted hepatic architecture that allows antigenic substances absorbed from the intestine to bypass the normal hepatic sinusoidal filtering process. Increased portal vein pressure (portal hypertension) and impaired lymphatic drainage lead to accumulation of fluid in the peritoneal cavity (ascites). The consequent decrease in blood volume results in secondary hyperaldosteronism with dilutional hyponatraemia and often hypokalaemia (p. 54).

HEPATOCELLULAR FAILURE

Hepatocellular failure may occur during the course of severe hepatitis or decompensated cirrhosis, or following ingestion of liver toxins such as paracetamol. Jaundice is usually progressive although fulminating cases may die before it develops. Any or all of the biochemical abnormalities of hepatitis may be found, depending on the stage of the disease.

Other features may include:

severe electrolyte disturbances, particularly hypokalaemia, due to secondary hyperaldosteronism (in cirrhosis);

a prolonged prothrombin time and other coagulation defects;

a low plasma urea. Normally ammonia from deamination of amino acids is utilised in the synthesis of urea in the liver. Impairment of this process results in deficient urea production and accumulation of amino acids in the blood with consequent overflow aminoaciduria (p. 390). In many cases, however, renal failure develops and urea levels rise despite reduced production;

occasionally, hypoglycaemia (p. 214).

HEPATIC INVASION OR INFILTRATION

Hepatomegaly may be due to invasion or infiltration by carcinoma or lymphoma, or to infiltration by granulomata such as sarcoidosis. The results of biochemical tests may be normal; sometimes the only abnormal finding is a raised AST, or the ALT may also be elevated to a lesser extent. The picture may reflect cholestasis, with or without jaundice. Metabolic function is rarely demonstrably impaired. If a primary hepatocellular carcinoma develops, either in a cirrhotic liver or *de novo*, the transaminases and ALP usually rise rapidly, and plasma α-fetoprotein levels are usually very high: slightly raised levels are *not* diagnostic of primary hepatic malignancy (p. 477).

HAEMOLYTIC JAUNDICE

Jaundice due to haemolysis is not a liver disease, and is discussed more fully on p. 316. Jaundice is usually mild in the adult (bilirubin less than 70 μmol/l or 4 mg/dl) because of the large reserve hepatic secretory capacity. Erythrocytes contain a large amount of AST and LD (HBD) (p. 380); a rise in plasma levels of these enzymes after severe haemolysis should not be interpreted as evidence of myocardial damage.

In the newborn, the massive red-cell destruction occurring in haemolytic disease, together with the immature hepatic handling of bilirubin, can increase the level of unconjugated bilirubin to 400–500 μmol/l (25–30 mg/dl) or more. If the level of unconjugated bilirubin exceeds the plasma protein binding capacity, free unconjugated bilirubin may be deposited in the brain and cause *kernicterus*. Levels may be reduced by exposing the infant to ultraviolet light, which destroys bilirubin, or by exchange transfusion. In premature infants, too, the poorly developed conjugating mechanism may result in so-called "physiological" jaundice with markedly raised levels of unconjugated bilirubin, also necessitating treatment.

The course and severity of neonatal hyperbilirubinaemia may be influenced by drugs in three ways:

several drugs *displace bilirubin from plasma albumin* and increase the risk of deposition in the brain with resultant cerebral damage. These include salicylates and sulphonamides. The commonest of these is salicylates;

novobiocin *inhibits the glucuronyl transferase system* and aggravates unconjugated hyperbilirubinaemia;

any drug producing *haemolysis* aggravates the condition.

INHERITED HYPERBILIRUBINAEMIA

There is a group of disorders, apparently inherited, in which hyperbilirubinaemia (unconjugated or conjugated) is the only detectable abnormality.

Unconjugated Hyperbilirubinaemia

Gilbert's disease is a relatively common condition characterised by plasma bilirubin levels usually of 20–40 μmol/l and rarely, if ever, exceeding 80 μmol/l. Probably because of the non-specificity of the assay "conjugated bilirubin" levels may appear to be as high as 10 per cent of the total. Values fluctuate and rise during intercurrent illness or during fasting. The condition may be noted at any age. It is often discovered when bilirubin levels fail to return to normal after an attack of hepatitis, or during any mild illness which, because of the jaundice, may be misdiagnosed as hepatitis. The condition is harmless but must be differentiated from haemolysis and from hepatitis. Some patients do have shortened red-cell survival. The reason for the hyperbilirubinaemia is not clear and may be due to several factors.

The **Crigler-Najjar syndrome,** due to a deficiency of hepatic glucuronyl transferase, is more serious. It usually presents at birth and the plasma unconjugated bilirubin may increase to levels that exceed the binding capacity of plasma albumin, and so cause *kernicterus*. The defect may be complete (Type I) or partial (Type II). In the latter the plasma bilirubin may be reduced by drugs such as phenobarbitone that induce enzyme synthesis.

Conjugated Hyperbilirubinaemia

In two conditions there is defective excretion of conjugated bilirubin, but not of bile acids.

The **Dubin-Johnson syndrome** is characterised by mildly raised conjugated bilirubin levels that tend to fluctuate. Bilirubin is therefore present in the urine. Alkaline phosphatase levels are normal. There may be hepatomegaly and the liver is dark brown due to the presence in the cells of a pigment with the staining properties of lipofuscin. The condition is harmless and the diagnosis may be confirmed by the characteristic staining in the liver biopsy.

The pigment is not present in the liver in the clinically similar **Rotor syndrome.**

BILE AND GALL STONES

BILE ACIDS AND BILE SALTS

Four bile acids are produced in man. Two of these, *cholic acid* and *chenodeoxycholic acid*, are synthesised in the liver from cholesterol and are referred to as *primary bile acids*. These are secreted in the bile into the gut where bacterial action converts them to the *secondary bile acids*, *deoxycholic acid* and *lithocholic acid*, respectively. In human bile the bile acids are in the form of sodium salts, all conjugated with the amino acids glycine or taurine (*bile salts*). Some of the secondary bile salts are absorbed and re-excreted by the liver (enterohepatic circulation of bile salts). Bile therefore contains a mixture of primary and secondary bile salts.

Bile salts are important because they contain both polar and non-polar chemical groups and *form micelles*. In these micelles the non-polar groups are orientated towards the centre of the molecule and form a small pool of lipid solvent, while the polar (water-soluble) groups are on the outside.

Deficiency of bile salts in the intestinal lumen leads to impaired micelle formation, and malabsorption of fat (p. 295). Such deficiency may be caused by cholestatic liver disease (failure to reach the gut) or by ileal resection or disease (failure of reabsorption, with reduced bile-salt pool). The role of deficient bile salts in cholelithiasis is discussed below.

FORMATION OF BILE

About 1 to 2 litres of bile are produced by the liver daily. This *hepatic bile* contains bilirubin, bile salts, phospholipids and cholesterol as well as electrolytes in similar concentrations to those in plasma. Small amounts of protein are also present. In the gall bladder there is active reabsorption of sodium, chloride and bicarbonate, together with an isosmotic amount of water. The end result is *gall-bladder bile* which is ten times more concentrated than hepatic bile and in which sodium is the major cation and bile salts are the major anions. The concentrations of other non-absorbable molecules, conjugated bilirubin, cholesterol and phospholipids also increase.

GALL STONES

Although most gall stones contain all constituents of bile, there are several types differing in the main constituent. Only about 10 per cent of gall stones contain sufficient calcium to be radio-opaque (unlike renal calculi).

Cholesterol gall stones may be single or multiple and are not usually radio-opaque. They are white or yellowish and the cut surface has a crystalline appearance. They may be mulberry shaped.

Pigment gall stones consist largely of bile pigments with organic material and variable amounts of calcium. They are small multiple stones, dark green or black and are hard. Rarely they are radio-opaque.

Mixed gall stones, the commonest form, are, as the name suggests, composed of a mixture of cholesterol, bile pigments, protein and calcium. They are multiple and appear as faceted dark brown stones with a hard shell and softer centre. They may be radio-opaque.

CAUSES OF GALL STONE FORMATION

Pigment stones are found in *chronic haemolytic states* such as hereditary spherocytosis in which there is an increase in bilirubin formation and therefore biliary excretion. In all other forms of gall stones, however, the aetiology is uncertain.

Cholesterol-containing calculi are most likely to occur in bile already supersaturated with this steroid. Precipitation forms a nucleus of crystals on which more precipitation causes progressive enlargement. However, not all patients with a high biliary cholesterol concentration form calculi. Predisposing factors may include a low total concentration, or a change in relative concentrations of different bile salts.

Gall stones are said to be commonest in multiparous women, and the incidence is increased by oral contraceptive preparations: this may suggest a hormonal influence.

There is *no* association between hypercholesterolaemia and the formation of cholesterol gall stones.

Routine chemical analysis of gall stones is of little value.

CONSEQUENCES OF GALL STONES

Gall stones may remain silent for an indefinite length of time and be discovered only at laparotomy for an unrelated condition, or on x-ray of the abdomen. They may, however, lead to several clinical consequences:

acute cholecystitis, due to obstruction of the cystic duct by a gall stone with chemical irritation of the gall bladder mucosa by trapped bile and secondary bacterial infection;

chronic cholecystitis;

common bile duct obstruction, if a stone lodges in the bile duct. This may present as biliary colic, obstructive jaundice (usually intermittent) or acute pancreatitis if the pancreatic duct is also occluded;

extremely rarely, carcinoma of the gall bladder occurs.

SUMMARY

LIVER DISEASE

1. The liver has a central role in many metabolic processes.

2. Bilirubin derived from haemoglobin is conjugated in the liver and excreted in bile. Conversion to stercobilinogen (faecal urobilinogen) takes place in the bowel. Some reabsorbed faecal urobilinogen is excreted in the urine.

3. Bilirubin metabolism may be assessed by measuring plasma levels of bilirubin, and by visual inspection of the stool and urine.

4. Jaundice is due to a raised plasma bilirubin level. It may be due to a raised unconjugated bilirubin only, or, in adults, most commonly to an increase in both fractions.

5. The results of initial biochemical tests may be characteristic of one or more of three basic pathological abnormalities:

liver cell damage (high transaminases);

cholestasis (high alkaline phosphatase);

reduced functioning tissue mass (low albumin or prolonged prothrombin time).

Jaundice may or may not be present with any of these processes.

6. Further tests, whether biochemical or not, are selected on the basis of the clinical and preliminary biochemical findings.

7. Unconjugated hyperbilirubinaemia is usually due to haemolysis. In the newborn it may, if severe enough, exceed the plasma protein-binding capacity, and free bilirubin may enter brain cells and cause kernicterus.

8. There is a group of inherited conditions characterised by hyperbilirubinaemia. Most are relatively harmless, but the Crigler-Najjar syndrome may cause kernicterus.

9. Drugs may produce jaundice in several ways and a drug history is important in the assessment of the jaundiced patient. In the newborn, drugs may increase the risk of kernicterus.

BILE AND GALL STONES

Bile secreted by the liver is concentrated in the gall bladder before passing into the gut. Gall stones form when cholesterol crystals precipitate from supersaturated bile. Initiating factors are unknown. Pigment stones may occur in chronic haemolytic states.

FURTHER READING

SHERLOCK, S. (Ed.) (1980). Virus hepatitis. *Clin. Gastroenterol.*, **9**, 1–224. (An issue devoted to the clinical and laboratory features of viral hepatitis.)

CZAJA, A. J. (1979). Serologic markers of hepatitis A and B in acute and chronic liver disease. *Mayo Clin. Proc.*, **54**, 721–732.

CZAJA, A. J. and DAVIS, G. L. (1982). Hepatitis non-A, non-B. Manifestations and implications of acute and chronic disease. *Mayo Clin. Proc.*, **57**, 639–652.

WHITCOMB, F. F. Jr. (1979). Chronic active liver disease. Definition, diagnosis and management. *Med. Clin. N. Amer.*, **63**, 413–422.

READ, A. E. (1978). Some clinical features of liver cell failure: an appraisal of their causes. *Gut*, **19**, 543–548.

ZIMMERMAN, H. J. (1979). Intrahepatic cholestasis. *Arch. Intern. Med.*, **139**, 1038–1045.

WARNES, T. W. (1982). Investigation of the jaundiced patient. *Brit. J. Hosp. Med.*, **28**, 385–391.

INVESTIGATION OF SUSPECTED LIVER DISEASE

The most commonly available laboratory tests for liver disease include measurement of plasma levels of:

bilirubin	excretory function
transaminases (AST and/or ALT)	cell damage
alkaline phosphatase	cholestasis
albumin and/or prothrombin time	synthetic function

The selection of investigations may depend on the clinical features. The earlier part of this chapter considered the biochemical changes and clinical course of specific disorders. Different diseases may however present in a similar way and a single disorder may present in more than one way. In this section we will consider the differential diagnosis of clinical problems.

Jaundice as a Presenting Feature

While a patient with chronic liver disease may present with jaundice, the differential diagnosis of jaundice with bilirubinuria (due to conjugated bilirubin) in a previously well patient is usually between acute hepatocellular damage and cholestasis.

1. Pay special attention to a history of:

   recent exposure to hepatitis or infectious mononucleosis;
   recent receipt of blood or blood products;
   associated symptoms such as pain, pruritis or weight loss, or anorexia and nausea;
   drug and alcohol intake;
   changes noted in the colour of urine (dark) or faeces (pale);
   the presence of hepatomegaly.

2. (a) Request transaminases and alkaline phosphatase assays.

   (b) If suggested by the clinical history, if there is a predominant elevation of transaminases, and especially if the ALT is higher than the AST, request serological tests for hepatitis and infectious mononucleosis.

3. If there is predominant elevation of ALP determine if there is bile-duct dilatation by ultrasound or radiological tests:

   (a) if bile ducts are dilated, there is an obstruction that may require surgical intervention;

   (b) if dilated ducts are not demonstrated, there is probably intrahepatic cholestasis. Request serological tests for hepatitis, for smooth muscle, mitochondrial and anti-nuclear antibodies, and immunoglobulin levels.

If these do not provide a diagnosis, and if the prothrombin time is normal, a liver biopsy may be indicated.

4. If acute alcoholic hepatitis is suspected, contributory evidence may be the finding of a GGT activity which is disproportionately high compared with that of transaminases, and of macrocytosis, hypertriglyceridaemia and hyper-uricaemia. *None of these findings is diagnostic.*

Suspected Chronic Liver Disease (With or Without Jaundice)

1. Relevant points in the clinical evaluation are:

previous hepatitis;
alcohol intake;
presence or history of other autoimmune disorders;
pruritis or features of malabsorption.

2. Request transaminases and alkaline phosphatase assays.

(a) An ALT level higher than the AST may be due to reversible alcoholic hepatitis, to chronic persistent hepatitis, or to early chronic active hepatitis;

(b) An AST level higher than that of ALT may be due to cirrhosis or severe chronic active hepatitis;

(c) A high ALP suggests cholestasis.

3. A high blood alcohol level, despite denial of drinking by the patient, suggests, *but does not prove*, an alcoholic aetiology.

4. Detectable mitochondrial or smooth-muscle antibodies suggest a non-alcoholic cause.

5. Electrophoresis and immunoglobulin assay may help. A high IgG and IgA, causing β-γ fusion on the strip, suggests cirrhosis. A high IgG and normal IgA suggests chronic active hepatitis. A high IgM suggests biliary cirrhosis.

6. If macrocytosis, hyperuricaemia and hypertriglyceridaemia are all present they suggest *but do not prove* a chronic high alcohol intake: any or all of them may be abnormal for other reasons, and any or all of them may be normal in a known alcoholic.

Hepatic Invasion or Infiltration

1. Significant infiltration of the liver by tumour cells or granulomata, such as sarcoidosis, may occur without biochemical abnormality, or any of the changes described on p. 325 may be present. The most sensitive of the transaminases is the AST, the level of which may be high despite a normal ALT.

2. If primary hepatocellular carcinoma is suspected request α-fetoprotein assay.

3. Radionuclide scans or other imaging procedures, or a liver biopsy, may be indicated.

Unconjugated Hyperbilirubinaemia

Jaundice in which conjugated bilirubin levels are normal or less than about 10 per cent of the total, and in which there is therefore little or no bilirubinuria, may be due to:

increased bilirubin load. If there is no obvious cause, such as extensive bruising, haematological tests for haemolysis are indicated;
inherited defects of bilirubin uptake or conjugation (p. 326). Other tests for liver disease are usually normal and the diagnosis is made on the history and by the exclusion of haemolysis.

DRUGS AND THE LIVER

Drugs, or their metabolites, may damage the liver or produce changes that mimic liver disease clinically and biochemically. It is important that this cause be recognised, as normal function is usually restored on stopping the drug. *A drug history is an essential part of the assessment of a patient with liver disease.*

The mechanism of the drug effect is not known in many cases and may be due to individual hypersensitivity. Others will produce damage in any patient if safe dosage limits are exceeded. Table XXVII lists some of the drugs that may affect the liver. The following reference contains a more comprehensive list:

LUDWIG J. (1979). Drug effects on the liver. A tabular compilation of drugs and drug-related hepatic diseases. *Dig. Dis. Sci.*, **24**, 785–796.

TABLE XXVII

SOME DRUG EFFECTS ON THE LIVER.
An asterisk indicates that damage is dose-dependent and predictable.

	Hepatic necrosis	Hepatitis-like reaction	Chronic hepatitis	Cholestasis
α-methyldopa	+	+	+	+
aspirin	+*			
carbamazepine				+
chlorambucil		+		+
chlordiazepoxide		+		
chlorpromazine				+
chlorpropamide				+
chlortetracycline	+			
cytotoxic drugs	+*			
dantrolene	+	+	+	
erythromycin				+
ferrous sulphate	+*			
halothane	+	+		+
indomethacin				+
isoniazid		+	+	
MAO inhibitors		+		
methotrexate			+	
nitrofurantoin		+	+	+
oxyphenisatin		+	+	
para-amino salicylic acid		+		+
paracetamol	+*		+	
phenothiazines				+
phenylbutazone		+		+
phenytoin		+		
propylthiouracil	+			
17 α-alkylated steroids (in oral contraceptives)				+
tolbutamide				+
valproate	+			+

HANDLING OF BLOOD SAMPLES FROM PATIENTS WITH POSSIBLE HEPATITIS

Samples from all patients with viral hepatitis, undiagnosed jaundice, or positive HB_s Ag tests, as well as at-risk patients from dialysis units, should be considered *infective*. Anyone handling such a sample (medical and nursing staff, porters and laboratory staff) is at risk. It is *the duty of the clinician sending the blood to identify it clearly as potentially dangerous*. The sample should be *sent to the laboratory in leak-proof tubes in a sealed plastic bag*.

TEST PROTOCOLS

Rapid Tests for Urinary Bile Constituents

Urine containing bilirubin is usually dark yellow or brown, whereas fresh urine containing urobilinogen only is initially of normal colour.

Reagent strips (Ames or similar products) are available for the detection of these substances. *Fresh* urine is essential.

Ictostix (incorporated in Multistix and Bili-Labstix) includes stabilised diazotised 2, 4-dichloraniline, which reacts with *bilirubin* to form azobilirubin. The test will detect about 3 μmol/l (0·2 mg/dl) of bilirubin. Drugs (such as large doses of chlorpromazine) may give *false positive* reactions.

Urobilistix includes paradimethylaminobenzaldehyde, which reacts with *urobilinogen*. It does *not* react with porphobilinogen.

This test will detect urobilinogen in urine from some normal subjects. *False positive* results may occur with drugs such as *p*-aminosalicylic acid and certain sulphonamides. Positive results should be confirmed by a test for *urobilin* (below).

Both these reagent strips must be stored, and the test performed, strictly according to the instructions of the manufacturers.

Test for Urobilin

This test is preferable to the above if the urine is not fresh, because urobilinogen is converted to urobilin on standing.

1. Mix 5 ml of urine with 2 drops of alcoholic iodine solution (converts urobilinogen to urobilin).

2. Add 5 ml of zinc acetate suspended in absolute alcohol, mix and allow to settle.

3. A greenish fluorescence in the supernatant fluid is due to a zinc-urobilin complex. This is best seen in darkened surroundings by shining the light from a pencil torch through the fluid.

Chapter XIV

PROTEINS IN PLASMA AND URINE

PLASMA PROTEINS

PLASMA contains a mixture of proteins which differ in origin and function.

Metabolism of Plasma Proteins

The plasma levels of proteins depend on the balance between their synthesis and their catabolism or loss from the body.

Many plasma proteins are *synthesised* in the *liver*, but the *plasma cells* and *lymphocytes* of the immune system synthesise immunoglobulins, and proteins of the complement system are synthesised by macrophages as well as hepatic cells.

Small proteins are passively *lost* through the *renal glomerulus* and *intestinal wall*. Some of these are reabsorbed, either directly by renal tubular cells or after digestion in the intestinal tract; some are catabolised by renal tubular cells. Most plasma protein is *catabolised* by capillary endothelial cells or mononuclear phagocytes after being taken up by pinocytosis.

Functions of Plasma Proteins

The following is an outline of the main functions of the plasma proteins.

Inflammatory response and control of infection.—The immunoglobulins and the complement proteins form part of the immune system, and the latter, together with the group of proteins known as "acute phase reactants", are involved in the inflammatory response.

Transport.—Albumin and specific binding proteins transport many hormones, vitamins, lipids, bilirubin, calcium, trace metals and some drugs. Combination with protein renders these substances soluble in plasma, or physiologically inactive.

Control of extracellular fluid distribution.—Distribution of water between the intra- and extravascular compartments is affected by the concentration of plasma proteins, especially albumin (p. 36).

Peptide hormones and *blood clotting factors* contribute quantitatively relatively small, but physiologically important amounts of plasma protein. Only a few circulating *enzymes* are funċtional: most originate from breakdown of cells.

This list is by no means complete. The function of many of the proteins which have been identified in plasma is unknown.

Methods of Assessing Plasma Proteins

Proteins may be quantitated, either as groups (for example total protein) or, more often, as individual proteins (most commonly albumin).

Changes in the relative proportions of groups of proteins may be visually assessed after electrophoresis of serum or plasma.

TOTAL PROTEIN

Albumin normally makes the largest single contribution to plasma total protein, and *hypoproteinaemia* is almost always due to *hypoalbuminaemia* (p. 342).

The *concentration*, but not the actual amount of protein rises if excessive amounts of protein-free fluid are lost from the vascular compartment, either due to dehydration, or to stasis during venepuncture (p. 498).

True hyperalbuminaemia probably does not occur, and *hyperproteinaemia* is usually due to a major *increase in one or more of the immunoglobulins* (p. 348). If the albumin concentration is normal or low and the total protein level high, electrophoresis and, if necessary, immunoglobulin assay should be carried out.

Total protein levels may be misleading, and may be normal in the face of quite marked changes in the constituent proteins. For example:

a fall in albumin may roughly be balanced by a rise in immunoglobulin levels. This is quite a common combination;

most individual proteins, other than albumin, make a relatively small contribution to total protein: quite a large *percentage* change in the concentration of one of them may not be detectable as a change in total protein.

ELECTROPHORESIS

Electrophoresis, using the differences in electrical charge on different proteins to separate them, is usually performed by applying a small amount of serum to a strip of cellulose acetate or agarose and passing a current across it for a standard time. In this way five main groups of protein may be distinguished after staining, and may be compared with those in a normal control serum.

Each of the five fractions, albumin and the α_1-, α_2-, β- and γ-globulins, includes many proteins (Fig. 29). Changes in electrophoretic pattern are most obvious when the level of a protein normally present in high concentration (for instance, albumin) is abnormal, or when there are parallel changes in several proteins in the same fraction.

The following description applies to the normal appearance, in adults, of cellulose acetate strips.

> **Albumin,** usually a single protein, makes up the most obvious band.
>
> The α_1-**globulin** band consists almost entirely of α_1-*antitrypsin*;
>
> The α_2-**globulin** band consists mainly of α_2-*macroglobulin* and *haptoglobin*;
>
> The β-**globulin** often separates into two bands: β_1 consists mainly of *transferrin*, with a contribution from *LDL*, and β_2 consists of the *C3 fraction of complement*;

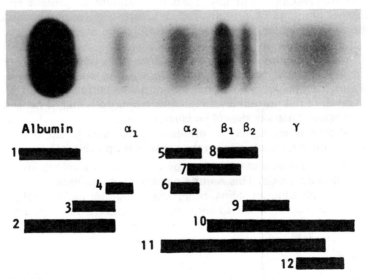

FIG. 29.—The plasma protein electrophoretic pattern and the twelve main plasma proteins (excluding fibrinogen).

In this example the β-globulin has separated into β_1 and β_2 fractions. This finding is not invariable, especially in stored specimens.

1. Albumin	5. α_2-macroglobulin	9. C_3 fraction of complement
2. α-lipoprotein (HDL)	6. haptoglobin	10. IgG
3. α_1-acid glycoprotein	7. β-lipoprotein (LDL)	11. IgA
4. α_1-antitrypsin	8. transferrin	12. IgM

The γ-globulins are *immunoglobulins*. Some immunoglobulins are also found in the α_2 and β regions.

If *plasma* rather than serum is used a sixth band, *fibrinogen*, appears in the β-γ region. This may make interpretation difficult, and clotted blood should be taken if electrophoresis is requested.

ELECTROPHORETIC PATTERNS IN DISEASE (Fig. 30)

Some abnormal electrophoretic patterns are characteristic, or suggestive, of a group of related clinical conditions. The finding of such a pattern may sometimes be an indication for further investigation.

Parallel Changes in All Fractions (not shown in Fig. 30)

This is a *normal pattern with an abnormal total protein concentration*. An *increase* in all protein fractions (including immunoglobulins) may be found in *dehydration* and due to *stasis* during venepuncture (p. 498), and a reduction in *overhydration* or in specimens taken from a "drip arm". A reduction also occurs in severe protein *malnutrition* and *malabsorption*, unless accompanied by infection.

The Acute-phase Pattern

Tissue damage of any kind triggers the sequence of biochemical and cellular events associated with inflammation (p. 343). The biochemical changes include stimulation of synthesis of the so-called acute-phase proteins, although secondary utilisation of synthesised complement causes a fall in its concentration: the plasma levels of these proteins reflect the activity of the inflammatory response, and their presence is responsible for the rise in the erythrocyte sedimentation rate (ESR) and the increased plasma viscosity characteristic of such a response.

Some of the individual acute-phase proteins are discussed on p. 344. They include those that are incorporated mostly in the α_1- and α_2-globulin fractions seen after routine electrophoresis.

An *increased density of the α_1- and α_2-fractions*, often with a reduced albumin level, is therefore commonly associated with such conditions as:

  infection;
  malignancy (especially the α_1 fraction);
  trauma, including surgical operation;
  autoimmune disease (for example, rheumatoid arthritis).

The electrophoretic pattern may suggest the need to monitor the levels of one or more of the individual acute-phase proteins.

NORMAL

NORMAL

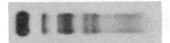

ACUTE TISSUE DAMAGE
(raised α_1, α_2)

NEPHROTIC SYNDROME
(low alb, raised α_2)

CHRONIC TISSUE DAMAGE
(raised α_1, α_2, γ)

α_1 ANTITRYPSIN DEFICIENCY
(absent α_1)

CIRRHOSIS ($\beta-\gamma$ fusion)

HYPOGAMMAGLOBULINAEMIA

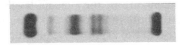

PARAPROTEIN (γ region)
(with immune paresis)

PARAPROTEIN (β region)
(with immune paresis)

FIG. 30.—Electrophoretic patterns in disease.

Note:
1. *Nephrotic syndrome.* In this typical example the level of all fractions except that of the α_2 globulin is low. Occasionally the γ fraction may be normal or increased.
2. *Paraproteinaemias.* Immune paresis is not invariable (p. 353).

Chronic Inflammatory States

In chronic inflammation immunoglobulin synthesis is increased, and may be evident after electrophoresis as a diffuse *rise in γ-globulin*. If there is still an active inflammatory reaction this increased density of γ-globulin will be associated with the increase of α_1- and α_2-fractions of the acute-phase response.

Cirrhosis of the Liver

The changes in plasma proteins in liver disease are considered more fully on p. 332. They are usually "non-specific", but in cirrhosis a characteristic pattern is sometimes seen. Albumin and often α_1-globulin levels are reduced and the γ-globulin concentration is markedly raised, with apparent fusion of the β and γ bands.

Nephrotic Syndrome

Plasma protein changes depend on the severity of the renal lesion (p. 358). In early cases a low albumin level may be the only abnormality, but the typical pattern in established cases is a reduced albumin, α_1- and γ-globulin with an increase in α_2-globulin. β-Globulin levels are usually normal. If the syndrome is due to systemic lupus erythematosus the γ-globulin may be normal or raised.

α_1-Antitrypsin Deficiency

Because the α_1-band seen after cellulose acetate electrophoresis consists almost entirely of α_1-antitrypsin, absence or an obvious reduction in density of this band suggests α_1-antitrypsin deficiency (p. 345).

Paraproteinaemia and hypogammaglobulinaemia are discussed in the section on immunoglobulins (pp. 352 and 351).

Although the changes in the electrophoretic pattern usually indicate disease, they are rarely pathognomonic.

Proteins of Clinical Importance

Concentrations of many of the individual plasma proteins may be measured by immunochemical or chemical methods. Of these the lipoproteins are discussed in Chapter X, transferrin in Chapter XVIII, and the immunoglobulins on p. 347. Some other proteins of clinical value are discussed here. Concentration changes in some, but not all, are visible on electrophoretic strips.

Albumin

Albumin, with a molecular weight of 65 000, is synthesised in the liver. In the plasma it normally has a biological half-life of about 20 days. Although about 60 per cent of extracellular albumin is in the interstitial compartment, much of which is subcutaneous, the plasma *concentration* is much higher because of the relative impermeability of the blood-vessel wall: this concentration gradient is important in maintaining plasma volume (p. 36).

We will discuss two rare groups of congenital abnormalities of albumin synthesis. There are several inherited variants of albumin synthesis including the *bisalbuminaemias*, in which two types of albumin are coded for; these are curiosities only, as there are usually no clinical consequences. In another *analbuminaemia*, there is deficient synthesis of the protein. Clinical consequences are slight, and oedema, although present, is surprisingly mild.

An abnormally high plasma albumin level is found only with dehydration or, artefactually, in a sample taken with prolonged venous stasis (p. 498), and only low albumin levels are of clinical importance.

Causes of hypoalbuminaemia.—A low plasma albumin level despite a normal total body albumin may be due to dilution by an excess of protein-free fluid, or to redistribution into the interstitial fluid due to increased capillary permeability. There may be true albumin deficiency due to a decreased rate of synthesis, or to an increased rate of catabolism or loss from the body.

Dilutional hypoalbuminaemia may be the result of:

> taking blood from the arm into which an infusion is flowing (artefactual p. 500);
> administration of an excess of protein-free fluid;
> fluid retention, usually in oedematous states or during late pregnancy.

Redistribution of albumin from plasma to interstitial fluid may be the result of:

> recumbency: albumin levels may be 5 to 10 g/l lower in the recumbent than the upright position, perhaps because of redistribution of water;
> increased capillary permeability: this is probably the cause of the sudden fall in plasma albumin level in "*shock*".

The slight fall in the albumin level found in even mild acute illness may be due to a combination of the above two factors.

Decreased synthesis of albumin.—Normally about 4 per cent of the body albumin is replaced each day by hepatic synthesis: impairment will cause hypoalbuminaemia. Synthesis depends on an adequate dietary supply of amino acids to replace the nitrogen lost, mainly as urinary urea, after protein catabolism. Hypoalbuminaemia may therefore be the result of:

chronic liver dysfunction;
malnutrition or malabsorption.

Increased catabolism of albumin.—Catabolism, and therefore nitrogen loss, is increased in many illnesses. This may aggravate hypoalbuminaemia due to other causes.

Increased loss of albumin from the body.—Because of its relatively low molecular weight, significant amounts of albumin are lost in conditions in which there is an increased permeability of membranes separating plasma from the outside of the body. The plasma concentration of albumin is higher than that of other low-molecular-weight plasma proteins, and its loss is therefore more obvious. Such protein-losing states include loss through:

the glomerulus in the nephrotic syndrome (p. 357);
the skin because of extensive burns or skin diseases such as psoriasis. A large proportion of the interstitial fluid is subcutaneous;
the intestinal wall in protein-losing enteropathy (p. 301).

Results of Hypoalbuminaemia

Fluid distribution.—Albumin is quantitatively the most important protein contributing to the plasma colloid osmotic pressure. *Oedema* can occur in severe hypoalbuminaemia.

Binding functions.—About half the plasma calcium is bound to albumin (p. 252) and hypoalbuminaemia is accompanied by *hypocalcaemia*. As this involves only the protein-bound (physiologically inactive) fraction, symptoms of tetany do not develop and calcium or vitamin D supplementation is contra-indicated.

Albumin also binds *bilirubin, free fatty acids*, and a number of *drugs* such as salicylates, penicillin and sulphonamides. The albumin-bound fractions are physiologically and pharmacologically inactive. A marked reduction in plasma albumin, by reducing the binding capacity, may increase free levels of these substances and may cause toxic effects if drugs are given in their normal dosage. Drugs which are albumin bound may, if administered together, compete for binding sites, also increasing free concentrations: an example of this is simultaneous administration of salicylates and the anticoagulant warfarin, with potentiation of the effect of the latter.

PROTEINS OF INFLAMMATION AND THE IMMUNE SYSTEM

The body responds to tissue damage, and to the presence of infecting organisms or other foreign substances, by a complex, interrelated series of cellular and chemical responses. The cells and humoral factors act together to initiate and control the inflammatory reaction, and so remove damaged tissue and foreign substances.

The inflammatory response, the ability to kill infecting organisms, or both, may be impaired if there is deficiency of either cellular or humoral components. Inappropriate responses may, by damaging host tissue, cause autoimmune disease.

In this chapter we shall confine ourselves to describing briefly some of the chemical responses to tissue damage, including the response of the complement system, and some of the changes in immunoglobulins (synthesised by B lymphocytes) which may be associated with antigenic challenge. Further details, and an account of the cellular response (polymorphonuclear leucocytes and T lymphocytes), will be found in textbooks of immunology, one of which is given at the end of the chapter.

THE PROTEINS ASSOCIATED WITH ACUTE INFLAMMATION

ACUTE-PHASE PROTEINS

We have already mentioned the rise in α_1- and α_2-globulins due to the acute-phase response. Most acute-phase reactants are synthesised in the liver, probably in response to peptide mediators (e.g., interleukin I) released from inflammatory cells. These reactants include:

activators of other inflammatory pathways such as *C-reactive protein* (so-called because it reacts with the C-polysaccharide of pneumococci). This protein combines with bacterial polysaccharides or phospholipids from damaged tissue to become an activator of the complement pathway.

inhibitors such as α_1-*antitrypsin* (see below), which control the inflammatory response and so minimise damage to host tissue.

haptoglobin, which binds haemoglobin released by local haemolysis during the inflammatory response.

Inflammation causes an increase in the levels of acute-phase proteins. The level of complement proteins may fall due to secondary utilisation; the complement system will be discussed separately.

The non-specific nature of the response means that measurement of individual acute-phase proteins is rarely helpful as an aid to diagnosis. However, their rates of synthesis and half-lives differ, and measurements of the relative changes in the levels of more than one may be useful to monitor the progress of inflammatory disease such as rheumatoid arthritis.

C-reactive protein estimation can be useful to detect acute inflammation, to which it responds rapidly. It is less reliable for monitoring chronic inflammation, for which measurements of various combinations of other acute-phase proteins have been suggested. *It is essential to contact your laboratory for advice before requesting such assays.*

Genetically low levels of some of the acute-phase proteins may produce clinical effects.

α_1-**Antitrypsin deficiency.**—The α_1-antitrypsins normally control the proteolytic action of enzymes from phagocytes (protease inhibitors; Pi).

There are about 30 genetic variants of α_1-antitrypsin, which are inherited as autosomal codominant alleles. The normal allele is PiM and the normal genotype MM.

The most important abnormal alleles which cause α_1-antitrypsin deficiency are called *null*, in which none of the protein is synthesised, and Z, in which the protein accumulates in the liver because it cannot be secreted after synthesis. The condition may be suspected after serum protein electrophoresis if the α_1 band is much reduced or absent (Fig. 30). The unopposed action of proteases from phagocytes in the lung may, by destroying elastic tissue, cause basal emphysema in young adults homozygous for either of these abnormal alleles; the condition may be exacerbated by cigarette smoking or infection. Hepatic damage occurs in 10 to 20 per cent of subjects, such as those with PiZZ, in whom protein cannot be secreted by hepatocytes; the condition may present as hepatitis in the neonatal period or as cirrhosis in children or young adults.

The diagnosis can usually be confirmed by demonstrating that the α_1-antitrypsin level is low, but sometimes the phenotype should also be identified. Blood relations should be investigated, and all those with abnormal levels advised not to smoke.

COMPLEMENT

The complement proteins are synthesised by macrophages or hepatocytes and, because of the presence of inhibitors in plasma, usually circulate in an inactive form. Sequential activation with utilisation of complement proteins during the inflammatory process reduces their plasma levels. The products of activation attract phagocytes to the area of inflammation (*chemotaxis*), and by increasing the permeability of the capillary wall to both cellular and chemical components allow them to reach affected cells; acting together with immunoglobulins they opsonise and lyse foreign cells. Other acute-phase proteins, synthesised in response to mediators released from the inflammatory cells, help to control and limit the response (see earlier).

Clinically the most important of the groups of complement proteins is C_3, activation of which results in chemotaxis. Two main pathways are concerned with activation of C_3 (Fig. 31). The names of these pathways reflect the order in which they were discovered, rather than their importance. Activation of either results in low C_3 levels.

The alternative pathway is the most important. C_3 is activated by IgA immune complexes, lipopolysaccharides on the surface of invading organisms, or by the products of activation of the classical pathway. C_3b is formed, which in turn activates more C_3. Other products of C_3b cause vasodilatation and cell lysis. The cyclical process is self-perpetuating and C_3 *levels fall.* If the initial stimulus is removed, inhibitors, such as those formed in the acute-phase reaction, control the reaction and C_3 levels become normal.

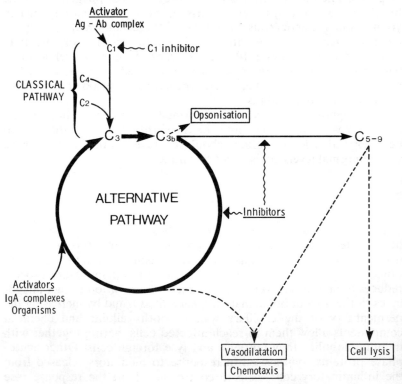

FIG. 31.—The complement pathways (much simplified).

In the classical pathway C_1 is usually activated by antigen-bound IgG or IgM (immune complexes). C_4 (and C_2) are used during the resultant sequence of events, which activates the alternative pathway. *C_3 and C_4 levels therefore fall.* Once the formation of immune complexes stops, C_3 and C_4 levels are restored.

In immune complex disease (for example systemic lupus erythematosus, SLE) circulating immune complexes persist. The products of C_3, released by continued activation of the classical pathway, may damage blood vessels, joints and kidneys. Both plasma C_3 and C_4 levels are low, and persistently low C_3 levels may help to distinguish chronic mesangiocapillary glomerulonephritis, with a poor prognosis, from the less serious and self-limiting acute streptococcal glomerulonephritis, in which C_3 levels become normal within a few months.

In hereditary angioneurotic oedema, C_1 inhibitor (Fig. 31) is deficient or non-functioning. The condition is characterised by episodically increased capillary permeability and consequent oedema of the subcutaneous tissues and mucous membranes of the upper respiratory or gastro-intestinal tract. Laryngeal oedema may be fatal. The level, or sometimes the activity, of the inhibitor may be measured.

This account is, of necessity, simplified. *Before requesting complement studies you should contact your laboratory for advice.*

IMMUNOGLOBULINS

It has long been known that the γ-globulin fraction of plasma proteins has antibody activity. Although the terms antibody and γ-globulin are often used synonymously, antibodies also occur in the β- and α_2-fractions (Fig. 29), and the term immunoglobulin (Ig) is preferable.

Structure

The immunoglobulin unit is a Y-shaped molecule depicted schematically in Fig. 32.

The following points should be noted:

Usually, four polypeptide chains are linked by disulphide bonds. There are two heavy (H) and two light (L) chains in each unit. The *H chains* in a single unit are similar and determine the immunoglobulin *class* of the protein. H chains γ, α, μ, δ and ε occur in IgG, IgA, IgM, IgD and IgE respectively. *L chains* are of two *types* κ or λ. In a single molecule the L chains are of the same type, although the Ig class as a whole contains both types.

There are two antigen-combining sites per unit, known as the F(ab)₂ piece. These lie at the ends of the arms of the Y: both

H and L chains are necessary for full antibody activity. The amino acid composition of this part of the chain varies in different units (variable region). When both of these sites combine with antigen, conformational changes are transmitted through the hinge region (Fig. 32) and the Fc portion of the molecule becomes activated.

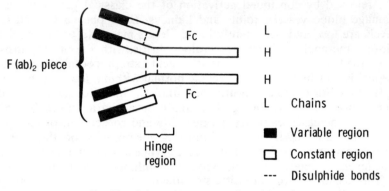

Fig. 32.—Schematic representation of Ig unit.

The rest of the H and L chains is less variable (constant or Fc region). The constant region of the H chains is responsible for such properties of the Ig unit as the ability to bind complement or actively to cross the placental barrier. The H chains are associated with a variable amount of carbohydrate; IgM has the highest content.

Some immunoglobulin molecules may contain more than one unit, usually bound together by "J" chains: for example, the IgM molecule consists of 5 units. Because of this variation in size and therefore in density, the classes can be separated by ultracentrifugation. They are classified according to their Svedberg coefficient (S), the S value of a protein increasing with increasing size.

The S values, most important functions and other properties of immunoglobulins are given in Table XXVIII. As we have seen, they act synergistically with the acute-phase proteins, including complement.

Normal Immunoglobulin Response to Infection

Immunoglobulins are synthesised by B lymphocytes.

IgM immunoglobulins are synthesised first in response to particulate

TABLE XXVIII

PROPERTIES AND FUNCTIONS OF PLASMA IMMUNOGLOBULINS

	IgG	IgA	IgM	IgE	IgD
Molecular weight	160 000	160 000 (polymers occur)	1 000 000	200 000	190 000
Sedimentation coefficient	7 S	7 S	19 S	8 S	7 S
% total plasma Ig	73	19	7	0·001	1
Able to activate complement	Yes	Yes (as complexes)	Yes	No	No
Able to cross placenta	Yes	No	No	No	No
Approximate mean normal adult concentration:					
g/l	10	2·5	1·0	0·0003	0·03
IU/ml	140	150	140	90	
Adult levels reached by:	3–5 years	15 years	9 months	? 15 years	? 15 years
Major function	Protects extra-vascular tissue spaces. Secondary response to antigen. Neutralises toxins.	Protects body surfaces as secretory IgA (11 S)	Protects blood stream. Primary response to antigen. Lyses bacteria.	Mast-cell bound antibodies of immediate hypersensitivity reactions.	Not known

antigens such as blood-borne organisms. Because of their large size they are almost confined to the intravascular compartment, and this fact, together with the speed of synthetic response, makes them the first line of defence amongst immunoglobulins against invading blood-borne organisms. The fetus can synthesise IgM, and high levels at birth usually indicate that there has been intra-uterine infection.

IgG immunoglobulin levels rise slightly later in response to soluble antigens such as bacterial toxins. Because of their relatively low molecular weight they can diffuse fairly freely into the interstitial fluid, and act in tissue against infection. Within a few weeks of the initial infection raised levels of all immunoglobulins may be demonstrated, and may be evident as a diffuse γ band on the electrophoretic strip.

Much **IgA** is synthesised submucosally and is present in intestinal and respiratory secretions, sweat, tears and colostrum. It is affected more than other immunoglobulins in diseases of the gastro-intestinal and respiratory tracts. Secretory IgA is a dimer in which the two subunits are joined by a peptide J chain and, in addition, have a "secretory piece" synthesised by epithelial cells. Circulating IgA does not contain the secretory piece.

Most infections produce a general immunoglobulin response. In such circumstances immunoglobulin estimation adds little to the observation of an increase in γ-globulin on routine electrophoresis. In certain conditions, some of which are listed in Table XXIX, one or more immunoglobulin classes predominate. Although there is considerable overlap, individual Ig estimation may help in the diagnosis of such cases.

TABLE XXIX

Predominant Ig	Examples of Clinical Conditions
IgG	Autoimmune diseases, such as SLE or chronic active hepatitis
IgA	Diseases of intestinal tract, for example Crohn's disease Diseases of respiratory tract, for example tuberculosis, bronchiectasis Early cirrhosis of the liver
IgM	Primary biliary cirrhosis Viral hepatitis Parasitic infestations, especially when there is parasitaemia At birth, indicating intra-uterine infection

The Immunoglobulin Response to Allergy

IgE is synthesised by plasma cells beneath the mucosae of the gastro-intestinal and respiratory tracts and by those in the lymphoid tissue of the nasopharynx. It is present in nasal and bronchial secretions. Circulating IgE is rapidly bound to cell surfaces, particularly to those of mast cells and circulating basophils, and plasma levels are therefore very low. Combination of antigen with this cell-bound antibody results in the cells releasing mediators and accounts for immediate hypersensitivity reactions such as occur in hay fever. Desensitisation therapy of allergic disorders aims at stimulating production of circulating IgG against the offending antigen, to prevent it reaching cell-bound IgE, and/or suppression of IgE synthesis. *Raised concentrations* are found in several diseases with an *allergic* (atopic) component such as some cases of eczema, asthma and parasitic infestations.

IMMUNOGLOBULIN DEFICIENCY

Susceptibility to recurrent infections may be the result of quantitative or functional deficiency of immunoglobulins, complement, or the cells of the immune system. It is important to remember that *normal serum immunoglobulin concentrations do not exclude an immune deficiency state.*

A severe reduction in immunoglobulin levels may cause obvious hypogammaglobulinaemia.

The effects of deficiency of individual immunoglobulins is related to their functions and distribution. In *IgM deficiency* septicaemia is common. *IgG deficiency* may result in recurrent pyogenic infections of tissue spaces, especially in the lungs and skin, by toxin-producing organisms such as staphylococci and streptococci. *IgA deficiency* may be symptomless, or may be associated with recurrent, mild respiratory-tract infections or intestinal disease.

Primary immunoglobulin deficiency is less common than deficiency secondary to other disease.

Several classifications have been proposed for these deficiencies. We will, very briefly, discuss three categories presenting with Ig deficiency.

Transient immunoglobulin deficiency.—In the newborn infant circulating IgG is derived from the mother by placental transfer. Levels decrease over the first 3 to 6 months of life, and then gradually rise as endogenous IgG synthesis increases. In some subjects onset of synthesis is delayed, and "physiological hypogammaglobulinaemia" may persist for several more months.

Most of the IgG transfer across the placenta takes place in the last

3 months of pregnancy. Severe deficiency may therefore develop in very premature babies, because the level of IgG derived from the mother falls before the endogenous level rises.

Primary immunoglobulin deficiency.—Primary *IgA deficiency* in plasma, saliva and other secretions is relatively common, with an incidence of about 1 in 500 of the population.

Several rare syndromes, usually familial, have been described. In one, *infantile sex-linked agammaglobulinaemia* (Bruton's disease), which occurs only in males, there is almost complete absence of B cells and circulating immunoglobulins, while cellular (T cell) immunity seems normal. Other syndromes have varying degrees of immunoglobulin deficiency and impaired cellular immunity, and can occur in either sex.

Secondary immunoglobulin deficiency.—Low serum immunoglobulin levels are most common in patients with malignant disease, particularly of the haemopoietic and immune systems, and are often precipitated by chemo- or radiotherapy; they are an almost invariable finding in patients with myelomatosis. In severe protein-losing states, such as the nephrotic syndrome, low immunoglobulin levels (especially of IgG) are partly due to the loss of relatively low-molecular-weight Ig, and partly due to increased catabolism.

B-CELL DISORDERS

The normal B-cell response to infection is *polyclonal*. Many different groups (clones) of B cells synthesise a range of different immunoglobulins which cause the appearance of diffuse hypergammaglobulinaemia in the electrophoretic pattern.

Paraproteinaemia

Each B cell is highly specialised and synthesises a single class and type of immunoglobulin. If, by contrast to the usual response to infection, one of these cells proliferates to form a clone, a single protein will be produced in excess. This *monoclonal* proliferation of B cells is often, but not always, malignant.

The term "paraprotein" refers to the appearance of a narrow, dense band on the electrophoretic strip, most commonly in the γ region, but which may be anywhere from the α_2 to the γ regions inclusive (Fig. 30). A paraprotein can often be shown to be monoclonal.

Immune Paresis

The synthesis of immunoglobulins from other clones may be suppressed by monoclonal proliferation. In such cases the γ band, other than the narrow paraprotein, is reduced or absent and levels of the other immunoglobulins can be shown to be reduced (Fig. 30).

Bence Jones Protein

Bence Jones protein (BJP) is found in the urine of many patients with malignancy of B cells, but can rarely be found without malignant disease. It consists of free monoclonal light chains, or fragments of them, which have been synthesised much in excess of H chains, implying a degree of dedifferentiation. Because of its relatively low molecular weight (20 000 to 40 000) the protein is filtered at the glomerulus and only accumulates in the plasma if there is glomerular failure or if it polymerises. BJP may damage renal tubular cells and may form large casts, producing the "myeloma kidney". It may also form amyloid deposits in tissues.

The coexistence of paraproteinaemia, immune paresis and Bence Jones protein is very suggestive of malignancy of B cells.

Causes of Paraproteinaemia

Although the presence of a paraprotein is strongly suggestive of a malignant process, this is not always so.

Paraproteins may be found in the following malignant conditions:

> myelomatosis (this accounts for most of the malignant para-proteinaemias);
> macroglobulinaemia;
> B-cell lymphomas (including chronic lymphatic leukaemia);
> heavy-chain diseases.

Results of Malignant B-cell Proliferation

Some of the clinical and laboratory findings are similar in all the malignant B-cell tumours. Whether or not they occur depends on the level of paraprotein, the presence or absence of immune paresis, and the presence or absence of BJP. It should be remembered that malignant tumours of B cells can exist without all, or rarely any, of these findings:

Consequences of the presence of paraprotein.—Unless the level of paraprotein is very high these findings are not present. Very high levels (which are very suggestive of malignancy) may be associated with a very high ESR and may cause:

> *in vivo* effects of increased plasma viscosity with sluggish blood flow in small vessels; these effects include *retinal-vein thrombosis* with impairment of vision, *cerebral thrombosis*, or even *peripheral gangrene* (hyperviscosity syndrome);

an obviously increased blood viscosity during venepuncture; the blood may clot in the syringe, and it may be difficult to prepare blood films;

(Hyperviscosity is most common in macroglobulinaemia, but may sometimes occur in myeloma);

a high plasma total protein despite a normal or low albumin level;

spurious hyponatraemia due to the space-occupying effect of the protein (p. 37).

Consequences of immune paresis.—Because of the abnormal spectrum of immunoglobulins there may be increased susceptibility to infection.

Effects of Bence Jones protein.—Cases with large amounts of Bence Jones protein are especially likely to be associated with:

renal failure, due to deposition of BJP in tubular cells;

amyloidosis.

Other common findings in cases with malignant B-cell tumours are:

normochromic, normocytic anaemia, a common presenting feature of any malignant disease;

small *haemorrhages*, perhaps due to complexing of coagulation factors by the paraprotein;

Raynaud's phenomenon if the paraprotein is a cryoglobulin (see p. 356).

Myelomatosis (multiple myeloma; plasma-cell myeloma)

Myelomatosis is a condition which becomes increasingly frequent after the age of 50; it is rare before the age of 30. It occurs equally in both sexes. In the commonest form there is malignant proliferation of plasma cells throughout the bone marrow. In such cases the clinical and laboratory features are due to:

malignant proliferation of plasma cells;

disordered immunoglobin synthesis and/or secretion from the cell.

Malignant proliferation of plasma cells.—*Bone pain*, which may be severe, is due to pressure from the proliferating cells. *X-rays* may show discrete *punched-out areas* of radiotranslucency, most frequently in the skull, vertebrae, ribs and pelvis. There may be generalised osteoporosis. Histologically, there is little osteoblastic activity around the lesion and plasma alkaline phosphatase activity is therefore normal.

Pathological fractures may occur.

The effects of paraproteinaemia, immune paresis and BJP (which occurs in about 70 per cent of cases) and the other possible findings, have been discussed above.

Other laboratory findings.—The raised plasma protein is usually IgG, less commonly IgA (about 2·5:1) and rarely Bence Jones protein (if renal failure is present). Occasionally IgD, IgM or IgE are found, the last two being very rare.

In about 20 per cent of cases, usually those with BJP production, no paraprotein can be detected in the plasma. Rarely, neither a paraprotein nor BJP can be found. In either case there is usually immune paresis.

In IgD myeloma an increase in γ globulin may not be detectable by routine electrophoresis.

Hypercalcaemia may be present (p. 266). High levels are usually suppressed by hydrocortisone (see steroid suppression test, p. 269) and steroids can be used to treat such hypercalcaemia.

As there is little osteoblastic activity, the *alkaline phosphatase level is normal* unless there is liver involvement, when the raised level can be shown to be of hepatic, not bony, origin. A normal alkaline phosphatase level in cases with bone lesions may help to differentiate the picture from bony metastases.

Bone marrow appearance.—The proportion of plasma cells in the bone marrow is increased, and many of these cells are atypical ("myeloma cells"). *Examination of a marrow film must be carried out before myelomatosis is diagnosed or excluded.*

Soft-tissue plasmacytoma.—Rarely myeloma involves soft tissues, without marrow changes (extramedullary plasmacytoma). Although the protein abnormalities of myeloma are often found in these cases, the behaviour and prognosis of the two conditions are different. Spread of soft-tissue plasmacytoma is slow and tends to be local. Local excision of a solitary tumour is often effective. The paraprotein should be typed, and the level monitored to follow progress.

Waldenström's Macroglobulinaemia

Like myeloma, macroglobulinaemia usually occurs in the older age group (between 60 and 80 years), but it is more common in males than in females. Also like myeloma, it is due to malignancy of B cells, but the malignant cells resemble lymphocytes rather than plasma cells. Symptoms of the "hyperviscosity syndrome" are commoner than in myeloma, probably because of the large size of the IgM molecule, but skeletal manifestations are rare. There is lymphadenopathy.

Laboratory findings and diagnosis.—*Serum protein changes.*—The paraprotein in the γ region can be identified as monoclonal IgM. Serum IgA concentration is usually reduced, but that of IgG *may* be raised.

The *bone marrow* aspirate or lymph node biopsy contains atypical lymphocytoid cells.

Heavy-Chain Diseases

This is a rare group of disorders characterised by the presence of an abnormal protein in plasma or urine identifiable as part of the H chain (α, γ or μ). The clinical picture is that of generalised lymphoma (γ-chain disease), intestinal lymphomatous lesions, with malabsorption (α-chain disease) or chronic lymphatic leukaemia (μ-chain disease). In some cases, a paraprotein is detectable in the serum.

Cryoglobulinaemia

Proteins which precipitate when cooled below body temperature are called cryoglobulins. They may be associated with diseases known to produce paraproteins. About half of them can be shown to consist of a monoclonal immunoglobulin (usually IgM or IgG). The patient usually presents with other symptoms of the underlying disease, and the cryoglobulin is found during investigation. Occasionally, especially if the concentration of protein is high and if it precipitates at temperatures above 22° C, intravascular precipitation may cause such skin lesions as purpura and Raynaud's phenomenon.

In some cases the protein can be shown to be polyclonal, and sometimes to include complement; these cases may be associated with immune complex disease (p. 347), but occasionally no underlying abnormality can be found (*essential cryoglobulinaemia*).

Paraproteinaemia Without Obvious Cause

If a paraprotein is found, investigation for one of the diseases discussed above should be initiated. In between 10 and 30 per cent of cases in hospital (and probably many more in the "well population") no cause can be found. The condition which may be transient, has been called "*benign*" or "*essential*" *paraproteinaemia* or *monoclonal gammopathy of undetermined significance*. The diagnosis should be made provisionally, and the patients followed up; they may later develop frank myeloma or macroglobulinaemia.

PROTEINURIA

The loss of most plasma proteins through the kidney is restricted by the size of the pores in, and possibly by a negative charge on the glomerular basement membrane that repels negatively-charged protein molecules. Alteration of either of these factors by glomerular disease may allow albumin and larger proteins to enter the filtrate. Low-molecular-weight proteins are filtered even under normal conditions. Most are absorbed and metabolised by renal tubular cells, and normal subjects excrete up to 0·08 g of protein a day in the urine, amounts undetectable by usual screening tests. Proteinuria of more than 0·15 g a day almost always indicates disease.

Proteinuria may be due to renal disease or more rarely may occur because large amounts of low-molecular-weight proteins are circulating.

It is important to remember that blood and pus in the urine give positive tests for protein.

Renal Proteinuria

Glomerular proteinuria is due to increased glomerular permeability (*nephrotic syndrome*): this is discussed more fully below. *Albumin* is usually the predominant protein in the urine.

"*Orthostatic (postural) proteinuria.*"—Most proteinuria is more severe in the upright than in the prone position. The term "orthostatic" or "postural" has been applied to proteinuria, often severe, which disappears at night. It appears to be glomerular in origin and is commonest in adolescents and young adults. Although often harmless, evidence of renal disease may occur after some years.

Tubular proteinuria may be due to renal tubular damage from any cause, especially pyelonephritis. If glomerular permeability is normal, proteinuria is usually less than 1 g a day, and consists mainly of the relatively low-molecular-weight α_2- *and* β-*microglobulins*.

Proteinuria with Normal Renal Function

Proteinuria can be due to production of *Bence Jones protein*, to severe haemolysis with *haemoglobinuria*, or to severe muscle damage with *myoglobinuria*. In the latter two cases the urine will be red or brown in colour.

Apparent proteinuria has occasionally been found because the patient has added egg white to the urine.

Bence Jones proteinuria can be inferred by inspection of the electrophoretic pattern of urinary proteins. BJP is the only protein of a molecular weight lower than albumin likely to be found in significant amounts in unconcentrated urine (in the absence of haemoglobinuria or myoglobinuria). The presence in the urine, especially if it is not present in the serum, of a band which is denser than that of albumin suggests BJP.

NEPHROTIC SYNDROME

In the nephrotic syndrome *increased glomerular permeability* causes a protein loss, by definition of *more than 6 g a day*, with consequent hypoalbuminaemia and oedema, and with hyperlipoproteinaemia. The renal disease may be primary, or secondary to other pathology. It has been reported:

> *in most types of glomerulonephritis*, usually due to deposition of circulating immune complexes in the glomerulus (about 80 per cent of nephrotics). In children "minimal change" glomerulonephritis is the commonest cause.

> *secondary to*:

> diabetes mellitus;
> systemic lupus erythematosus (SLE) (due to immune complexes);
> inferior vena caval or renal-vein thrombosis;
> amyloidosis;
> malaria due to *P. malariae* (due to immune complexes).

Laboratory Findings

Protein abnormalities.—*Proteinuria* in the nephrotic syndrome ranges from 6 to 50 g per day. The proportion of different proteins lost is an index of the severity of the glomerular lesion. In mild cases albumin (MW 65 000) and transferrin (MW about 80 000) are the predominant urinary proteins, and α_1-antitrypsin (MW 50 000) is also present. With increasing glomerular permeability IgG (MW 160 000) and larger proteins appear.

The *serum protein* electrophoretic picture reflects the pattern of loss and has been described on p. 341.

Electrophoresis of urine on cellulose acetate gives some idea of the severity of the lesion. The *differential protein clearance* is a more precise measure of the selectivity of the lesion. The clearance of a low MW protein, such as transferrin or albumin, is compared with that of a larger one, such as IgG. The result is usually expressed as a ratio, obviating the need for timed collections. A ratio IgG to transferrin or albumin clearance of less than 0·2 indicates high selectivity (predominant loss of small molecules), with a more favourable prognosis than those cases in which the ratio is higher: such cases usually respond well to steroid or cyclophosphamide therapy.

The consequences of the protein abnormalities are:

> *oedema* due to hypoalbuminaemia (p. 343);
> *reduction in the concentration of protein-bound substances* due to loss of carrier protein. It is important not to misinterpret low total levels of calcium, thyroxine, cortisol and iron.

Lipoprotein abnormalities.—In mild cases LDL increases, with consequent *hypercholesterolaemia*. In more severe cases a rise in

VLDL (*triglycerides*) may cause turbidity of the plasma. The cause of these changes is not fully understood.

Fatty casts may appear in the urine.

Renal function tests.—In the early stages glomerular permeability is high, and the plasma urea concentration normal. Later glomerular failure may develop, with uraemia. At this stage, protein loss is reduced and plasma levels of protein and lipid may revert to normal. In the presence of uraemia this does *not* indicate recovery.

The student should read the indications for protein estimations on p. 362.

SUMMARY

Plasma Proteins

1. Albumin is the plasma protein most often measured specifically. Its functions include control of fluid distribution between plasma and the extracellular compartment, and binding and consequent inactivation of many endogenous and exogenous substances.

2. Groups of proteins may be separated by electrophoresis. In normal serum five bands can be seen—albumin, and α_1-, α_2-, β- and γ-globulins. In plasma a sixth band, due to fibrinogen, is present.

3. Clinically helpful electrophoretic patterns may be found in the nephrotic syndrome, the paraproteinaemias, hypogammaglobulinaemia, α_1-antitrypsin deficiency, and some cases of cirrhosis of the liver.

4. The acute-phase reaction occurs in many inflammatory states and in tissue damage. It is characterised by an increased density of the α_1 and α_2 bands on electrophoresis.

5. Measurement of specific acute-phase proteins, such as α_1-antitrypsin, C-reactive protein and haptoglobin, is occasionally helpful in monitoring inflammatory disease. The level of the C_3 fraction of complement may fall in such disease.

6. The immunoglobulins, complement and the other acute-phase proteins act synergistically in the presence of invading organisms.

7. The immunoglobulins (Ig) are a group of proteins that are structurally related. Five classes are described. The main ones are IgG, IgA and IgM. Paraproteins are immunoglobulins and Bence Jones protein is related to them.

8. The functions and properties of the Ig classes and their response to antigenic stimuli differ. Estimation of IgG, IgA and IgM may be helpful in the diagnosis of a few conditions.

9. Immunoglobulin deficiency is only one aspect of immunological deficiency. *Normal Ig levels do not exclude immunological deficiency.*

Ig deficiencies may be primary or secondary and may involve one or all Ig classes.

10. A paraprotein is a narrow band, usually found in the γ-globulin region of the electrophoretic strip. It usually consists of monoclonal immunoglobulins, and usually, but not always, indicates malignant proliferation of B cells.

11. Paraproteins are most commonly associated with myeloma. Myeloma presents clinically in a variety of ways, reflecting bone marrow replacement and abnormal plasma protein levels. The laboratory diagnosis is made by bone marrow examination, and by finding protein abnormalities in serum and/or urine.

12. Bence Jones protein (usually found only in the urine) consists of monoclonal free light chains. Its presence usually indicates malignancy of B cells.

13. Cryoglobulins are proteins which precipitate when cooled below body temperature. They cause symptoms on exposure to cold. They may occur in any of the diseases associated with paraproteinaemia.

Proteinuria

1. Proteinuria may be due to glomerular or tubular disease. Glomerular proteinuria is the commoner form: massive proteinuria is always of glomerular origin.

2. Proteinuria may occur with normal renal function if abnormally large amounts of low molecular weight proteins are being produced.

3. The nephrotic syndrome is characterised by proteinuria of at least 6 g a day, with a low serum albumin, oedema and hyperlipoproteinaemia. The proteinuria is glomerular in type. The severity of the lesion may be assessed by differential protein clearance.

FURTHER READING

BRENNER, B. M., HOSTETTER, T. H. and HUMES, H. D. (1978). Molecular basis of proteinuria of glomerular origin. *New Engl. J. Med.*, **298**, 826–833.

HOBBS, J. R. (1980). Myeloma. *Medicine, (3rd series)*, No. **29**, 1507–1512.

KYLE, R. A. (1975). Multiple myeloma. Review of 869 cases. *Mayo Clin. Proc.*, **50**, 29–40.

LEVINSKY, R. J. (1978). Role of circulating soluble immune complexes in disease. *Arch. Dis. Childh.*, **53**, 96–99.

PARFREY, P. S. (1982). The nephrotic syndrome. *Brit. J. Hosp. Med.*, **27**, 155–162.

ROITT, I. M. (1980). *Essential Immunology*, 4th edit. Oxford: Blackwell Scientific Publications.

TOBIN, M. J. and HUTCHISON, D. C. S. (1982). An overview of the pulmonary features of α_1-antitrypsin deficiency. *Arch. Intern. Med.*, **142**, 1342–1348.

WHICHER, J. T. (1980). The interpretation of electrophoresis. *Brit. J. Hosp. Med.*, **24**, 348–360.

INDICATIONS FOR PROTEIN ESTIMATIONS

We have discussed the changes that may occur in plasma protein levels in disease. Demonstrating these changes does not always aid diagnosis. Before making a request, some assessment should be made as to whether the result of the estimation will aid diagnosis or treatment.

The following are the commonest indications for protein estimation.

To Assess Changes in Hydration

Changes in plasma *total protein or albumin* levels over short periods of time are almost certainly due to changes in hydration (p. 41).

To Evaluate Apparently Abnormal Levels, or Changes in Level, of a Protein-Bound Substance

Plasma *albumin* estimation must always accompany that of calcium (pp. 278 and 281). If changes in other protein-bound substances parallel those of the albumin, they are probably due to changes in protein levels.

To Investigate Oedema (see p. 55)

Very low plasma *albumin* levels suggest that hypoalbuminaemia is the cause of the oedema.

In the presence of hypoalbuminaemia a *serum electrophoretic pattern* typical of nephrotic syndrome or hepatic cirrhosis (Fig. 30) points to the cause. If oedema is due to causes other than hypoalbuminaemia the dilution leads to a fall in level of albumin and all the other protein fractions.

To Investigate Suspected Immune Complex Disease

Plasma C_3 determination is indicated if immune complex disease is suspected. If such disease is in an active phase, *levels will usually be low*: if it is high it suggests another cause for tissue damage.

The *electrophoretic pattern* may show a diffuse rise in γ-globulins in immune complex disease. It should be remembered that this finding may be due to a number of other types of disease.

To Investigate the Cause of Recurrent Infections

Serum immunoglobulin concentrations should be measured as part of the assessment of the adequacy of the immune system. Total haemolytic complement may be measured to detect inherited complement deficiencies. Note that these tests assess only *humoral* immunity.

Serum electrophoresis may indicate the cause of low immunoglobulins (for example, there may be a protein-loss pattern [p. 341] or a paraprotein).

Estimation of the 24-hour *urinary protein loss* may be indicated if this cause is suspected.

To Investigate Suspected Myeloma or Macroglobulinaemia

Electrophoresis should be carried out on *serum*, to detect a paraprotein, *and on urine*, to detect Bence Jones protein. The diagnosis must be confirmed by inspection of a bone marrow aspirate or biopsy.

Plasma immunoglobulin levels should be measured to assess the degree of immune paresis.

To Investigate Liver Disease

In chronic liver disease *serum albumin* levels may be low.

Serum electrophoresis may show β-γ fusion in cirrhosis or a diffuse increase in γ-globulin in chronic active hepatitis.

Serum immunoglobulin levels may help in differential diagnosis. In cirrhosis there is a predominant increase of IgA, in biliary cirrhosis of IgM, and in chronic active hepatitis of IgG.

Protein estimations are usually unhelpful in acute liver disease.

BLOOD SAMPLING FOR PROTEIN ESTIMATION

Blood for protein estimation, including that for immunoglobulins, should be taken with a *minimum of stasis*, or falsely high results may be obtained.

Clotted blood is essential for routine electrophoresis, because the presence of fibrinogen may mask, or be interpreted as, an abnormal protein.

Samples for *complement* estimations must be taken and processed so as to minimise *in vitro* activation. *It is essential to contact your laboratory for details before taking the blood.*

Blood for *cryoglobulin* measurement should be collected in a syringe *warmed to 37° C, and should be maintained at this temperature until it has been tested.* Failure to observe this precaution may result in false negative findings, because the cryoprecipitate is incorporated into the blood clot on cooling.

TESTING FOR URINARY PROTEIN

Several rapid screening tests are in routine use. There are two common *limitations:*

1. Most of the tests were developed to detect albumin and may be negative in the presence of even large amounts of other proteins such as BJP.

2. As these tests depend on protein *concentration*, very dilute urines may give negative results despite significant proteinuria.

It is *essential* that the samples should be *fresh*.

Albustix (Ames).—The test area of the reagent strip is impregnated with an indicator, tetrabromphenol blue, buffered to pH 3. At this pH it is yellow in the absence of protein; because protein forms a complex with the dye, stabilising it in the blue form, it is green or bluish-green if protein is present. The colour after testing is compared with the colour chart provided, which indicates the approximate protein concentration. The strips should be kept in the screwtop bottles, in a cool place. The instructions on the container should be carefully followed.

False positive results occur:

1. In strongly alkaline (infected or stale) urine, when the buffering capacity is exceeded. A green colour in this case is a reflection of the alkaline pH.

2. If the urine container is contaminated with disinfectants such as chlorhexidine.

False negative results occur if acid has been added to the urine as a preservative (for example, for estimation of urinary calcium).

Salicylsulphonic acid test.—The urine is filtered if turbid. 2 to 3 drops of 25 per cent salicylsulphonic acid are added to about 1 ml of clear urine. The appearance of turbidity or flocculation indicates the presence of protein.

False positive results may be due to:

radio-opaque media such as are used in intravenous pyelograms;
the presence of metabolites of tolbutamide;
administration of large doses of penicillin;
the presence of urate in high concentration.

False negative results may occur in very alkaline (stale) specimens.

Bradshaw's test for globulins, including Bence Jones protein.—*It must be stressed that the presence of BJP can only be confirmed or excluded in the laboratory.*

Urine is layered gently on a few ml of concentrated hydrochloric acid in a test tube. A thick, white precipitate at the interface indicates the presence of *globulin* in the urine (most commonly BJP). Although albumin does not give a positive reaction, in severe nephrotic syndrome the presence of globulins may cause a positive result. The test is only positive if Bence Jones proteinuria is gross.

Chapter XV

PLASMA ENZYMES IN DIAGNOSIS

MOST enzymes are present in cells at much higher concentrations than in plasma: some enzymes occur predominantly in certain types of cell. "Normal" plasma levels reflect the balance between the *synthesis and release of enzymes* during ordinary cell turnover and their *clearance* from the circulation. *Proliferation of cells*, an *increase in the rate of cell turnover, cell damage* or *enzyme induction* usually result in *raised plasma levels*; these are readily demonstrable, because very low concentrations produce easily measurable activity *in vitro*. Hence enzymes can be used as "markers" to detect and localise cell damage or proliferation.

Very occasionally plasma activities may be *lower than normal*, due either to *reduced synthesis* or to *congenital deficiency or variants*; cholinesterase deficiency is an example.

Assessment of Cell Damage and Proliferation

Plasma enzyme levels depend on:

> *the rate of release* from cells, in turn dependent on the rate of cell damage;
> *the extent of cell damage*, or the *degree of proliferation* or *induction of enzyme synthesis*.

These two factors are balanced against
> *the rate of enzyme clearance from plasma*.

Rapid damage of relatively few cells as occurs, for example, in many cases of viral hepatitis, may result in very high enzyme levels, which fall as the condition improves. By contrast, the liver may be much more extensively involved in advanced cirrhosis, but the rate of active cell damage is often low and enzyme levels may be only slightly raised or may even be "normal". In very severe liver disease, if many cells have been destroyed, fewer are left to undergo further damage and plasma enzyme levels may even *fall* terminally.

The mechanism by which most enzymes are removed from the plasma is unknown. The relatively small peptide enzymes, such as α-amylase, can be cleared by the kidney, but most are large proteins and may be catabolised by proteases. Each enzyme has a fairly constant biological half-life which differs from that of other enzymes. This knowledge may help in assessing the stage of an *acute* illness. After a myocardial infarction, for example, plasma levels of creatine kinase and aspartate transaminase (with short half-lives) fall to normal before those of lactate dehydrogenase (with a longer half-life). If there is circulatory failure the half-life may be prolonged.

If the fall of levels during recovery depends largely on renal excretion, as in the case of α-amylase, renal disease may delay this fall; it may even be the cause of high levels in the absence of pancreatic disease.

For these reasons it is unwise to predict the extent of cell damage from single or serial enzyme estimations without taking the clinical picture and other findings into account.

Localisation of Damage

Most enzymes of clinical importance are common to nearly all cells, although some are associated predominantly with certain tissues. Even if the enzyme is relatively specific for a particular tissue, elevated levels indicate only cell abnormality, and not the type of disease process. For example, following *major surgical procedures* or extensive *trauma*, levels of lactate dehydrogenase, aspartate transaminase and creatine kinase may be raised as the enzymes leak out of damaged cells, mainly of skeletal muscle. If *circulatory failure* is present due to cardiac failure or shock, and especially after cardiac arrest, very high levels of several enzymes may occur because of generalised hypoxic damage to cells and reduced rates of clearance; raised levels of "cardiac" enzymes do not necessarily mean that myocardial infarction caused the arrest. When unexplained raised levels (especially of lactate dehydrogenase) are found, the possibility of *malignancy* should be considered: some malignant cells contain very high concentrations of such enzymes. *Haemolysis* is often associated with increases in transaminases and other enzymes liberated from damaged erythrocytes.

Diagnostic precision may be improved in two ways:

by isoenzyme determination.—Measured enzyme activity may be due to closely related but slightly different molecular forms of the enzyme. These different forms (isoenzymes) can be identified by physical or chemical means and assume clinical significance when isoenzymes have different tissues of origin: for instance, different lactate dehydrogenase isoenzymes predominate in heart and liver;

by estimation of more than one enzyme.—Although many enzymes are widely distributed, their *relative concentrations in different tissues* may vary. For instance, both alanine and aspartate transaminases are abundant in liver, while in heart muscle there is much more aspartate than alanine transaminase. This difference is reflected in the markedly higher plasma aspartate than alanine transaminase level after myocardial infarction, whereas both are usually elevated if there is liver involvement.

The *distribution of enzymes within the cell* may differ. Some, such as alanine transaminase and lactate dehydrogenase, occur only in the *cytoplasm*. Others are found only in *mitochondria*, while aspartate transaminase, for example, occurs in both of these cellular compartments. Different relative levels may indicate different types of disease (p. 318).

Non-specific Causes of Raised Enzyme Levels

Before attributing enzyme changes to a specific disease process, it is advisable to consider more generalised causes. Some of these have already been mentioned.

Small rises in aspartate transaminase are common as a non-specific finding in a variety of illnesses, some of which may be minor. Moreover, even only moderate exercise or large intramuscular injections may elevate the plasma levels of muscle enzymes such as creatine kinase.

Other causes which should be considered are:

physiological

newborn.—The levels of some enzymes such as aspartate transaminase are moderately raised in the neonatal period;

childhood.—Alkaline phosphatase levels are high until after puberty;

pregnancy.—In the last trimester levels of alkaline phosphatase are usually raised. During and immediately after labour there are moderate rises in several enzymes, such as transaminases and creatine kinase, probably due to the uterine contractions.

enzyme induction by drugs

Some drugs, particularly phenytoin and barbiturates may stimulate the production of microsomal enzymes (induction). Raised levels of, for example, γ-glutamyltransferase in patients taking such drugs do not necessarily imply a pathological process.

artefactual elevations of enzyme levels

haemolysed or old samples are usually *unsuitable* for enzyme estimation. Even if the red cell does not contain the enzyme being measured, other released enzymes may interfere with the estimation to produce a falsely elevated result.

MEASUREMENT OF ENZYMES

The total plasma enzyme concentration (as protein) is less than 1 g/l. Results of assays are not expressed as concentrations, but as *activities*. *In vivo* and *in vitro* activity of enzymes depends not only on their concentrations, but also on the presence of activators and inhibitors.

Several analytical factors may affect the *in vitro* enzyme activity. For this reason results of the assay of the same enzyme, apparently in the same units, may vary between different laboratories. Similarly, numerical ratios of one enzyme to another may not have the same diagnostic implications in two laboratories. Important factors are the substrate and buffer used and the temperature of the assay. Because *the definition of "international units" does not take these factors into account, results from different laboratories, apparently in the same units, may not be directly comparable.*

Enzyme activities must be interpreted in relation to the "normal range" of the issuing laboratory, and reference ranges will not be given in this chapter.

ALTERED PLASMA ENZYME LEVELS

In this section, we will consider individual enzymes of clinical importance. Their application in defined clinical situations is considered later.

TRANSAMINASES

The transaminases are enzymes that are involved in the transfer of an amino group from an α-amino to an α-oxo acid. Pyridoxal phosphate is the cofactor for this reaction. Transaminases are widely distributed in the body.

Aspartate Transaminase (AST)

AST (serum glutamate oxaloacetate transaminase, SGOT) is widely distributed, with high concentrations in the heart, liver, skeletal muscle, kidney and erythrocytes, and damage to any of these tissues may cause raised levels.

Causes of Raised AST

Artefactual: *in vitro* haemolysis, or delayed separation of plasma from cells.

Physiological: newborn (approximately 1½ times adult "normal").

Markedly raised levels (10–100 times normal):

Myocardial infarction (p. 377);
Viral hepatitis (p. 320);
Toxic liver necrosis (p. 324);
Circulatory failure, with "shock" and hypoxia.

Moderately raised levels:

Cirrhosis (up to twice normal);
Cholestatic jaundice (up to 10 times normal);
Malignant infiltration of the liver;
Skeletal muscle disease (p. 379);
After trauma or surgery (especially cardiac surgery);
Severe haemolytic anaemia;
Infectious mononucleosis (liver involvement).

Alanine Transaminase (ALT)

ALT (serum glutamate pyruvate transaminase, SGPT) is present in high concentration in liver and to a lesser extent in skeletal muscle, kidney and heart.

Causes of Raised ALT

Markedly raised levels (10–100 times normal):

Viral hepatitis (p. 320):
Toxic liver necrosis (p. 324);
Circulatory failure, with "shock" and hypoxia.

Moderately raised levels:

Cirrhosis (up to twice normal) (p. 324);
Cholestatic jaundice (up to 10 times normal) (p. 322);
Liver congestion secondary to cardiac failure;
Infectious mononucleosis (liver involvement);
Extensive trauma and muscle disease (much less than AST).

LACTATE DEHYDROGENASE (LD)

LD catalyses the reversible interconversion of lactate and pyruvate. It is widely distributed, with high concentrations in the heart, skeletal muscle, liver, kidney, brain and erythrocytes, and measurement of total LD is therefore a non-specific index of cell damage.

Causes of Raised LD Levels

Artefactual: in vitro haemolysis, or delayed separation of plasma from cells.

Marked increase (more than five times normal):

Myocardial infarction (p. 377);

Some haematological disorders.

In blood diseases such as pernicious anaemia and leukaemias, very high values (up to 20 times normal) may be found. Lesser increases occur in other states of abnormal erythropoiesis, such as thalassaemia, myelofibrosis and haemolytic anaemia.

Circulatory failure, with "shock" and hypoxia;

Renal infarction, or sometimes rejection of a renal transplant.

Moderate increase:

Viral hepatitis;

Any malignancy;

Skeletal muscle disease;

Pulmonary embolism;

Infectious mononucleosis;

Cerebral infarction ⎱
Renal disease ⎰ occasionally.

Isoenzymes of LD

Five isoenzymes are normally detectable by electrophoresis and are referred to as LD_1 to LD_5. LD_1, the fraction that moves fastest towards the anode, predominates in heart muscle, erythrocytes and the kidney. The slow moving (cathodal) LD_5 isoenzyme is the most abundant form in liver and skeletal muscle. While in many conditions there is an increase in all fractions, the finding of certain patterns is of diagnostic value:

predominant elevation of LD_1 and LD_2 (LD_1 greater than LD_2) occurs in myocardial infarction, megaloblastic anaemia, and renal infarction;

predominant elevation of LD_2 and LD_3 occurs in acute leukaemia; LD_3 is the main isoenzyme elevated in many other malignancies;

elevation of LD_5 occurs after damage to liver or skeletal muscle.

A rise in LD_1 is most significant in the diagnosis of myocardial infarction. LD_1 (and, to a lesser extent, LD_2 and LD_3) can use β-hydroxybutyrate as well as lactate as substrate, whilst LD_4 and LD_5 cannot. Some laboratories make use of this fact, and assay hydroxybutyrate dehydrogenase (HBD) as an index of LD_1. Immunological methods for the specific measurement of LD_1 are now available but are expensive.

CREATINE KINASE (CK)

CK (CPK) is found mainly in heart muscle, brain and skeletal muscle, but also occurs in such other tissues as smooth muscle.

Causes of Raised CK Levels

Artefactually high levels occur in haemolysed samples with most of the commonly used methods of analysis.

Physiological:

Newborn—slightly raised;
Parturition—for a few days.

Marked increase:

Myocardial infarction (p. 377);
Muscular dystrophies (p. 379), and rhabdomyolysis (breakdown of skeletal muscle);
"Shock" and circulatory failure.

Moderate increase:

Muscle injury;
After surgery (for about a week);
Physical exertion. There may be a significant rise in plasma levels after only moderate exercise, or even muscle cramp;
After intramuscular injection;
Hypothyroidism (thyroxine apparently influences the catabolism of the enzyme);
Alcoholism (possibly due partly to alcoholic myositis);
Acute psychotic episodes;
Some cases of cerebrovascular accident and head injury;
Some patients predisposed to malignant hyperpyrexia (p. 398).

Isoenzymes of CK

The three CK isoenzymes are combinations of the protein subunits, M and B.

MM is the predominant *muscle* isoenzyme (*skeletal and cardiac*), and is detectable in normal plasma.

The *myocardium* also contains the *MB* form; this is normally undetectable in the plasma, and its presence suggests myocardial damage. The uses and limitations of CK-MB estimation are considered on p. 379.

BB is present in the *brain* and in the *smooth muscle* of the gastro-intestinal and genital tracts. Raised plasma levels may occur during parturition, and have been described after brain damage and in association with malignant tumours of bronchus, prostate and breast, but its measurement is not of proven diagnostic value for these conditions. In malignant disease, total CK activity is not always elevated.

α-AMYLASE (AMS)

α-Amylase is the enzyme concerned with the breakdown of dietary starch and glycogen to maltose. It is present in pancreatic juice and saliva as well as in the liver, fallopian tubes and muscle. The enzyme is excreted in the urine. The amylase in normal plasma is derived from both the pancreas and the salivary glands.

Estimation of total amylase is of most use for the diagnosis of acute pancreatitis (p. 297), during which the activity may be very high. Moderately raised levels are found in a number of conditions and are not of diagnostic significance.

Causes of Raised Plasma Amylase Levels

Marked increase (5–10 times normal):

> Acute pancreatitis;
> Severe glomerular dysfunction;
> Severe diabetic ketoacidosis.

Moderate increase (up to 5 times normal):

> Other acute abdominal disorders:
>> Perforated peptic ulcer;
>> Acute cholecystitis;
>> Intestinal obstruction;
>> Abdominal trauma;
>> Ruptured ectopic pregnancy;
> Salivary gland disorders:
>> Mumps—not usually required for diagnosis;
>> Salivary calculi;
>> After sialography;
> Morphine administration (spasm of sphincter of Oddi);
> Severe glomerular dysfunction (may be markedly raised);
> Myocardial infarction (occasionally);
> Acute alcoholic intoxication (transient);
> Diabetic ketoacidosis (may be markedly raised).

Sustained high levels may be associated with a pancreatic pseudocyst.

Macroamylasaemia. In some subjects a high plasma amylase activity is accompanied by low renal excretion of the enzyme, despite normal renal function. The condition is symptomless, and it is thought either that the enzyme is bound to a high molecular weight plasma component, such as a protein, or that the amylase molecules themselves form high molecular weight polymers. This harmless condition may be confused with other causes of hyperamylasaemia.

Low levels of plasma amylase may be found in infants up to one year.

α-Amylase is a relatively small molecule, rapidly cleared by the kidneys. Raised *urinary* levels occur whenever plasma levels are raised, unless these are due to renal failure or macroamylasaemia

Isoenzymes of Amylase

It is usually necessary to measure only total amylase activity if the diagnosis of acute pancreatitis is suspected (p. 297), although levels are not invariably raised in this condition. Pancreatic and salivary isoenzymes can now be distinguished by simple methods, which may be helpful to detect:

high levels of pancreatic isoenzyme in pancreatitis, when, for example, coexistent mumps or renal failure complicate the interpretation of a high total level;

low levels of pancreatic isoenzyme in chronic pancreatic disease.

Alkaline Phosphatase (ALP)

The alkaline phosphatases are a group of enzymes which hydrolyse phosphates at an alkaline pH. The activity measured by routine methods includes that of several isoenzymes. They are found in bone, liver, kidney, intestinal wall, lactating mammary gland and placenta. In bone the enzyme is found in *osteoblasts* and is probably important for normal bone formation.

In *adults*, the normal levels of alkaline phosphatase are derived largely from the *liver*. In *children*, in whom osteoblasts are very active, there is an additional contribution from *bone*, and this accounts for the higher total levels at this age; during the *pubertal growth spurt* activities rise even higher. In the elderly the levels of the bone isoenzyme may again increase slightly. In both adults and children there is a variable contribution from the intestine. *Pregnancy* raises the "normal" range because of the production of a heat-stable alkaline phosphatase by the *placenta*.

A placental-like isoenzyme (the *"Regan isoenzyme"*, named after the first patient in whom it was found) may be produced by malignant tumours, especially of the bronchus.

Causes of Raised Plasma ALP

Physiological:

Children—until about the age of puberty (up to 2–2½ times adult normal);

Puberty—during the pubertal growth spurt levels may be 5 or 6 times the adult normal;

Pregnancy—in the last trimester.

Bone disease:

Osteomalacia and rickets (p. 259);

Primary hyperparathyroidism *with bone disease* (p. 260);

Paget's disease of bone (may be very high);

Secondary carcinoma in bone;

Some cases of osteogenic sarcoma.

Liver disease:

Intra- or extrahepatic cholestasis (p. 319);

Space-occupying lesions—tumours, granulomata, infiltrations.

Malignancy—bone or liver involvement, or direct production.

Low Levels of Plasma ALP

Arrested bone growth:
Achondroplasia;
Cretinism;
Vitamin C deficiency.
Hypophosphatasia.

Isoenzymes of Alkaline Phosphatase

Bone disease (with increased osteoblastic activity) or liver disease (with involvement of the biliary tract) are the commonest causes of a raised plasma alkaline phosphatase activity.

If the cause of a raised plasma alkaline phosphatase is not clinically apparent further tests may be necessary. The isoenzymes originating from bone, liver, the intestine and the placenta may be separated by electrophoresis, but interpretation can be difficult, especially if both bone and liver fractions are elevated, or if the total ALP activity is only marginally raised. Differentiation can be facilitated by repeating electrophoresis after heating the plasma: the hepatic isoenzyme is more heat stable than that from bone, and the reduced level of bone isoenzyme may be evident when the first and second strips are compared. The placental isoenzyme is even more heat stable, and heat inactivation methods may be used during pregnancy to distinguish whether bone and/or liver fractions are also raised.

ACID PHOSPHATASE (ACP)

Acid phosphatase is found in the prostate, liver, red cells, platelets and bone. The main use of the estimation is in the diagnosis of prostatic carcinoma.

Normally acid phosphatase in prostatic secretions drains via the prostatic ducts and very little appears in the blood. In prostatic carcinoma, particularly if it has metastasised, plasma acid phosphatase levels rise, probably because of increased numbers of prostatic cells secreting the enzyme into the systemic circulation. If the tumour is confined to the prostate or if it is too undifferentiated to secrete acid phosphatase, plasma levels may be normal.

Many methods have been devised to measure only the prostatic fraction in plasma, without complete success. One of the best chemical methods uses l-tartrate to inhibit prostatic acid phosphatase: the assay is performed with and without tartrate and the difference between the two results is the prostatic "tartrate-labile" acid phosphatase. It is now possible to measure the enzyme protein, rather than its activity, by radioimmunoassay. The technique is oversensitive because it cannot distinguish early malignancy from benign prostatic enlargement: it is also non-specific, and the result may be abnormal in the presence of non-prostatic malignancy. The clinical value remains to be proved, and it should not be used as a screening test for carcinoma of the prostate.

Recent papers have suggested that rectal examination never increases plasma acid phosphatase levels. It is true that the effect is relatively rare, especially if the examination is performed by an experienced clinician who knows that the estimation is to be requested. However, in our experience the level does rise to two or three times the upper limit of "normal" in some cases, and then only falls to its basal level after some days. An example is given and a sampling procedure, taking this possibility into account, is suggested on p. 498. It is obvious that prostatectomy will release large amounts of the enzyme, and the assay should never be requested for at least a week after this operation.

Acid phosphatase is unstable, and specimens must be sent to the laboratory without delay.

Causes of Raised Plasma Acid Phosphatase

Tartrate-labile:

Prostatic carcinoma with extension;
Following rectal examination;
Acute retention of urine;
Passage of a catheter.

Total:

Artefactually raised levels occur in a haemolysed specimen;
Paget's disease of bone and some cases of metastatic malignant
disease (only if the ALP levels are very high, and therefore of
no diagnostic significance);
Gaucher's disease (probably from Gaucher cells);
Occasionally in thrombocythaemias.

γ-GLUTAMYLTRANSFERASE (GGT)

γ-Glutamyltransferase (γ-glutamyltranspeptidase) occurs mainly
in liver, kidney and pancreas. Raised plasma levels, however, are
almost always of hepatic origin and occur in two situations:

cholestatic liver disease, when they usually parallel the increase
in ALP. In cholestatic jaundice of pregnancy, GGT levels do
not increase;
induction by drugs. Many drugs, the most common being anticonvulsants and alcohol, induce synthesis of the enzyme.

Because of this, mildly and moderately raised (2–3 times normal)
levels are particularly difficult to interpret. The earlier assumption
that such levels are due to alcohol abuse is probably unfair to the
patient involved. *Markedly elevated* GGT levels, out of proportion to
those of transaminases suggest:

alcoholic hepatitis or gross alcohol abuse;
induction by anticonvulsants.

ENZYME PATTERNS IN DISEASE

MYOCARDIAL INFARCTION

The enzyme estimations of greatest value in the diagnosis of
myocardial infarction are AST, LD (or HBD) and CK. The choice of
estimation depends on the time interval after the suspected infarction.
A guide to the sequence of changes is given in Table XXX.

TABLE XXX

Enzyme	Starts to rise (hours)	Time after infarction Peak elevation (hours)	Duration of rise (days)
Total CK	4–8	24–48	3–5
AST	6–8	24–48	4–6
LD (HBD)	12–24	48–72	7–12

All enzyme levels may be normal until at least 4 hours after infarction: *blood should not be taken for diagnostic enzyme assay until this time after the onset of chest pain.*

Enzyme levels are raised in about 95 per cent of cases of myocardial infarction and are sometimes very high. The height of the rise is a very rough index of the extent of damage and as such is of some value in prognosis. It is, however, only one of many factors (p. 367). A second rise in enzyme levels after their return to normal indicates extension of the infarction, or the development of congestive cardiac failure; in the latter HBD and CK do not increase. Levels in angina are usually normal. If a patient is first seen after the total LD has returned to normal, diagnosis may still be possible on the basis of the finding of a raised heart isoenzyme (LD_1) as detected by HBD estimation or isoenzyme electrophoresis.

After even a small myocardial infarction there is usually some hepatic congestion due to right-sided heart failure, and therefore a slight rise in ALT levels. This is not usually a diagnostic problem because the AST is relatively much more elevated than the ALT, and the HBD (LD_1) is unequivocally high. If there is primary hepatic dysfunction, congestive cardiac failure without infarction, or pulmonary embolism (which usually causes some hepatic congestion), there is an approximately equal rise in ALT and AST, with normal HBD levels.

Measurement of total creatine kinase activity is rarely helpful in the diagnosis of myocardial infarction. If it is high when the AST and LD_1 (HBD) are normal, the elevation is most likely to be due entirely to the MM isoenzyme derived from skeletal muscle, perhaps because of an intramuscular injection, recent exercise or surgery. Detection of CK-MB by isoenzyme electrophoresis may occasionally help to detect a very recent infarction, but in most cases nothing is lost by taking a later sample for AST and LD_1 (HBD). Any patient suspected of having had a myocardial infarction should be kept under observation, and the delay of a few hours will not be detrimental to his management. Most of the CK released after myocardial infarction is the MM isoenzyme, common to skeletal and myocardial muscle; the MM has a longer half-life than the MB fraction and, *after about 24 hours*, plasma elevation of MM with undetectable MB does *not* exclude myocardial damage as a cause of elevated total CK levels: at this time the transaminase and HBD pattern is usually characteristic. *In most cases of suspected myocardial infarction, AST and HBD (LD₁) estimations, together with the clinical and ECG findings, are adequate to make a diagnosis.*

LIVER DISEASE

Enzyme changes in liver disease are discussed in Chapter XIII.

MUSCLE DISEASE

In the muscular dystrophies, probably because of increased leakage of enzymes from the diseased cells, plasma levels of muscle enzymes are increased. The relevant enzymes are CK and the transaminases. Of these CK estimation is the most valuable. Points to consider in interpretation are:

> levels are highest (up to 10 times normal or more) in the early stages of the disease. Later, when much of the muscle has wasted, they are lower and may even be normal;
>
> levels are higher following activity immediately after rest (release of CK built up in muscle during rest) than after prolonged activity;
>
> in detecting affected newborns, it must be remembered that levels at this age are higher than in adults.

Similar but less marked changes are found in many subjects with myositis.

Carriers of the Duchenne type of muscular dystrophy can often be detected by raised CK levels. Because elevations are only moderate, non-specific causes of raised enzyme activity (p. 367) must be excluded.

The specimen should be collected:

> at a time when normal levels would be expected to be maximum, that is *late in the day after ordinary physical activity* (levels may be normal in known carriers in the morning);
>
> *NOT during pregnancy* (levels may be low in early pregnancy);
>
> at a time when release from skeletal muscle is not abnormally high, that is:
>> *NOT for 48 hours after severe or prolonged exercise;*
>>
>> *NOT for 48 hours after an intramuscular injection.*

The assay must be performed on a fresh or carefully preserved specimen, and the laboratory should be informed *before* blood is taken. Preferably the estimation should be performed on three separate occasions, to minimise errors of interpretation due to random variations in plasma levels.

Although enzyme levels are usually *normal* in *neurogenic muscular atrophy*, the number of exceptions makes such tests an unreliable index in the differential diagnosis from primary muscle disease.

Enzymes in Malignancy

1. Levels of prostatic acid phosphatase rise in some cases of malignancy of the gland.

2. Malignancy anywhere in the body may be associated with a non-specific increase in LD, HBD and occasionally transaminases.

3. For follow-up of treated cases of malignancy, alkaline phosphatase estimations are of value. Raised levels may indicate secondary deposits in bone or liver. Liver deposits may, in addition, produce an increase in transaminases or LD.

4. Tumours may produce a number of enzymes, such as ALP, LD (HBD) and CK-BB.

Haematological Disorders

The extreme elevation of LD and HBD in megaloblastic anaemia and leukaemia has been mentioned (p. 371). Similar elevations may be found in other conditions with abnormal erythropoiesis. Typically there is much less change in the level of AST than in that of LD and HBD. Severe haemolysis produces changes in AST and LD (HBD).

PLASMA CHOLINESTERASE AND SUXAMETHONIUM SENSITIVITY

There are two cholinesterases, one found predominantly in erythrocytes and nervous tissue (acetylcholinesterase) and the other in plasma. This latter enzyme, cholinesterase ("pseudocholinesterase") is synthesised chiefly in the liver and is the one routinely measured.

Causes of Decreased Plasma Cholinesterase
Hepatic parenchymal disease:
hepatitis;
cirrhosis;
Anticholinesterases (organophosphates);
Inherited abnormal cholinesterase;
Myocardial infarction.

Causes of Increased Plasma Cholinesterase

Recovery from liver damage;
Nephrotic syndrome.

Suxamethonium Sensitivity

The muscle relaxant suxamethonium (Scoline) is normally broken down by cholinesterase and this limits its period of action. In certain subjects administration is followed by a prolonged period of apnoea. Rarely no enzyme is detectable, but most patients have been found to have low levels of plasma cholinesterase, and the enzyme present is qualitatively different from the normal. The abnormal cholinesterase may be classified by the degree of enzyme inhibition produced by dibucaine (a spinal anaesthetic) or by fluoride. Results are expressed as *dibucaine* and *fluoride numbers*. For example, normal cholinesterase is 80 per cent inhibited by dibucaine (dibucaine number 80) while in subjects homozygous for the defective gene the dibucaine number is about 20. Heterozygotes have intermediate numbers.

Ten possible combinations have been described. In general, marked sensitivity is found only in homozygotes. Discovery of a patient with this abnormality indicates the need to investigate the whole family and to issue approriate warning cards to all affected individuals.

SUMMARY

1. Enzyme activities in cells are high. Natural decay of cells releases enzymes into the plasma, where activities are measurable, but usually low.

2. Plasma enzyme estimations are most useful to detect raised levels due to cell damage.

3. Assays of few enzymes are specific for any one tissue, but isoenzyme studies may increase specificity. In general, patterns of enzyme changes together with clinical findings are used for interpretation.

4. Non-specific causes of raised plasma enzyme activity include peripheral circulatory insufficiency, trauma, malignancy and surgery.

5. Enzyme estimations are of value in:

(*a*) Myocardial infarction—AST, LD and isoenzymes (HBD), and sometimes CK.

(*b*) Liver disease—transaminases, ALP, and sometimes GGT.

(*c*) Bone disease—ALP.

(*d*) Prostatic carcinoma—tartrate-labile ACP.

(*e*) Acute pancreatitis—α-amylase.

(*f*) Muscle disorders—CK.

6. Artefactual increases may occur in haemolysed samples.

FURTHER READING

WILKINSON, J. H. (1976). *The Principles and Practice of Diagnostic Enzymology*. London: Edward Arnold. (The chapters dealing with myocardial infarction (8), enzymes in disease of skeletal muscle (9) and enzyme tests in diseases of the liver and hepatobiliary tract (10) are particularly useful.)

ROMAS, N. A., ROSE, N. R. and TANNENBAUM, M. (1979). Acid phosphatase: new developments. *Hum. Path.*, **10**, 501–512.

IRVIN, R. G., COBB, F. R. and COE, C. R. (1980). Acute myocardial infarction and MB creatine phosphokinase. *Arch. Intern. Med.*, **140**, 329–334.

WAGNER, G. S. (1980). Optimal use of serum enzyme levels in the diagnosis of acute myocardial infarction. *Arch. Intern. Med.*, **140**, 317–319.

PENN, R. and WORTHINGTON, D. J. (1983). Is γ-glutamyltransferase a misleading test? *Brit. Med. J.*, **286**, 531–535.

Chapter XVI

INBORN ERRORS OF METABOLISM

THE chemical make-up of an individual is determined by about 50 000 gene pairs transmitted from generation to generation on the chromosomes. Random selection and recombination during meiosis, as well as occasional mutation, introduce individual variations. Such variations may at one extreme be incompatible with life or, at the other, produce biochemical differences detectable only by special techniques, if at all. In the latter category are the genetic variations of plasma proteins such as the transferrins or haptoglobins: such differences are useful in population and inheritance studies but do not necessarily impair function. Between the two extremes there are many variations that do produce functional abnormalities. It is to these variations that the term "inborn errors of metabolism" applies.

GENERAL PRINCIPLES

Because the sequence of bases making up the DNA strands in the genes codes via RNA for protein structure, it is not surprising that most, if not all, inherited biochemical abnormalities can be explained by defective synthesis of a single peptide. This abnormality may be due to an abnormal *structural* gene with production of an abnormal protein, or to an abnormal *control* gene which alters the level of function of one or more structural genes and consequently the amounts of one or more structurally normal proteins. In most of the examples discussed in this chapter, the affected protein is an enzyme.

Possible Metabolic Consequences

Deficiency of a single enzyme in a metabolic chain may produce its effects in several ways. Let us suppose that substance A is acted on by enzyme X to produce substance B, and that substance C is on an alternative pathway.

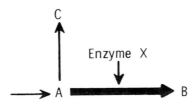

The consequences of a deficiency of X may be due to:

deficiency of the products of the enzyme reaction (B). Examples of this are cortisol deficiency in congenital adrenal hyperplasia (p. 152), and the hypoglycaemia of some forms of glycogenosis (p. 215);

accumulation of the substance acted on by the enzyme (A) (for instance, phenylalanine in phenylketonuria, p. 392). In many lysosomal storage diseases, substances normally enzymatically hydrolysed in lysosomes accumulate there because one of the enzymes is deficient;

diversion through an alternative pathway; some product of the latter may produce effects (C). The virilisation due to androgens in congenital adrenal hyperplasia falls into this group.

The effects of the last two types of abnormality will be aggravated if the whole metabolic pathway is controlled by feedback from the final product. For instance, in congenital adrenal hyperplasia, cortisol deficiency stimulates steroid synthesis and therefore the accumulation of androgens, and consequent virilisation, is accentuated.

The clinical effects of some inborn errors may be modified by, or depend entirely on, environmental factors. For example, because women lose iron during menstruation and pregnancy, those with idiopathic haemochromatosis accumulate less iron than men with the same condition. Patients with cholinesterase variants develop prolonged paralysis only if the muscle relaxant suxamethonium is administered (p. 380); in some subjects with glucose-6-phosphate dehydrogenase deficiency, haemolysis occurs only if they ingest drugs such as primaquine. Such cases appear "normal" without the intervention of modern therapeutics.

When to Suspect an Inborn Error of Metabolism

The possibility of an inherited metabolic defect should be considered in the presence of any bizarre clinical or biochemical picture in infancy or childhood, especially if more than one baby in the family has been affected. The following, *without obvious cause*, are particularly suggestive:

failure to thrive;
vomiting;
hypoglycaemia;
a peculiar smell or staining of napkins;
hepatosplenomegaly;
jaundice;
retarded mental development; fits; spasticity;
metabolic acidosis;

renal calculi;
rickets which is refractory to treatment.

Clinical Importance of Inborn Errors of Metabolism

The recognition of many inborn errors of metabolism is of academic interest only, because the abnormality produces no clinical effect. If no effective treatment is available it may be important to make a diagnosis so that genetic counselling can be undertaken. There is a group of diseases in which recognition in early infancy is vital, since *treatment may prevent irreversible clinical consequences or death.* Some of the more important of these are:

> phenylketonuria (p. 392);
> galactosaemia (p. 216);
> maple syrup urine disease (p. 395).

Examples of conditions which should be sought *in relatives of affected patients,* either because *further ill effects may be prevented,* or because a *precipitating factor should be avoided,* are:

> cholinesterase abnormalities (p. 380);
> glucose-6-phosphate dehydrogenase deficiency (p. 397);
> acute porphyrias (p. 434);
> haemochromatosis (p. 424);
> cystinuria (p. 391);
> Wilson's disease (p. 396).

Other conditions can be *treated symptomatically.* Examples are:

> hereditary nephrogenic diabetes insipidus (p. 51);
> congenital disaccharidase deficiency (p. 300);
> Hartnup disease (p. 391).

Some inborn errors of metabolism are completely, or almost completely, harmless. Their importance lies in the fact that they produce *effects which may lead to misdiagnosis* or which may alarm the patient. Examples of these are:

> renal glycosuria (p. 224);
> alkaptonuria (p. 395);
> Gilbert's disease (p. 326).

Finally, the clinical effects of some inborn errors of metabolism may not appear until after child-bearing age is reached: in these cases *genetic counselling* of blood relations is desirable. An example of this type of disease may be:

> Wilson's disease (p. 396).

Laboratory Diagnosis of Inborn Errors of Metabolism

The enzyme deficiency is usually demonstrated *indirectly* by detecting a high concentration of the substance normally metabolised by the deficient enzyme, for example phenylalanine in phenylketonuria. *Direct* enzyme assay is available only in special centres but, if possible, all cases should be confirmed by this method. *Prenatal diagnosis* of some metabolic defects is possible by studying cells cultured from the amniotic fluid in early pregnancy.

Screening for Inborn Errors

Because of the importance of early treatment many countries have instituted programmes of screening all newborn infants for inherited metabolic disorders, particularly for phenylketonuria. Tests may be performed on blood (obtained from a heel prick) or on urine. Blood, or more rarely urine, is often collected on filter paper, which is easy to transport to the laboratory.

The timing of the sample collection is important to avoid false negative results. Substances on the metabolic pathway before the enzyme block (for example, phenylalanine in phenylketonuria or galactose in galactosaemia) only accumulate once the infant begins to ingest the precursor (such as protein or milk products). Blood samples for screening apparently well infants are usually collected on the 6th to 9th day of life. Abnormal metabolites may not be found in the urine before 4 to 6 weeks after birth if the "renal threshold" is relatively high.

A positive result on screening should be confirmed by quantitative analysis or by repeat testing. Many abnormalities detected ,are transient.

The many pitfalls in the interpretation of the results of screening the newborn for inborn errors, and the responsibility of those who undertake to do so, are reviewed in the reference quoted at the end of the chapter.

Treatment of Inborn Errors

Some inborn errors of metabolism are amenable to treatment by supplying the missing metabolite or by limiting the dietary intake of precursors in the affected metabolic pathway. Accumulated products may occasionally be removed (for example, iron in haemochromatosis).

PATTERNS OF INHERITANCE

The following account is intended for quick revision only. Details are available in books on genetics.

Every inherited characteristic is governed by a pair of genes on homologous chromosomes (one received from each of the parents). Different genes governing the same characteristic are called alleles. If an individual has two identical alleles he is *homozygous* for that gene or inherited characteristic; if he has two different alleles he is *heterozygous*. Genes may be carried on the sex chromosomes (X and Y) or on the autosomes (similar in both sexes) and the patterns of inheritance differ.

Autosomal Inheritance

1. Let us suppose that one parent (Parent 1 in the example below) carries an "abnormal" gene (A). If N is a normal gene, the possible gene combinations in the offspring are shown in the square.

	Parent 2	
	N	N
Parent 1 A	*AN*	*AN*
N	NN	NN

It will be seen that, on a *statistical basis*, half the offspring will carry one abnormal gene (*AN*): they are *heterozygous* for this gene, like Parent 1. None will be homozygous for the abnormal gene (*AA*).

2. If both parents are heterozygous, a quarter of the offspring (in a large series) will be homozygous (*AA*) and half heterozygous.

	Parent 2	
	A	N
Parent 1 A	*AA*	*AN*
N	*AN*	NN

3. If one parent is homozygous and the other "normal" all offspring will be heterozygous.

Since most genes producing clinical abnormalities are rare, example 1 above has the highest statistical likelihood. With consanguineous marriages example 2 becomes more probable, since blood relatives are more likely to carry the same abnormal genes than two unrelated people.

The *consequences* of the carriage of the abnormal gene depend on its potency compared to that of the normal one.

A *dominant* gene produces abnormality in both heterozygotes and homozygotes although homozygotes may be more severely affected. Thus, in example 1, Parent 1 and half the offspring would be affected, and in example 2, both parents and 75 per cent of the offspring would be affected. Characteristically, cases appear in *successive generations*.

A *recessive* gene produces the abnormality only in homozygous individuals. Thus, in example 1, neither parents nor offspring would be affected, and in example 2 the parents would appear normal but 25 per cent of the offspring would be affected. Characteristically, one or more cases appear in a *single generation* with apparently normal parents.

The terms "dominant" and "recessive" are relative. A dominant gene may fail to manifest itself (*incomplete penetrance*) and so appear to skip a generation. A gene may vary in its *degree of expression*, and therefore in the degree of abnormality it produces. Finally, a recessive gene, which produces disease only in the homozygote, may nevertheless be detectable by biochemical tests in the heterozygote.

Sex-linked Inheritance

Some abnormal genes are carried only on the sex chromosomes, almost always on the X chromosome.

X-linked recessive inheritance.—Females carry two X chromosomes and males one X and one Y. In X-linked recessive inheritance an abnormal X chromosome (Xa) is latent when combined with a normal X chromosome, but active when combined with a Y. If the mother carries Xa she will appear to be normal, but, statistically, half the sons will be affected (*YXa*). Half the daughters will be carriers (*XXa*), but all daughters will appear clinically normal.

	Mother		
	Xa	X	
Father X	*XXa*	XX	←Daughters
Y	*YXa*	YX	←Sons

If the father is affected and the mother carries two normal genes, none of the sons will be affected, but all daughters will be carriers.

	Mother		
	X	X	
Father Xa	*XXa*	*XXa*	←Daughters
Y	XY	XY	←Sons

Inherited disease manifesting only in male offspring, but carried by females, is typical of X-linked recessive inheritance. The female is only clinically affected in the extremely rare circumstance when she is homozygous for the abnormal gene. This would only occur if there were an affected father and carrier mother.

Haemophilia is the classical example of X-linked recessive inheritance.

X-linked dominant inheritance.—In this type of inheritance both XXa and YXa (females and males) are affected. An example is familial hypophosphataemia (p. 267).

Multiple Alleles

Occasionally there may be several alleles governing the same characteristic. Different pair combinations may then produce different disease patterns (for example, some haemoglobinopathies), or the variation may only be detectable by biochemical testing (for example, plasma protein variants).

DISEASES DUE TO INBORN ERRORS OF METABOLISM

Only a very small number of the known inborn errors of metabolism will be discussed. Those selected illustrate some of the principles mentioned earlier. The choice must be biased by what the authors feel to be important, and others might disagree. On p. 385 some of the more clinically important abnormalities have been listed, and on p. 400 a fuller (but by no means complete) list is included under systematic headings; this includes the mode of inheritance, when known. Many of these conditions have been mentioned briefly in the relevant chapters. A few remain, which the authors feel to be of relative importance.

Incidence

All the inborn errors of metabolism are very rare. The approximate incidence of disorders has been established by screening programmes in several countries. Of the disorders discussed below phenylketonuria, Hartnup disease, cystinuria, familial iminoglycinuria and histidinaemia are the commonest (1 in 10 000–20 000); maple syrup urine disease is much rarer (about 1 in 350 000).

AMINOACIDURIA

Because disturbances of amino acid metabolism or excretion occur in many inborn errors of metabolism, aminoaciduria is one of the first manifestations to be sought in suspected cases. It will be discussed briefly before considering specific diseases.

Amino acids are normally filtered at the glomerulus, and reach the proximal renal tubule at concentrations equal to those in plasma: almost all are actively reabsorbed at this site. Aminoaciduria may therefore be of two types:

overflow aminoaciduria in which, because of raised plasma levels, amino acids reach the proximal tubule at concentrations higher than the reabsorptive capacity of the cells;

renal aminoaciduria in which plasma levels are low because of urinary loss due to defective tubular reabsorption.

A further subdivision may be based on the pattern of the excreted amino acids:

specific aminoaciduria is the excessive excretion of either a single amino acid or a group of related amino acids. It may be overflow or renal in type and is almost always due to a genetic defect.

non-specific aminoaciduria is the excessive excretion of a number of unrelated amino acids. It is almost always due to an acquired lesion. It may be overflow in type, as in severe hepatic disease, when failure of deamination of amino acids causes raised plasma levels; more commonly renal aminoaciduria results from non-specific proximal tubular damage from any cause (p. 19). In the latter, known as the *Fanconi syndrome*, other substances reabsorbed in the proximal tubule are lost in excessive amounts (phosphoglucoaminoaciduria): its occurrence in inborn errors of metabolism is much more commonly due to secondary tubular damage by the substance not metabolised normally (for instance, copper in Wilson's disease) than to a direct primary genetic defect. Acquired lesions will not be discussed further in this chapter.

INHERITED ABNORMALITIES OF TRANSPORT MECHANISMS

Groups of chemically similar amino acids are often reabsorbed in the renal tubule by shared or interrelated mechanisms. In several cases similar group-specific mechanisms are involved in intestinal absorption and defects involve both the renal tubule and intestinal mucosa. Inborn errors involving the following group pathways have been identified:

the dibasic amino acids (with two amino groups), cystine, ornithine, arginine and lysine (COAL is a useful mnemonic) (*cystinuria*);

many neutral amino acids (with one amino and one carboxyl group) (*Hartnup disease*);

the imino acids, proline and hydroxyproline, probably shared with glycine (*familial iminoglycinuria*).

Cystinuria

Cystinuria is due to an inherited abnormality of tubular reabsorption of the dibasic amino acids, cystine, ornithine, arginine and lysine, resulting in excessive urinary excretion of these four amino acids. A similar transport defect is present in the intestinal mucosa, but, although cystine absorption is diminished, it is produced in the body, and deficiency does not develop. The failure of renal tubular reabsorption results in a high urinary excretion and because cystine is relatively insoluble it may precipitate in the renal tract with calculus formation and its attendant complications. The solubility of cystine is such that only in homozygotes do urinary concentrations reach levels at which precipitation may occur and cause crystalluria and stone formation, although increased excretion can be demonstrated in heterozygotes.

The *diagnosis* of cystinuria is made by the demonstration of excessive urinary excretion of cystine and the other characteristic amino acids. The demonstration of the latter is necessary to distinguish the stone-forming homozygotes from heterozygous cystine-lysinurias and from cystinuria occurring as part of a generalised aminoaciduria.

The *management* of cystinuria is aimed at the prevention of the formation of calculi by a high fluid intake *day and night*, thus reducing the urinary cystine concentration. Alkalinising the urine also increases the solubility of cystine. If these measures are inadequate, administration of D-penicillamine, which produces a more soluble compound, may be tried.

The several genetic forms of cystinuria follow an autosomal recessive pattern of inheritance. Many cases are asymptomatic.

The relatively harmless condition described above must not be confused with **cystinosis**. This is a very rare inherited disorder of cystine metabolism characterised by accumulation of intracellular cystine in many tissues. In the kidney this produces tubular damage and consequently the Fanconi syndrome. The aminoaciduria is nonspecific and of renal origin. Death occurs at an early age.

Hartnup Disease

Hartnup disease, named after the first described patient, is a rare but interesting disorder in which there is a renal and intestinal transport defect involving neutral amino acids.

Most, if not all, the clinical manifestations can be ascribed to the reduced intestinal absorption and increased urinary loss of *tryptophan*. The amino acid is normally partly converted to nicotinamide, this conversion being especially important if the dietary intake of nicotinamide is marginal (p. 448), an example of environmental modification of an inherited defect. The clinical features of Hartnup disease are intermittent and resemble those of pellagra, namely:

a red, scaly rash on exposed areas of skin;
reversible cerebellar ataxia;
mental confusion of variable degree.

The thesis that nicotinamide deficiency is the cause of the clinical picture is supported by the response to administration of the vitamin and the fact that the features of the disease are frequently preceded by a period of dietary inadequacy.

In spite of the generalised defect of amino acid absorption there is no evidence of protein malnutrition: this may possibly be due to absorption of intact peptides by a different pathway.

An associated chemical feature is the excretion of excessive amounts of *indole* compounds in the urine. These originate in the gut from the action of bacteria on the unabsorbed tryptophan.

Hartnup disease has a recessive mode of inheritance.

Diagnosis is made by demonstrating the characteristic amino acid pattern in the urine. Heterozygotes are not readily detectable by present techniques.

Familial Iminoglycinuria

An abnormality of the transport mechanism of the imino acids leads to increased urinary excretion of proline, hydroxyproline and glycine, but with normal plasma levels. The condition is apparently harmless but must be distinguished from other, more serious causes, of iminoglycinuria. It is inherited as an autosomal recessive trait.

DISORDERS OF AMINO ACID METABOLISM

A number of inborn errors of amino acid metabolism have been described. Most are characterised by raised levels of the relevant amino acid(s) in the blood, with overflow aminoaciduria. Only a few of the better known conditions are described here.

Disorders of Aromatic Amino Acid Metabolism

The main chemical reactions of this pathway are outlined in Fig. 33, together with the site of known enzyme defects. It will be seen that *tyrosine*, normally produced in the body from phenylalanine, is the precursor of several important substances.

The inherited defects in thyroid hormone synthesis are considered on p. 185.

Phenylketonuria.—This condition is caused by an abnormality of the *phenylalanine hydroxylase system*, usually of the enzyme itself,

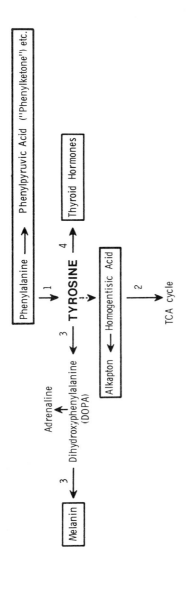

Enzyme Affected

1. Phenylalanine Hydroxylase (Phenylketonuria)
2. Homogentisic acid oxidase (Alkaptonuria)
3. Tyrosinase (Albinism)
4. Thyroid enzymes (Thyroid Dyshormonogenesis)

Squares indicate some of the substances which may be shown to be present in abnormal amounts in the diseases discussed

Fig. 33.—Some inborn errors of the aromatic amino acid pathway.

but occasionally synthesis of a cofactor, tetrahydrobiopterin. This shows that several different inherited defects may have very similar biochem- ical and clinical consequences. As phenylalanine hydroxylase catalyses the conversion of phenylalanine to tyrosine the former accumulates in the blood and is excreted in the urine together with its derivatives, such as phenylpyruvic acid: the disease acquires its name from the recognition of this (phenylketone) in the urine.

Extensive screening of newborn infants is carried out in many countries to detect cases early. The problems arising from such surveys are discussed later.

The clinical features are:

> irritability, feeding problems, vomiting and fits in the first few weeks of life;
> mental retardation developing at between 4 and 6 months with unusual psychomotor irritability;
> generalised eczema in many cases;
> a tendency to reduced melanin formation. Many patients have a pale skin, fair hair and blue eyes.

Diagnosis.—The *phenylalanine concentration* may be measured in *blood* taken from a heel-prick. The technique is suitable for mass screening and tests are best performed in special centres. The timing of the test is critical (p. 386). It is recommended that tests be performed between 6 and 10 days after birth (just before the infant leaves hospital).

Phenylpyruvic acid in the urine reacts with ferric chloride or Phenistix (Ames), but these tests may not be positive until about six weeks after birth, by which time there may be permanent cerebral damage.

Since the introduction of screening tests it has become apparent that raised blood levels of phenylalanine occur in conditions other than phenylketonuria, particularly in premature babies, when it is possibly due to delayed maturation of enzyme systems. These cases may be distinguished by repeated blood testing and by estimation of blood tyrosine levels (not raised in phenylketonuria, but raised in many of the other conditions).

A variant, persistent hyperphenylalaninaemia, without mental retardation has been described.

Babies exposed *in utero* to the high phenylalanine levels of unrecognised or untreated phenylketonuric mothers may be mentally retarded and show other congenital abnormalities, even though they themselves are not phenylketonuric (*maternal phenylketonuria*).

Management.—The aim of management is to lower blood phenyl-alanine levels by giving a low phenylalanine diet. Such treatment is difficult, expensive and tedious for the patient and parents, and requires careful biochemical monitoring of blood phenylalanine levels. Phenylalanine deficiency itself has deleterious effects. Tyrosine must be included in the diet as it is the precursor of many important metabolites (Fig. 33).

Phenylketonuria is inherited as an autosomal recessive. *Heterozygotes* are clinically normal, but may be detected by biochemical tests.

Alkaptonuria.—An inherited deficiency of *homogentisic acid oxidase* results in alkaptonuria. Homogentisic acid accumulates in blood, tissues and urine. Oxidation and polymerisation of this substance produces the pigment "alkapton", in much the same way as polymerisation of DOPA (see Fig. 33) produces melanin. Deposition of alkapton in cartilages, with consequent darkening, is called *ochronosis*: this may cause *arthritis* in later life and may be clinically visible as darkening of the ears. Conversion of homogentisic acid to "alkapton" is speeded up in alkaline conditions and the most obvious abnormality in alkaptonuria is passage of *urine which is black*, or which darkens as it becomes more alkaline on standing. This finding may, however, be absent in a significant number of cases. The condition is often first noticed by the mother who is worried by the black nappies, which only become blacker when washed in alkaline soaps or detergents. The condition is compatible with a normal life span and treatment is unnecessary, but arthritis in middle and later life is common. Homogentisic acid is a *reducing substance* and reacts with Clinitest tablets (p. 225).

Alkaptonuria is inherited as an autosomal recessive trait. Heterozygotes are not detectable by clinical or by biochemical findings.

Albinism.—A deficiency of *tyrosinase* in melanocytes causes one form of albinism and is inherited as a recessive character. The patient lacks pigment in the skin, hair and iris (the eyes appear pink) and the condition is especially obvious in negroes. Acute photosensitivity occurs because of pigment lack in the skin and iris. The tyrosinase involved in catecholamine synthesis is a different enzyme, controlled by a different gene. Adrenaline metabolism is normal in albinos.

Disorders of other Amino Acids

Maple syrup urine disease.—In maple syrup urine disease there is deficient decarboxylation of the oxoacids resulting from deamination of the three *branched-chain amino acids*, leucine, isoleucine and valine. These accumulate in the blood and are excreted in the urine together with their corresponding oxoacids. The odour of the urine, which resembles that of maple syrup, gives the disease its name.

The disease presents in the first week of life and if it is not treated severe neurological lesions develop with death in a few weeks or months. If, on the other hand, the condition is recognised and a diet low in branched-chain amino acids is given, normal development seems possible.

Diagnosis is made by demonstrating the raised levels of branched-chain amino acids in blood and urine. It may be confirmed by demonstrating the enzyme defect in the leucocytes.

The condition has a recessive mode of inheritance.

Histidinaemia.—Histidinaemia is associated with deficiency of the enzyme *histidase* which is required for the normal metabolism of histidine. Blood levels of histidine are raised and histidine and a metabolite, *imidazole pyruvic acid*, appear in increased amounts in the urine. Like phenylpyruvic acid (excreted in phenylketonuria) imidazole pyruvic acid reacts with ferric chloride to give a blue-green colour with ferric chloride or Phenistix (Ames). About half the cases described have shown mental retardation and speech defects but the rest appear normal. The results of dietary therapy are as yet inconclusive.

The condition is probably inherited as an autosomal recessive trait.

DEFECTS OF METAL METABOLISM

Two inherited disorders are associated with an abnormal accumulation of metals in the body. Iron overload (*idiopathic haemochromatosis*) is discussed in Chapter XVIII. Copper accumulates in Wilson's disease.

Wilson's Disease (Hepatolenticular Degeneration)

Some plasma copper is loosely bound to albumin, but most is normally incorporated in the protein *caeruloplasmin*. Copper is mainly excreted in the bile.

There are two defects of copper metabolism in Wilson's disease:

impaired biliary excretion leads to *copper deposition in the liver*;

deficiency of caeruloplasmin results in *low plasma copper* levels: most of this copper is in a loosely bound form, and is *deposited in tissues* and *filtered at the glomerulus* more readily than normal. Urinary copper excretion is increased.

Excessive deposition of copper in the basal ganglia of the brain, liver, renal tubules and the eye produces:

neurological symptoms due to basal ganglia degeneration;

liver damage leading to cirrhosis;

renal tubular damage with any or all of the biochemical features of this condition, including aminoaciduria (Fanconi syndrome);

Kayser-Fleischer rings at the edges of the cornea due to copper deposition in Descemet's membranes.

Diagnosis.—The vast majority of patients have low plasma caeruloplasmin and copper concentrations.

When interpreting caeruloplasmin concentrations it is important to remember that *low levels* may also occur during the *first few months of life*, with *malnutrition* and in the *nephrotic syndrome* due to loss in the urine. *Raised levels* are found in *active liver disease* (this may account for the rare finding of "normal" levels in patients with Wilson's disease), in the *last trimester of pregnancy*, in women taking *oral contraceptives* and non-specifically when there is *tissue damage* (for example, inflammation and neoplasia).

In some cases copper estimation on a liver biopsy specimen may be needed for diagnosis.

The clinical condition has a recessive mode of inheritance, but heterozygotes may have reduced caeruloplasmin levels. Distinction between presymptomatic homozygotes and heterozygotes is important, because the former should be treated.

Treatment with agents chelating copper, such as D-penicillamine, aims to reduce tissue copper concentration.

DRUGS AND INHERITED METABOLIC ABNORMALITIES

The variation in individual response to identical drugs may be partly due to genetic variation. There are, however, a number of well-defined inherited disorders that are aggravated by, or which only become apparent after, administration of certain drugs. These may be classified into two groups.

Disorders Resulting from Deficient Metabolism of a Drug

The muscle relaxant suxamethonium (succinylcholine) normally has a very brief action as it is rapidly broken down by plasma cholinesterase. In *suxamethonium sensitivity* (p. 380) an abnormal cholinesterase is present, breakdown of the drug is impaired and prolonged respiratory paralysis may occur.

Two other inherited disorders are characterised by defective metabolism of the drugs *isoniazid* and *phenytoin* respectively. In both, toxic effects appear more frequently and at lower dosages than in normal individuals.

Disorders Resulting from Abnormal Response to a Drug

A form of haemolytic anaemia, common in many parts of the world, is due to a deficiency of *glucose-6-phosphate dehydrogenase* (G-6-PD) in the erythrocyte. This is the first enzyme in the hexose monophosphate shunt and is required for the formation of NADPH. This, in turn, is probably essential for the maintenance of an intact red cell membrane. Numerous variants of G-6-PD deficiency have been described. In many cases haemolysis is precipitated by drugs, notably certain antimalarial drugs such as primaquine, and sulphonamides and vitamin K analogues.

In the inherited *hepatic porphyrias* (p. 434) acute attacks may be precipitated by several drugs, particularly barbiturates.

Some people react to general anaesthetics (most commonly halothane with suxamethonium) with a rapidly rising temperature, muscular rigidity and acidosis; the majority die as a result (*malignant hyperpyrexia*). Many, but not all, susceptible subjects in affected families have an elevated creatine kinase (CK) level.

This short section should remind readers to consider the possibility of an inborn error when an abnormal reaction to a drug is encountered. A reference to this growing field of *pharmacogenetics* is given at the end of the chapter.

SUMMARY

1. Inborn errors of metabolism may cause diseases due to inherited defects of protein synthesis. Most cases presenting with clinical symptoms are due to abnormalities of enzyme synthesis.

2. Inborn errors of metabolism may produce no clinical effects, may only produce them under certain circumstances (for example, cholinesterase variants) or, at the other extreme, may produce severe disease. Some are incompatible with life.

3. Recognition of some inherited abnormalities is of academic interest only. Diagnosis is important if the condition is serious but treatable, if precipitating factors can be avoided, or if confusion with other diseases is possible.

4. Inheritance may be autosomal or sex-linked, dominant or recessive. In diseases producing severe clinical effects inheritance is most commonly autosomal recessive, and they are most common in the offspring of consanguineous marriages.

5. In many cases in which the clinical disease is inherited in a recessive manner, lesser degrees of the abnormality can be detected by chemical testing.

6. Some inborn errors of metabolism not mentioned elsewhere in the book are discussed in this chapter.

FURTHER READING

HOLTON, J. B. (1982). Diagnosis of inherited metabolic diseases in severely ill children. *Ann. Clin. Biochem.*, **19**, 389–395.

HOLTZMAN, N. A. (1978). Newborn screening for inborn errors of metabolism. *Pediat. Clin. N. Amer.*, **25**, 411–421.

SCRIVER, C. R. and CLOW, C. L. (1980). Phenylketonuria: epitome of human biochemical genetics. *New Engl. J. Med.*, **303**, 1336–1342 and 1394–1400.

VESELL, E. S. (1979). Pharmacogenetics: multiple interactions between genes and environment as determinants of drug response. *Amer. J. Med.*, **66**, 183–187.

REFERENCE

STANBURY, J. B., WYNGAARDEN, J. B. FREDRICKSON, D. S., GOLDSTEIN, J. L. and BROWN, M. S. (Eds.). (1983). *The Metabolic Basis of Inherited Disease*, 5th edit. New York: McGraw-Hill.

SOME INBORN ERRORS AND THEIR INHERITANCE

The following list of inborn errors of metabolism is far from complete. It is meant for reference only, and the student should not attempt to learn it. Most of the abnormalities have been discussed in this book, and a page reference is given. Where it is known the mode of inheritance is given, unless the heading applies to a group of diseases of different modes of inheritance.

D = Autosomal Dominant
R = Autosomal Recessive
X-linked D = X-linked Dominant
X-linked R = X-linked Recessive

	Inheritance	Page
I. DISORDERS OF CELLULAR TRANSPORT		
Most of these are recognised as renal tubular transport defects, and in some defective intestinal transport can also be demonstrated.		
Amino Acids		
Dibasic amino acids—Cystinuria	R	391
Neutral amino acids—Hartnup disease	R	391
Familial iminoglycinuria	R	392
Glucose		
Renal glycosuria	D	224
Water (failure to respond to ADH)		
Hereditary nephrogenic diabetes insipidus	X-linked R	51
Sodium (failure to respond to aldosterone)		
Pseudo-Addison's disease	?	53
Potassium (all cells)		
Familial periodic paralysis	D	74
Calcium (failure to respond to PTH)		
Pseudohypoparathyroidism	X-linked D	264
Phosphate		
Familial hypophosphataemia	X-linked D	267
Hydrogen Ion		
Renal tubular acidosis	D	111
Bilirubin (liver cells)		
Congenital hyperbilirubinaemias		
Crigler-Najjar syndrome Type 1	R	326
Crigler-Najjar syndrome Type II	D	326
Dubin-Johnson syndrome	?R	326
Gilbert's disease	?	326
Rotor syndrome	?R	327
II. DISORDERS OF AMINO ACID METABOLISM		
Aromatic Amino Acids		
Phenylketonuria	R	392
Alkaptonuria	R	395
Albinism	R	395
Thyroid dyshormonogenesis	All R	185

	Inheritance	Page
Sulphur Amino Acids		
Cystinosis	R	391
Homocystinuria	R	—
Branched-Chain Amino Acids		
Maple syrup urine disease	R	395
Histidinaemia	R	396
III. DISORDERS OF CARBOHYDRATE METABOLISM		
Glycogenoses	R	215
Galactosaemia	R	216
Hereditary fructose intolerance	R	216
Essential pentosuria	R	226
Essential fructosuria	R	225
Diabetes mellitus	?	204
IV. DISORDERS OF LIPID METABOLISM		
Hyperlipoproteinaemias	—	242/3
Hypolipoproteinaemias	R	243
Plasma LCAT deficiency	R	244
V. ABNORMALITIES OF PLASMA PROTEINS		
Immunoglobulin deficiencies	—	352
Carrier protein abnormalities		
Transferrin deficiency	?R	423
Thyroxine-binding globulin deficiency	X-linked	180
Cholinesterase variants	R	380
α_1-Antitrypsin deficiency		345
Bisalbuminaemia	D	342
Analbuminaemia	R	342
VI. DISORDERS OF METAL METABOLISM		
Haemochromatosis	?	424
Wilson's disease	R	396
VII. ERYTHROCYTE ABNORMALITIES (see haematology textbooks)		
Haemoglobinopathies		—
Glucose-6-phosphate dehydrogenase deficiency	X-linked	397
NADP methaemoglobin reductase deficiency	R	
VIII. DISORDERS OF PORPHYRIN METABOLISM		
Porphyrias	—	434
IX. DISORDERS OF STEROID METABOLISM		
Congenital adrenal hyperplasias	All R	152

Chapter XVII

PURINE AND URATE METABOLISM

HYPERURICAEMIA AND GOUT

HYPERURICAEMIA may be asymptomatic, or may give rise to the clinical syndrome of gout. Unequivocally high urate levels, even if symptomless, should be treated because the relatively insoluble urate may, like calcium, precipitate in tissues. If this takes place in the kidney renal damage can result. Hyperuricaemia may be due to a primary lesion of purine metabolism or be secondary to a variety of other conditions. The primary syndrome has a familial incidence.

At the pH of plasma most urate is ionised at position 8 (Fig. 34). This anionic group is associated with the predominant extracellular cation, sodium: monosodium urate, although relatively poorly soluble, is more so than the less ionised uric acid formed if the pH falls below about 6, as it may in the urine.

NORMAL URATE METABOLISM

Urate is the end product of purine metabolism in primates, including man. In most other mammals it is further broken down to the soluble compound, allantoin, and it is because of the poor solubility of urates that man is prone to clinical gout and renal damage by urate. The purines, adenine and guanine, are constituents of both types of *nucleic acid* (DNA and RNA). The purines used by the body for nucleic acid synthesis may be derived from the breakdown of ingested nucleic acid (mainly in meat which is rich in cells), or may be synthesised in the body from small molecules *de novo*.

Synthesis of Purines

The synthetic pathway of purines is complex, and involves the incorporation of many small molecules into the relatively complex purine ring. The upper part of Fig. 34 summarises some of the more important steps in this synthesis. Cytotoxic drugs inhibit various stages in this pathway, so preventing DNA formation and cell growth.

The following stages in Fig. 34 should be especially noted.

The first step in purine synthesis is condensation of pyrophosphate

PURINE SYNTHESIS

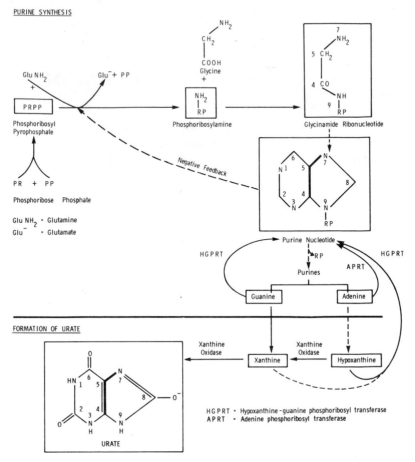

Glu NH₂ = Glutamine
Glu⁻ = Glutamate

HGPRT = Hypoxanthine-guanine phosphoribosyl transferase
APRT = Adenine phosphoribosyl transferase

FIG. 34—Summary of purine synthesis and breakdown to show steps of clinical importance.

with phosphoribose to form phosphoribosyl pyrophosphate (PRPP).

The amino group of glutamine is then incorporated into the ribose phosphate molecule and pyrophosphate is released. Amidophosphoribosyl transferase catalyses this *rate-limiting* or controlling step in purine synthesis. It is subject to feedback inhibition from increased levels of purine nucleotides: thus the rate of synthesis is slowed when its products increase. This step may be at fault in primary gout.

Glycine is next added to phosphoribosylamine. Labelled glycine can be used to study the rate of purine synthesis. The atoms in the glycine molecule have been numbered in the diagram to correspond with those of the purine and urate molecules, and the heavy lines further indicate the final position of the amino acid in these molecules. By the use of labelled glycine it has been shown that purine synthesis is increased in primary gout.

After many complex steps purine ribonucleotides (purine ribose phosphates) are formed and, as has already been stated, the levels of these control the first step in the synthetic pathway. Ribose phosphate is split off, thereby releasing the purines.

Fate of Purines

Purines synthesised in the body, those derived from the diet, and those liberated by endogenous breakdown of nucleic acids may follow one of two pathways.

They may be:

reused for nucleic acid synthesis;
oxidised to urate.

Formation of urate from purines.—As shown in the lower part of Fig. 34, some of the adenine is oxidised to hypoxanthine, which is further oxidised to xanthine. Guanine can also form xanthine. Xanthine, in turn, is oxidised to form urate. The oxidation of both hypoxanthine and xanthine is catalysed by the liver enzyme *xanthine oxidase*. Thus the formation of urate from purines depends on xanthine oxidase activity, a fact of importance in the treatment of gout.

Reutilisation of purines.—Some xanthine, hypoxanthine and guanine can be resynthesised to purine nucleotides by pathways involving, amongst other enzymes, hypoxanthine–guanine phosphoribosyl transferase (HGPRT) and adenine phosphoribosyl transferase (APRT).

Excretion of urate.—75 per cent of the urate leaving the body is excreted in the urine and 25 per cent passes into the intestine, where it is broken down by intestinal bacteria (*uricolysis*). The urate filtered at the renal glomerulus is probably almost completely reabsorbed in the tubules and most urinary urate is derived from active tubular secretion: urinary excretion is slightly lower in males than females: this may contribute to the higher incidence of hyperuricaemia in men. Some drugs used to treat hyperuricaemia may enhance renal secretion of urate.

Renal excretion of urate is inhibited by such organic acids as lactic and oxoacids.

CAUSES OF HYPERURICAEMIA

Figure 35 summarises the factors which may contribute to hyperuricaemia. These are:

    increased rate of urate formation;
        increased synthesis of purines (Step a);
        increased intake of purines (Step b);
        increased turnover of nucleic acids (Step c);
    reduced rate of urate excretion (Step e).

Abnormalities of steps (b), (c) and (e) are causes of secondary hyperuricaemia. Increased synthesis is probably the most important mechanism causing primary hyperuricaemia.

DANGERS OF HYPERURICAEMIA

The solubility of urate in plasma is limited. Precipitation in tissues may be favoured by a variety of local factors of which the most important are probably tissue pH and trauma. Crystallisation in *joints*, especially those of the foot, produces the classical picture of gout, first described by Hippocrates in 460 B.C. It is thought that local inflammation due to urate precipitation produces an increase in leucocytes in the area, and that lactic acid production by these cells lowers the pH locally: this reduces the solubility of urate and sets up a vicious circle in which further precipitation occurs. It should be noted that in attacks of acute gouty arthritis local factors are of more importance than the plasma urate levels; the latter may even be normal during the attack.

Precipitation can also occur in other tissues, and the subcutaneous collections of urate, which are especially common in the ear and in the olecranon and patellar bursae and tendons, are called *gouty tophi*.

Attacks of gout are extremely painful and unpleasant and may lead to permanent joint deformity. Tophi are disfiguring but harmless. A serious effect of hyperuricaemia is due to precipitation of urate in the kidney, leading to *renal damage*. For this reason it has been recommended that even asymptomatic cases with plasma urate levels consistently above 0·6 mmol/l (10 mg/dl) should be treated.

PRIMARY HYPERURICAEMIA AND GOUT

Familial Incidence

In A.D. 150 Galen said that gout was due to "debauchery, intemperance and an hereditary trait". "Intemperance", as we shall see, may aggravate the condition. The striking familial incidence of

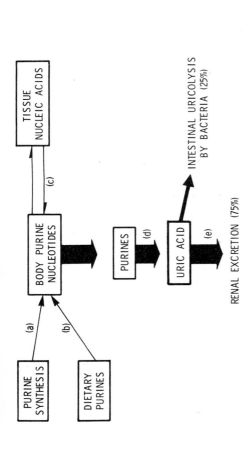

TISSUE NUCLEIC ACIDS

(c)

BODY PURINE NUCLEOTIDES

(a)

(b)

PURINE SYNTHESIS

DIETARY PURINES

PURINES

(d)

URIC ACID

(e)

INTESTINAL URICOLYSIS BY BACTERIA (25%)

RENAL EXCRETION (75%)

Causes of Hyperuricaemia

Pathway (a) Increased in Primary Hyperuricaemia

Pathway (b) Affected by diet

Pathway (c) Increased in malignancy, infection, cytotoxic therapy, psoriasis etc.

Pathway (e) Decreased in renal failure, thiazide diuretic therapy, some cases of primary hyperuricaemia and acidosis

Treatment of Hyperuricaemia

Pathway (d) Reduced by Xanthine Oxidase inhibitors (eg. Allopurinol)

Pathway (e) Increased by Uricosuric Drugs (eg. Probenecid)

Fig. 35.—Origin and fate of urate in normal subjects.

hyperuricaemia confirms that there is probably "an hereditary trait", but in this respect we know little more than Galen did, because the mode of inheritance (like that of diabetes mellitus) is still not fully established.

Sex and Age Incidence

Gout and hyperuricaemia are very rare in children, and rare in women of child-bearing age. The difference in incidence between males and females is not due to sex-linked inheritance, because it can be transmitted by males. Plasma urate levels are low in children and rise in both sexes at puberty, more so in males than females. Women become more prone to hyperuricaemia and gout in the post-menopausal period (compare plasma cholesterol and iron levels). It may be that renal excretion of urate is affected by sex-hormone levels.

Precipitating Factors

The classical image of the gouty subject is the red-faced, good-living, hard-drinking squire depicted in novels and paintings of the 18th century. Galen mentioned "debauchery and intemperance" as causes of gout. Two factors probably account for the high incidence of clinical gout in this type of subject.

> *Alcohol* has been shown to decrease renal excretion of urate. This may be because it increases lactic acid production, which inhibits urate excretion.
>
> A *high meat diet* contains a high proportion of *purines*.

Neither of these factors is likely to precipitate gout in a normal person, but may do so in a subject with a hyperuricaemic trait.

For reasons which are not clear, hyperuricaemia, and even clinical gout, are relatively common in patients with hypercalcaemia from any cause, and in patients with recurrent calcium-containing renal calculi, even when not accompanied by hypercalcaemia.

Biochemical Lesion of Primary Hyperuricaemia

Use of labelled glycine has shown that *purine synthesis is increased* in about 25 per cent of cases of primary hyperuricaemia. There may be overactivity of the enzyme controlling the formation of phosphoribosylamine (Fig. 34). This could be due to failure of normal feedback suppression by nucleotides.

Reduced renal secretion of urate has also been demonstrated in other cases of primary hyperuricaemia. Many subjects may have both increased synthesis and decreased excretion.

Principles of Treatment of Hyperuricaemia

Treatment may be based on:

reducing purine intake (Step (b) Fig. 35).—This is not very effective by itself;

increasing renal excretion of urate with *uricosuric drugs*, such as probenecid and salicylates in large doses (Step (e) Fig. 35). These drugs are very effective if renal function is normal, but are useless in the presence of renal failure. Fluid intake must be kept high. *Low doses* of most uricosuric drugs *reduce* urate secretion;

reducing urate production by drugs which inhibit xanthine oxidase (Step (d) Fig. 35), such as *allopurinol* (hydroxypyrazolopyrimidine). This compound is structurally similar to hypoxanthine and acts as a competitive inhibitor of the enzyme. *De novo* synthesis may also be decreased by this drug;

colchicine, which has an anti-inflammatory effect in acute gouty arthritis, does not affect urate metabolism.

Juvenile Hyperuricaemia (Lesch-Nyhan Syndrome)

This is an exceedingly rare inborn error, probably carried on an X-linked recessive gene, in which severe hyperuricaemia occurs in young male children. A deficiency of the enzyme *hypoxanthine-guanine phosphoribosyl transferase* (*HGPRT*) has been demonstrated in affected subjects. Hypoxanthine and other purines cannot be recycled to form purine nucleotides, and urate production from them is probably increased. The syndrome is associated with mental deficiency, a tendency to self-mutilation, aggressive behaviour, athetosis and spastic paraplegia.

Glucose-6-Phosphatase Deficiency (p. 215)

The tendency to hyperuricaemia in patients with glucose-6-phosphatase deficiency may be related directly to the inability to convert glucose-6-phosphate to glucose. More G-6-P is available for metabolism through intracellular pathways, including:

the pentose-phosphate pathway, thus increasing ribose phosphate (phosphoribose) synthesis. This may accelerate the first step in purine synthesis, with consequent *urate overproduction*;

glycolysis, thus increasing lactic acid production (p. 215). Lactic acid may *reduce renal urate excretion*.

Secondary Hyperuricaemia

High plasma urate levels may result from:

increased turnover of nucleic acids ((c) in Fig. 35);
  rapidly growing malignancy, especially leukaemias and poly-
    cythaemia vera;
  treatment of malignant tumours;
  psoriasis;
  increased tissue breakdown in:
    acute starvation
    tissue damage;
reduced excretion of urate (Step (e) in Fig. 35);
  glomerular failure;
  diuretic therapy;
  acidosis.

Increased turnover of nucleic acids in malignancy can cause hyperuricaemia. *Treatment* of large tumours by radiotherapy or cytotoxic drugs can cause massive release of urate and has been known to cause acute renal failure due to tubular blockage by crystalline uric acid. During such treatment allopurinol should be used, and, if renal function is good, fluid intake should be kept high.

Starvation and tissue damage.—Endogenous tissue breakdown is increased due to acute starvation or tissue damage. Increased amounts of urate are produced. In both these conditions acidosis is probably present (due to ketosis in starvation, and due to tissue catabolism in both) and these acids probably inhibit renal excretion of urate, aggravating the hyperuricaemia. Levels may reach 0·9 mmol/l (15 mg/dl) or more in complete starvation.

Glomerular failure causes retention of urate as well as retention of other waste products of metabolism. When estimating plasma urate, urea should always be estimated on the same specimen to exclude renal glomerular failure as a cause of hyperuricaemia. It has already been mentioned that hyperuricaemia may *cause* renal failure and, in the presence of uraemia, it may be difficult to know which is cause and which is effect. As a rough guide the plasma urate concentration would be expected to be about 0·6–0·7 mmol/l (10–12 mg/dl) at a urea level of about 50 mmol/l (300 mg/dl); if it is much higher than this, hyperuricaemia should be suspected as the primary cause of the renal failure. Clinical gout is rare in secondary hyperuricaemia due to renal failure.

Increased intestinal secretion and bacterial uricolysis have been claimed to occur in renal failure and may account for the fact that, although plasma urea and urate levels rise in parallel, the molar rise in urate is less than that of urea.

Clinical gout is a rare complication of therapy with *diuretics*, although hyperuricaemia is relatively common. These drugs inhibit renal excretion of urate.

PSEUDOGOUT

Pseudogout, while not a disorder of purine metabolism, produces a similar clinical picture to gout. Calcium pyrophosphate precipitates in joint cavities and calcification of the cartilages is visible radiologically. The crystals may be identified under a polarising microscope. The plasma urate is normal.

HYPOURICAEMIA

Hypouricaemia is rare, except as a result of treatment of hyperuricaemia. It is an unimportant finding associated with proximal renal tubular damage, when reabsorption of urate is decreased.

Xanthinuria is a very rare inborn error in which there is a deficiency of liver xanthine oxidase. Purine breakdown stops at the xanthine-hypoxanthine stage. Plasma and urinary urate levels are very low. The increased urinary excretion of xanthine may lead to the formation of xanthine stones (the reason why this does not happen during therapy with xanthine oxidase inhibitors is not clear, but may be due to reduction of synthesis). The mode of inheritance is probably autosomal recessive.

SUMMARY

1. Urate is the end product of purine metabolism.
2. Hyperuricaemia may be the result of:

   increased nucleic acid turnover (malignancy, tissue damage, starvation);
   increased synthesis of purines (primary gout);
   reduced rate of renal excretion of urate (glomerular failure, thiazide diuretics, acidosis).

3. Hyperuricaemia may be aggravated by:
   high purine diets;
   acidosis or a high alcohol intake.

4. Primary hyperuricaemia and gout have a familial incidence and are rare in women of child-bearing age.
5. Because severe hyperuricaemia may cause renal damage it should be treated, even if asymptomatic.

6. Hypouricaemia is rare and usually unimportant. It occurs in the very rare inborn error, xanthinuria.

FURTHER READING

CAMERON, J. S. and SIMMONDS, H. A. (1981). Uric acid, gout and the kidney. *J. Clin. Path.*, **34**, 1245–1254.

EMMERSON, B. T. (1983). *Hyperuricaemia and Gout in Clinical Practice.* Bristol: John Wright & Sons.

Chapter XVIII

IRON METABOLISM

In man the circulating iron-containing pigment, haemoglobin, carries oxygen from the lungs to metabolising tissues, and in muscle myoglobin increases the local supply of oxygen. The ability to carry oxygen depends, among other factors, on the presence of ferrous iron in the haem molecule; iron deficiency is associated with deficient haem synthesis, and the symptoms of anaemia are due to tissue hypoxia. Cytochromes and some enzymes necessary for electron-transfer reactions also contain iron: it is doubtful whether clinical iron deficiency, unless very severe, affects these.

NORMAL IRON METABOLISM

DISTRIBUTION OF IRON IN THE BODY

Figure 36 represents diagrammatically the distribution of iron in the body. The total body iron is about 50 to 70 mmol (3 to 4 g).

About 70 per cent of the total iron is circulating in erythrocyte *haemoglobin*.

Up to 25 per cent of the body iron is stored in the reticulo-endothelial system, in the liver, spleen and bone marrow. This storage iron is complexed with protein to form *ferritin* and *haemosiderin*. The iron in ferritin is more easily released from protein than that in haemosiderin. Haemosiderin (which may be aggregated ferritin) can be seen in unstained sections examined by light microscopy. Both ferritin and haemosiderin (but not iron incorporated in haem) stain with potassium ferrocyanide (Prussian blue reaction): this staining characteristic may be used to assess the size of iron stores.

Iron deficiency only becomes evident when *no stainable iron* is detectable in the *reticulo-endothelial cells* in *bone marrow* films: this is the iron normally drawn on for haemoglobin synthesis.

Iron overload is likely when *stainable iron* is demonstrable in *liver* biopsy specimens: such parenchymal deposition occurs when storage capacity is exceeded.

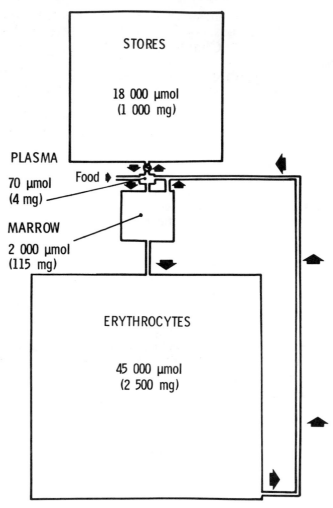

Fig. 36.—Body iron compartments.

Histological assessment is more reliable for detecting iron deficiency than iron overload (p. 429).

Only about 50 to 70 µmol (3 to 4 mg, or about 0·1 per cent) *of the total body iron is circulating in the plasma*, bound to protein. This is the fraction measured in *plasma iron* estimations.

The remainder of the body iron is incorporated in myoglobin, cytochromes and iron-containing enzymes.

Iron can only cross cell membranes by active transport in the ferrous (Fe^{++}) form: it is in this reduced state in both oxyhaemoglobin and "reduced" haemoglobin. In ferritin and haemosiderin, and when bound to transferrin, it is in the ferric state (Fe^{+++}).

The *control* of iron distribution in the body is poorly understood. *Plasma iron concentrations can vary by 100 per cent or more for purely physiological reasons, and are also affected by a variety of pathological factors other than the amount of iron in the body.* These variations in plasma iron are probably due to redistribution between stores and plasma.

IRON BALANCE

As shown in Fig. 36, iron, once in the body, is in a virtually closed system.

Iron Excretion

There appears to be no control of iron excretion, and loss from the body probably depends on the iron content of cells lost by desquamation, mostly into the intestinal tract and from the skin. The total daily loss by these routes is about 18 μmol (1 mg). Negligible amounts appear in the urine, reflecting the fact that all circulating iron is protein bound.

In women the mean monthly menstrual loss of iron is about 290 μmol (16 mg) and averages 10–18 μmol (0·5–1 mg) a day over the month above the basal 18 μmol (1 mg), although it can be much higher in those with menorrhagia, who may become iron deficient. During pregnancy the mean extra daily loss to the fetus and placenta is about 27 μmol (1·5 mg).

It should be noted for comparison that a male blood donor losing a unit (pint) of blood every 4 months averages an extra loss of 36 μmol (about 2 mg) daily above the basal loss of 18 μmol (1 mg) (Table XXXI

Iron Absorption

The control of the body iron content depends upon control of absorption.

Iron is absorbed by an active process in the upper small intestine. It can cross cell membranes, including those of intestinal cells, only in the ferrous form. Within the intestinal cell some of the iron is combined with the protein, apoferritin, to form ferritin: this, like ferritin elsewhere, is a storage compound. Ferritin is lost into the intestinal tract when the cell desquamates.

Normally about 18 μmol (1 mg) of iron is absorbed each day and

TABLE XXXI

COMPARISON OF IRON LOSSES

	Source of loss	Extra loss	Daily extra loss	Daily total loss
Men and non-menstruating women	Desquamation	—	—	18 μmol (1 mg)
Menstruating women (Mean value)	Desquamation + menstruation	290 μmol (16 mg)/month	9 μmol (0·5 mg)	27 μmol (1·5 mg)
Pregnancy	Desquamation + loss to fetus and in placenta	7000 μmol (380 mg)/9 months	27 μmol (1·5 mg)	45 μmol (2·5 mg)
Male blood donors	Desquamation + 1 unit blood	4500 μmol (250 mg)/4 months	36 μmol (2·0 mg)	54 μmol (3·0 mg)

this just replaces loss. The percentage absorption of dietary iron depends to some extent on the food with which it is taken, but is usually about 10 per cent.

Iron absorption appears to be influenced by any or all of the following factors:

> oxygen tension;
> marrow erythropoietic activity;
> the size of iron stores.

From the clinical point of view it is important to note that *iron absorption is increased in many anaemias which are not due to iron deficiency.*

Most normal women taking an adequate diet probably absorb slightly more iron than men, and so replace their greater losses.

The iron requirements for growth in children and adolescents are similar to, or slightly higher than, those of menstruating women. This need, too, can be met by increased absorption from a normal diet.

Normal iron loss is so small, and normal iron stores are so large, that it would take about 3 years to become iron-deficient on a completely iron-free diet. This period is much shorter if there is any blood loss.

IRON TRANSPORT IN PLASMA

Iron is carried in the plasma in the ferric form, attached to the specific binding protein, *transferrin*, at a concentration of about $18\,\mu mol/l$ ($100\,\mu g/dl$). The protein is normally capable of binding about $54\,\mu mol/l$ ($300\,\mu g/dl$) of iron, and is therefore about a third saturated. Transferrin-bound iron is carried to stores and to bone marrow: in stores it is laid down as ferritin and haemosiderin, and in the marrow some may pass directly from transferrin into the developing erythrocyte to form haemoglobin.

Free iron is toxic, and is probably absent in normal subjects. In intestinal cells and stores it is bound to protein in ferritin and haemosiderin, in the plasma it is bound to transferrin, and in the erythrocyte it is incorporated in haemoglobin.

FACTORS AFFECTING PLASMA IRON LEVELS

Plasma iron estimation is frequently requested, but is rarely of clinical value, and results are often misinterpreted. As we have seen, plasma iron, like plasma potassium, represents a very small proportion of the total body content, and for this reason alone is likely to be a poor index of it: plasma potassium levels are important to the body and are normally relatively constant, but those of plasma iron (which is only a protein-bound transport fraction) are much less so and can vary greatly, even under physiological conditions.

PHYSIOLOGICAL FACTORS AFFECTING PLASMA IRON LEVELS

The causes of physiological changes in plasma iron concentrations are not well understood, but they are very rapid, and almost certainly represent shifts between plasma and stores. The following factors are known to affect levels and the first three may cause changes, in an individual subject, of 100 per cent or more.

Cyclical Variations

Circadian (diurnal) rhythm.—Plasma iron is higher in the morning than in the evening. If subjects are kept awake at night this difference may be less marked or absent, and it is reversed in night workers.

Monthly variations in women.—Plasma iron may reach very low levels just before or during the menstrual period. This reduction is probably due to hormonal factors rather than to blood loss.

Random Variations

Very large day-to-day variations (occasionally as much as threefold) occur in plasma iron, and these usually overshadow cyclical changes. They may sometimes be associated with physical or mental stress, but more usually a cause cannot be found.

Effect of Pregnancy and Oral Contraceptives

In women taking some oral contraceptives the plasma iron rises to levels similar to those found in men. A similar rise occurs in the first few weeks of pregnancy.

Sex and Age Differences

Plasma iron levels, like those of haemoglobin and the erythrocyte count, are higher in men than in women. This difference is probably hormonal in origin. It is first evident at puberty, before significant menstrual iron loss has occurred, and disappears at the menopause. Androgens tend to increase plasma iron concentration and oestrogens to lower it.

PATHOLOGICAL FACTORS AFFECTING PLASMA IRON LEVELS

Iron deficiency and iron overload usually cause low and high plasma iron levels respectively.

Iron deficiency is associated with a hypochromic microcytic anaemia, and with reduced amounts of stainable marrow iron.

Iron overload is associated with increased amounts of stainable iron in liver biopsy specimens.

Any illness, whether acute or chronic, causes hypoferraemia: even a bad cold can lead to a fall in plasma iron concentration. Chronic conditions such as malignancy, renal failure, rheumatoid arthritis, and chronic infections are often associated with normocytic, normochromic anaemia. Iron stores are normal or even increased, and the anaemia does not respond to iron therapy.

In rheumatoid arthritis, iron deficiency is especially likely to be superimposed on the anaemia of chronic illness: salicylates, or other anti-inflammatory drugs used to treat the primary condition, may cause low-grade bleeding from small gastric erosions. The finding of hypochromic erythrocytes, *not* a low plasma iron level, is the most sensitive index of this complication.

In conditions in which the *marrow cannot utilise iron*, either because it is hypoplastic, or because of the absence of some other factor necessary for erythropoiesis (such as vitamin B_{12} or folate), plasma iron levels are often high. Blood and marrow films may show a typical picture, but, for instance, in pyridoxine-responsive anaemia and thalassaemia, the findings in the blood film may be somewhat similar to that of iron deficiency; in the latter two conditions iron stores are increased, and this can be shown on the marrow film.

In *haemolytic anaemia* the iron from the haemoglobin of broken down erythrocytes is released into the plasma and reticulo-endothelial system. Plasma iron may be high during the haemolytic episode and is usually normal during quiescent periods. Marrow iron stores are usually increased in chronic haemolytic conditions.

In *acute liver disease* disruption of cells may release ferritin iron into the blood stream and cause a transient rise of plasma iron. In *cirrhosis* plasma iron levels may sometimes be high. Although the reasons for this are not clear, it may sometimes be due to increased iron absorption associated with a high iron intake.

TRANSFERRIN AND TOTAL IRON-BINDING CAPACITY (TIBC)

It will be seen that plasma iron levels by themselves give no useful information about the state of iron stores. In the rare situations in which there is still doubt after haematological investigation, diagnostic precision may be improved by measuring the transferrin level or iron-binding capacity of the plasma at the same time as the plasma iron. It is a waste of time and money to estimate only plasma iron.

The transferrin level is often assessed by measuring the *iron-binding capacity* of the plasma. An excess of inorganic iron is mixed with the plasma and any which is not bound to transferrin is removed, usually with a resin. The remaining iron is estimated on the plasma sample and the result expressed as a total iron-binding capacity (TIBC). Direct transferrin assay is becoming more widely available.

PHYSIOLOGICAL CHANGES IN TRANSFERRIN LEVELS

The transferrin level is less labile than that of plasma iron. However, it rises rapidly in subjects on some *oral contraceptive* preparations, and this should be remembered when interpreting results in women. It also increases after about the 28th week of *pregnancy*, even in those women with normal iron stores.

PATHOLOGICAL CHANGES IN TRANSFERRIN LEVELS

The *transferrin level and TIBC rise in iron deficiency* and *fall in iron overload*.

The *transferrin level and TIBC fall* in *chronic illnesses associated with a low plasma iron concentration*. In the nephrotic syndrome both may be low because transferrin is lost in the urine.

The transferrin level and TIBC are unchanged in acute infection.

Thus the low plasma iron of uncomplicated iron deficiency is associated with a high transferrin level and TIBC. That of anaemia not due to iron deficiency is associated with a low TIBC.

If iron deficiency coexists with the anaemia of chronic infection the opposing effects of the two conditions on the transferrin level makes interpretation of this, as well as that of iron (p. 419), difficult.

PERCENTAGE SATURATION OF TIBC

The statement that the TIBC is normally a third saturated with iron is a very approximate one: physiological variations of plasma iron level are rarely associated with much change in TIBC and the saturation of the protein varies widely; the percentage saturation is, of course,

$$\frac{\text{Plasma Iron Concentration} \times 100}{\text{TIBC}}$$

It has been claimed that percentage saturation is a better index of iron stores than plasma iron concentration alone, and that, if it is below 16 per cent, iron deficiency is likely to be present. The first part of the statement is obviously true, because the low plasma iron of iron deficiency is accompanied by a high TIBC; this will result in a lower percentage saturation than with the same level of plasma iron in other conditions. However, saturation as low as 16 per cent can be found, for instance, premenstrually and in acute infections, with no change in TIBC or in iron stores. It is probably more useful to take account of the results of both plasma iron and TIBC rather than to calculate the percentage saturation.

The findings in conditions which may affect plasma iron and transferrin levels and TIBC are summarised in Table XXXII.

PLASMA FERRITIN LEVELS

Normal plasma ferritin levels are about 100 μg/l. Circulating ferritin is in equilibrium with that in stores, and while levels less than 10 μg/l probably indicate iron deficiency, the result rarely contributes to making the diagnosis. The high levels found in iron overload or liver disease are of more significance.

The student should read the section on "Investigation of Anaemia" on p. 428.

IRON THERAPY

Because iron excretion is not actively controlled, and because body content is controlled by absorption, *parenteral iron therapy* may easily lead to iron overload. Repeated blood transfusions carry the same danger, because a unit of blood contains 4·5 mmol (250 mg) of iron. In anaemias other than that of true iron deficiency stores are normal or even increased, and parenteral iron should not be given unless the diagnosis of iron deficiency is beyond doubt: even when this is so, the oral is preferable to the parenteral route. Repeated blood transfusion may be necessary to correct severe anaemia in, for instance, chronic renal disease and hypoplastic anaemia, but, in such cases, the danger of overload should be remembered and blood should not be given indiscriminately.

Anaemia increases the rate of iron absorption even in the presence of increased iron stores. Treatment of, for instance, chronic haemolytic anaemia with *oral iron supplements* may occasionally cause iron overload; it does not improve the anaemia, which occurs because the rate of breakdown of erythrocytes exceeds the rate of production, and

TABLE XXXII

CHANGES IN PLASMA IRON (FE) AND TRANSFERRIN OR TOTAL IRON-BINDING CAPACITY (TIBC)

	Fe	Transferrin or TIBC	% Saturation	Marrow stores
Low Iron Levels Iron deficiency	→	↑	↓↓	↓ or absent
Chronic illnesses (e.g. infection and malignancy)	→	→	Variable	Normal or ↑
Acute illnesses (e.g. infection)	→	Normal	→	Normal
Premenstrual	→	Normal	→	Normal
High Iron Levels Iron overload	←	→	↑↑	←
Oral contraceptives and late pregnancy	↑ (to male level)	←	Normal	Normal
Early pregnancy	↑ (to male level)	Normal	←	Normal
Hepatic cirrhosis	Variable. May be ↑	→	↑ to ↑↑	May be ↑
Failure of marrow utilisation and haemolysis	←	Normal or ↓	←	←

not because of iron deficiency. The released iron stays in the body. In other non-iron-deficient anaemias the danger is similar.

Although there is some control of iron absorption, it is inefficient and may rarely be "swamped" by large loads, even in the absence of anaemia. Iron overload has been reported in a non-anaemic woman who continued to take oral iron (against medical advice) over a matter of years.

Iron therapy is potentially dangerous: it should be prescribed with care, and only when iron deficiency is proven.

IRON OVERLOAD

As emphasised on p. 415, there is virtually no excretion of iron from the body. Iron absorbed from the gastro-intestinal tract, or administered parenterally, in excess of daily loss, accumulates in body stores. If such "positive balance" is maintained over long periods, iron stores may exceed 350 mmol (20 g) (about five times the normal amount).

Causes of Iron Overload

Increased intestinal absorption:

idiopathic haemochromatosis;
anaemia with increased, but ineffective, erythropoiesis;
liver disease (rare);
dietary excess;
inappropriate oral iron therapy.

Parenteral administration:

multiple blood transfusions;
inappropriate parenteral iron therapy.

A very rare cause of iron overload is an inherited deficiency of transferrin.

Consequences of Iron Overload

The effect of the accumulated iron depends on its distribution in the body. This in turn is influenced partly by the route of entry. Two main patterns are seen at post-mortem examination or in biopsy specimens.

Parenchymal iron overload occurs in idiopathic haemochromatosis and in patients with ineffective erythropoiesis. Iron accumulates in the parenchymal cells of the liver, pancreas, heart and other organs. There is usually associated functional disturbance or tissue damage.

Reticulo-endothelial iron overload is seen after excessive *parenteral administration of iron* or *multiple blood transfusions*. The iron accumulates initially in the reticulo-endothelial cells of the liver, spleen and bone marrow. There are few harmful effects but under certain circumstances (p. 425) the distribution may change so that parenchymal damage occurs.

In dietary iron overload both hepatic reticulo-endothelial and parenchymal overload may occur, associated with scurvy and osteoporosis (p. 426). Whatever the cause of *massive* iron overload, there may be parenchymal accumulation and tissue damage.

Two commonly used terms should be defined:

> *Haemosiderosis* is defined as an increase in iron stores as haemosiderin and is a histological definition. It does not necessarily mean that there is an increase in total body iron; for example, in many types of anaemia there is reduced haemoglobin iron (less haemoglobin) but increased storage iron;
>
> *Haemochromatosis* describes the clinical disorder due to parenchymal iron-induced damage.

Syndromes of Iron Overload

Idiopathic Haemochromatosis

Idiopathic haemochromatosis is a genetically determined disease in which increased intestinal absorption of iron over many years produces large iron stores of parenchymal distribution. It presents, usually in middle age, as cirrhosis with diabetes mellitus, hypogonadism and increased skin pigmentation. Because of the darkening of the skin, due to an increase in melanin rather than to iron deposition, the condition has been referred to as "bronzed diabetes", although the colour is grey rather than bronze. Cardiac manifestations may be prominent, particularly in younger patients, many of whom die in cardiac failure. In about 10 to 20 per cent of cases hepatocellular carcinoma develops.

Idiopathic haemochromatosis has an autosomal recessive mode of inheritance. The gene for this disorder is closely associated with the HLA gene locus, a fact of importance in family studies. Those members of a family with an HLA haplotype identical with that of the patient are likely to develop the disorder.

Factors such as alcohol abuse may hasten the accumulation of iron and development of liver damage.

It may be difficult to distinguish between idiopathic haemochromatosis and alcoholic cirrhosis. In both conditions diabetes mellitus and hypogonadism may occur, and although the incidence of these complications is higher in idiopathic haemochromatosis this does not help in diagnosis of the individual case. Examination of liver biopsy specimens may further confuse the issue. The liver in cases of alcoholic cirrhosis not infrequently contains increased stainable iron. Not only do some alcoholic drinks, notably wines, contain significant amounts of iron, but there is evidence that in cirrhosis there may be increased iron absorption due possibly to an effect of alcohol. However, the majority of patients with cirrhosis do not have increased iron stores as shown by chemical testing, and the liver biopsy specimen shows that the iron is mainly present in the portal tracts. The clinical, histological and biochemical changes of cirrhosis are usually more obvious than those of iron accumulation: this contrasts with the picture in haemochromatosis with an apparently equivalent iron

Rarely, patients with cirrhosis or a portocaval shunt may have true iron overload and the distinction from idiopathic haemochromatosis may be extremely difficult. A family history, or investigation of near relatives, may help in diagnosis. The treatment of iron overload is the same in either case.

The diagnosis of idiopathic haemochromatosis *must* be followed by investigation of other members of the family, and treatment of those in whom increased iron stores are found.

Treatment of Iron Overload

The excess iron is usually removed from patients with idiopathic haemochromatosis by weekly venesection until the haemoglobin level falls. Each unit of blood removes 4·5 mmol (250 mg) of iron.

Anaemia and Iron Overload

Several types of anaemia may be associated with iron overload. In some, such as aplastic anaemia and the anaemia of chronic renal failure, the cause is multiple blood transfusions, and the iron initially accumulates in the reticulo-endothelial system. With massive overload (over 100 units of blood), parenchymal overload and haemochromatosis may develop.

In anaemias characterised by erythroid marrow hyperplasia, but with ineffective erythropoiesis, such as thalassaemia major and sideroblastic anaemia, there is, in addition, increased absorption of iron. Haemochromatosis develops at a lower transfusion load than in aplastic anaemia.

Iron overload associated with anaemia can obviously not be treated by venesection, since this would aggravate the anaemia by removing more haemoglobin. The tendency for transfusion further to aggravate overload can be minimised by giving the iron-chelating agent, desferrioxamine, each time: this can be excreted in the urine with any administered non-haemoglobin iron.

Dietary Iron Overload

Increased iron absorption due to excessive intake is rare. One well-described form is, however, relatively common in the black population of Southern Africa. The source is a beer brewed in iron containers. Usually the excess iron is confined to the reticulo-endothelial system and the liver (both portal tracts and parenchymal cells), and there is no tissue damage. In a small number of cases, deposition in the parenchymal cells of other organs occurs and the clinical picture may resemble that of idiopathic haemochromatosis: it may be distinguished by the high concentrations of iron in the reticulo-endothelial system seen in bone marrow and spleen (at autopsy).

Scurvy and osteoporosis may occur in this form of iron overload. The ascorbate deficiency may be due to its irreversible oxidation in the presence of excessive amounts of iron, and osteoporosis sometimes accompanies scurvy. Ascorbate deficiency also interferes with normal mobilisation of iron from the reticulo-endothelial cells; plasma iron levels may be low, and the response to chelating agents poor, despite iron overload.

The student should read the section on "Investigation of suspected iron overload" on p. 429.

SUMMARY

1. No significant excretion of iron can occur from the body. Control of body stores is by control of absorption. For this reason parenteral iron therapy should be given with care.

2. Absorption of iron is increased by anaemia even in the absence of iron deficiency. For this reason oral iron therapy should not be given in anaemia other than that due to iron deficiency.

3. Plasma iron levels vary considerably under physiological circumstances.

4. Plasma iron levels fall in many cases of anaemia not due to iron deficiency.

5. For these two reasons (3 and 4) plasma iron levels alone are a very poor indication of body iron stores.

6. Iron is carried in the plasma bound to the protein transferrin. Transferrin is often measured indirectly by measuring the total iron-binding capacity (TIBC) of plasma.

7. The TIBC rises in iron deficiency and falls in iron overload.

8. The TIBC falls in many cases of anaemia associated with a low plasma iron, but not due to iron deficiency.

9. A low plasma iron concentration with a high TIBC is more suggestive of iron deficiency than a low plasma iron alone.

10. The quickest, cheapest and most informative tests for iron deficiency are simple haematological ones. These should be performed before requesting plasma iron estimation.

11. The factors governing the distribution of excessive iron are not fully understood. A feature common to all forms of parenchymal overload is a high percentage saturation of transferrin.

12. Iron overload may be the result of excessive intestinal absorption or of parenteral iron administration, usually as blood. The distribution of the iron in the body differs in the various forms of iron overload.

13. Iron overload can be demonstrated by the response to repeated venesection or to chelating agents. The diagnosis of idiopathic haemochromatosis can only be made if massive iron overload is present.

14. Virtually all cases of parenchymal iron overload show a high plasma iron concentration with a high percentage saturation of transferrin.

15. Although plasma ferritin levels are often raised in iron overload, results can be misleading.

FURTHER READING

FINCH, C. A. and HUEBERS, H. (1982). Perspectives in iron metabolism. *New Engl. J. Med.*, **306**, 1520–1528.

HALLIDAY, J. W. and POWELL, L. W. (1982). Iron overload. *Semin. Hemat.*, **19**, 42–53.

INVESTIGATION OF DISORDERS OF IRON METABOLISM

Investigation of Anaemia

Anaemia may be due to iron deficiency, or to a variety of other conditions. The subject of the diagnosis of anaemia is covered more fully in textbooks of haematology. However, so that we may see the value of plasma iron estimations in perspective, it is worth considering the order in which anaemia may usefully be investigated.

1. The clinical impression of anaemia should be confirmed by *haemoglobin* estimation. Iron deficiency can, however, exist with haemoglobin levels within the "normal" range.

2. The *mean corpuscular haemoglobin (MCH)* and *mean corpuscular volume (MCV)* should be noted and, if necessary, a *blood film* should be examined. Iron deficiency anaemia is hypochromic and microcytic in type, and hypochromia may be evident before the haemoglobin level has fallen below the accepted normal range. Normocytic, normochromic anaemia is a non-specific finding, usually associated with other disease; it is not due to iron deficiency unless there has been very recent blood loss. Typical appearances of other types of anaemia may be seen on the blood film.

In most cases of anaemia, consideration of these findings together with the clinical picture will give the cause. Anaemias such as those of thalassaemia and of the pyridoxine-responsive type, although rare, are most likely to confuse the picture, since they too are hypochromic, but are not due to iron deficiency.

3. A *marrow film* may be required to confirm the diagnosis (for example, of megaloblastic anaemia). If such a film is available, staining with potassium ferrocyanide is by far the best indicator of the state of iron stores.

If marrow puncture is not felt to be justified, and *in the rare cases* in which diagnosis is not yet clear, biochemical investigations may occasionally help. If these are necessary, plasma iron *and* TIBC or transferrin should be estimated. Plasma iron estimation alone is uninformative.

Iron absorption test.—If a patient with *proven iron deficiency* shows no response to oral iron within a few weeks he is probably not taking the tablets: normal coloured, rather than black, faeces (detectable, if necessary, by rectal examination) confirm this suspicion.

If iron *has* been taken, malabsorption, usually part of an obvious general absorption defect, is occasionally the cause. If there is no evidence for malabsorption (Chapter XII), the adequacy of iron absorption may be assessed by measuring the increment of plasma iron after a standard (300 mg) oral dose of iron elixir. This increment will be very large (up to 40 μmol/l) if absorption is normal *in the patient with proven iron-deficiency anaemia*, but it will be very small:

if absorption is deficient;
in the non-anaemic subject (who does not normally absorb much iron);
in the patient with anaemia not due to iron deficiency (the absorbed iron is rapidly deviated to stores).

To avoid misinterpretation the diagnosis of iron deficiency must be made before the test is performed.

Investigation of Suspected Iron Overload

Initial tests.—Measure plasma iron, percentage saturation of transferrin and plasma ferritin. The *plasma iron* concentration is almost invariably high in idiopathic haemochromatosis, often above 36 μmol/l (200 μg/dl). This is associated with a reduced transferrin level (as shown by a lowered TIBC) and the *percentage saturation* is usually over 80 per cent, and often 100 per cent: in the presence of infection or malignancy, however, the plasma iron level and percentage saturation may be lower than expected; the TIBC remains low. The lowering effect of ascorbate deficiency on plasma iron levels has already been mentioned.

Plasma ferritin levels are high in most patients with iron overload (whether reticulo-endothelial or parenchymal). Only a few families with idiopathic haemochromatosis despite normal plasma ferritin levels have been described.

If all three values are normal, it is unlikely that the patient has iron overload.

Demonstration of increased iron stores.—The diagnosis of iron overload can only be made after proof has been obtained of increased iron stores.

Response to venesection.—The lack of response of the patient to a therapeutic course of venesection offers the most convincing proof of increased iron stores, albeit retrospectively. Removal of a unit of blood (4·5 mmol or 250 mg iron), repeated weekly, produces a rapid fall in plasma iron, soon followed by iron deficiency anaemia in a subject with normal iron stores. In patients with idiopathic haemochromatosis, however, 350 mmol (20 g) or more of iron may be removed in this way before evidence of iron deficiency develops.

Use of chelating agents.—An alternative method uses the chelating action of substances, such as desferrioxamine, which bind iron and are subsequently excreted in the urine. Following administration of desferrioxamine, subjects with increased iron stores excrete more iron in the urine than do normal people.

Plasma ferritin levels are high in most cases of reticulo-endothelial iron overload, but *normal concentrations have been found in cases of idiopathic haemochromatosis*. Raised levels may also occur in many hepatic diseases, including cirrhosis.

Liver biopsy specimens contain large amounts of stainable iron, which may be mainly in parenchymal, or mainly in reticulo-endothelial cells. Chemical estimation of iron is more reliable than histochemical evaluation.

Marrow iron content is *usually normal* in haemochromatosis, in which overload is predominantly parenchymal, but may be greatly increased in reticulo-endothelial overload. A similar loading of the reticulo-endothelial cells is found when there is deficient utilisation of marrow iron for haemoglobin synthesis, as in many haematological, neoplastic and chronic inflammatory diseases.

Family Studies

Relatives of a patient found to have idiopathic haemochromatosis should be investigated. Plasma iron and percentage saturation of transferrin are the most sensitive tests. Plasma ferritin may be normal if iron stores are not significantly increased. If available, HLA typing (p. 424) will identify those most likely to have inherited the disorder.

Chapter XIX

THE PORPHYRIAS

THE porphyrias are a group of disorders, usually inherited, of haem synthesis. Although most of them are uncommon and some are very rare, it is important to recognise them and to investigate relatives of known cases. For example, drugs which may precipitate acute attacks, sometimes with fatal consequences, must be avoided in the inherited hepatic porphyrias. It is equally important not to confuse the commoner, secondary, causes of abnormal porphyrin excretion with true porphyria.

PHYSIOLOGY

Haem is synthesised in most tissues of the body. In the bone marrow it is incorporated into *haemoglobin*. In other cells it is used for the synthesis of *cytochromes* and related compounds. The cytochromes are constituents of the electron transport chain, which harnesses the energy of metabolic processes. Impaired cytochrome synthesis can therefore have serious consequences. The liver is quantitatively the largest non-erythropoietic haem-producing organ.

Biosynthesis of Haem

The main steps are outlined below and in Fig. 37.

5-Aminolaevulinate (ALA) is formed by condensation of glycine and succinate. The reaction requires pyridoxal phosphate and is catalysed by the enzyme *ALA synthase*.

Two molecules of ALA condense to form a monopyrrole, *porphobilinogen* (PBG).

Four molecules of PBG combine to form a tetrapyrrole, *uroporphyrinogen* (Fig. 38). Two isomers are formed, I and III. The major pathway involves the III isomer and leads to the formation of haem by the successive production of *coproporphyrinogen* and *protoporphyrin*, followed by the incorporation of iron.

Each step is controlled by a specific enzyme. Normally *the rate-limiting step* is that catalysed by *ALA synthase*, and this step is regulated by *feed-back inhibition by the final product, haem*.

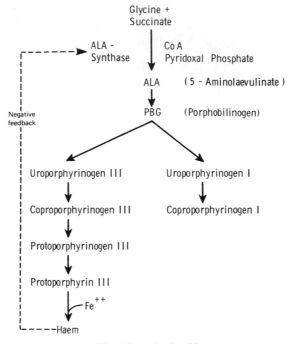

FIG. 37.—Biosynthesis of haem.

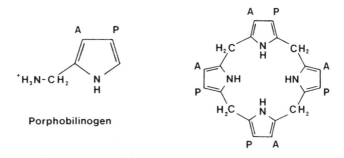

FIG. 38.—Porphobilinogen, and the tetrapyrrole uroporphyrinogen III which incorporates four porphobilinogen units. Uroporphyrinogen I differs only in the order of side-chains on one of the rings.

(Side-chains: A = acetate; P = propionate.)

The *porphyrinogens* and their precursors, *ALA* and *PBG*, are *colourless* compounds. Porphyrinogens, however, oxidise spontaneously to the corresponding *porphyrins* which are *dark red* in colour and which *fluoresce* in ultraviolet light. PBG, too, may spontaneously form uroporphyrin when exposed to air and light. A urine specimen containing large amounts of porphyrinogens or their precursors will gradually darken if left standing.

Excretion

Any excess of the intermediates on the haem pathway is excreted. ALA, PBG and uroporphyrin(ogen) are water-soluble and appear in the *urine*. Protoporphyrin is excreted in bile and appears in the *faeces*. Coproporphyrin(ogen) may be excreted by either route. Renal excretion of coproporphyrinogen, like that of urobilinogen, rises with increasing urinary alkalinity.

Normal urine contains ALA, PBG and porphyrin at concentrations undetectable by screening tests (see later). Faeces may contain sufficient porphyrin to impart a slight fluorescence to extracts.

THE PORPHYRIAS

The porphyrias are due to a deficiency of one of the enzymes on the haem pathway: haem production is therefore impaired. *Reduced feedback inhibition of ALA synthase may maintain adequate haem levels but at the expense of overproduction of porphyrins or their precursors.*

The *symptoms* of porphyria correlate well with the biochemical abnormalities.

Skin lesions, varying from mild photosensitivity to severe blistering, occur when *porphyrins* are produced in excess. The lesions typically occur in exposed areas where sunlight activates porphyrin in the skin to release energy which damages tissue.

Neurological disturbances, such as *peripheral neuritis, abdominal pain*, or both in the serious *acute attack*, occur only in those porphyrias in which the *precursors*, ALA and PBG, are produced in excess. It is not known whether the neurological damage is due to haem deficiency in the nervous system or to a direct toxic effect of ALA or PBG.

The porphyrias are usually classified according to whether the main site of porphyrin accumulation is in the liver or the erythropoietic system. This does not necessarily mean that the enzyme defect is confined to that system.

Hepatic Porphyrias

Acute intermittent porphyria;
Porphyria variegata;   } acute porphyrias
Hereditary coproporphyria;

Porphyria cutanea tarda;
  genetic predisposition;
  acquired.

Erythropoietic Porphyrias

Congenital erythropoietic porphyria;
Erythrohepatic protoporphyria.

The features of the porphyrias are outlined in Table XXXIII and discussed briefly below.

THE ACUTE HEPATIC PORPHYRIAS

Acute intermittent porphyria
Hereditary coproporphyria
Porphyria variegata

These three *dominantly* inherited disorders have latent and acute phases. The symptoms and biochemical abnormalities of the *latent phases* differ and reflect the nature of the enzyme defect. In the *acute phases*, however, the biochemical and clinical picture characteristic of excessive ALA and PBG excretion, associated with neurological and abdominal symptoms, develops. The similarities and differences are best explained by referring to a simplified scheme of the haem synthetic pathway (Fig. 39).

Latent Phase

The enzyme defect tends to reduce haem levels which in turn increase ALA synthase activity by the negative feedback loop. Increased ALA synthase activity has been demonstrated in all the porphyrias. This maintains haem levels at the expense of accumulation and excretion of the substance immediately before the block. The main biochemical abnormalities are predictable (Fig. 39).

Acute intermittent porphyria increased *urinary ALA and PBG*
                                    (not detectable in all patients)
Hereditary coproporphyria   increased *faecal coproporphyrin*
Porphyria variegata   increased *faecal protoporphyrin*

In acute intermittent porphyria the latent phase is usually asymp-

TABLE XXXIII

THE MAJOR CLINICAL AND BIOCHEMICAL FEATURES OF THE PORPHYRIAS

	Hepatic porphyrias							Porphyrias involving the erythropoietic system	
	Acute intermittent porphyria		Porphyria variegata		Hereditary coproporphyria		Porphyria cutanea tarda	Congenital erythropoietic porphyria	Erythrohepatic protoporphyria
	acute	latent	acute	latent	acute	latent			
Clinical Features									
Abdominal and neurological symptoms	+	−	+	−	+	−			
Skin lesions	−	−	+	+	rarely	−	+	+	+
Chemical Abnormalities									
Urine PBG and ALA	+ +	+	+ + +	−	+	−	−	−	−
Urine porphyrins	+	−	+	+	+ + +	+	+	+ +	−
Faecal porphyrins	−	−	+ + +	+ +	+ + +	+	−	+	+
	Acute attacks precipitated by many drugs (for instance barbiturates, oestrogens, sulphonamides)						Symptoms may be relieved by venesection	Erythrocyte porphyrins increased	

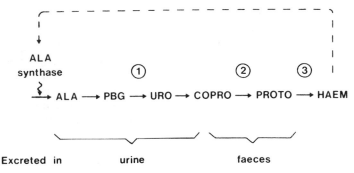

FIG. 39.—Sites of enzyme deficiencies in:
1. acute intermittent porphyria;
2. hereditary coproporphyria;
3. porphyria variegata.

tomatic. In porphyria variegata, and less commonly in hereditary coproporphyria, the increase in porphyrins may produce skin lesions.

Acute Phase

ALA synthase activity may be further increased by a number of drugs, particularly barbiturates, oestrogens, sulphonamides and griseofulvin, and by acute illness: this may be a direct effect or may be due to an increased demand for haem. It results in a marked increase in ALA and PBG excretion. In acute intermittent porphyria this increase is due to the block imposed by an inherited deficiency of *uroporphyrinogen-1-synthase*; in hepatic coproporphyria and porphyria variegata, this enzyme becomes rate-limiting and is unable to respond normally to the increased demand. *The increase in urinary ALA and PBG is the hallmark of the acute porphyric attack* and, in hereditary coproporphyria and porphyria variegata, is superimposed on the other biochemical abnormalities. The accelerated activity of the pathway and the spontaneous conversion of the precursors to porphyrin lead to increased urinary porphyrin excretion.

In the acute attack, colicky abdominal pain (due to involvement of the autonomic nervous system) and neurological symptoms ranging from peripheral neuritis to quadriplegia are usually the presenting features. Death may result from respiratory paralysis. Hyponatraemia is a common finding. Acute attacks occur only after puberty and are commoner in women than in men.

The acute attack closely resembles serious acute intra-abdominal conditions, and if the diagnosis is not made the patient may be subjected to surgery, with the use of a barbiturate anaesthetic; barbiturates and the stress of operation may aggravate the condition.

Diagnosis of Latent Porphyria

It is essential to investigate the family of any known porphyric. Screening tests for excess urinary PBG and ALA are inadequate to diagnose latent acute intermittent porphyria, and even quantitative estimation may fail to detect all carriers. The activity of the enzyme, *uroporphyrinogen-1-synthase*, should be measured in erythrocytes (the most easily available cells). There is often overlap in levels between affected and normal people, but within a family carriers can be shown to have levels about half those of unaffected members. This test will also detect affected children.

Carriers of porphyria variegata and hereditary coproporphyria can be identified, after puberty, by the demonstration of clearly increased *faecal porphyrin* excretion. Positive screening tests (see p.440) should be confirmed by quantitative estimation.

PORPHYRIA CUTANEA TARDA
(Cutaneous hepatic porphyria)

By contrast with acute intermittent porphyria, porphyria cutanea tarda does not present with acute exacerbations, nor are symptoms related to barbiturate ingestion.

A symptomatic mild coproporphyrinuria may occur in patients with liver disease, but less commonly excessive production and excretion of *uroporphyrin* (mostly of Type I) is associated with severe photosensitivity and skin lesions (porphyria cutanea tarda). Hirsutism and hyperpigmentation occur in chronic cases. It is probable that most patients have a dominantly inherited deficiency of the enzyme uroporphyrinogen decarboxylase, which converts uroporphyrinogen to coproporphyrinogen. Clinical disease, however, occurs only when an additional factor aggravates the deficiency; such factors are liver disease associated with alcohol abuse or iron overload, or high-dose oestrogen therapy. Skin sensitivity diminishes when they are removed. A similar picture may be produced by liver toxins, such as hexachlorobenzene, which may directly inhibit enzyme activity.

ERYTHROPOIETIC PORPHYRIAS

Two rare inherited disorders are associated with accumulation of porphyrins in erythrocytes. Acute porphyric attacks do not occur and ALA and PBG excretion are normal.

Congenital erythropoietic porphyria, unlike all the other porphyrias discussed, is inherited as a *recessive* characteristic. Usually from infancy onwards *erythrocyte* and *plasma uroporphyrin I* levels are very high and there is severe *photosensitivity*. Porphyrin is also deposited in bones and teeth, which fluoresce in ultraviolet light, and the teeth may be a brownish pink colour. Hirsutism, especially of the face, also occurs and there is a haemolytic anaemia.

Urinary porphyrin levels are grossly increased, and faecal levels less so.

Erythrohepatic protoporphyria.—In this *dominantly* inherited disorder *protoporphyrin* levels are increased in *erythrocytes* and *faeces*. There is mild photosensitivity, and hepatocellular damage may lead to liver failure.

OTHER CAUSES OF EXCESSIVE PORPHYRIN EXCRETION

Porphyria is not the only cause of disordered porphyrin metabolism and positive screening tests must be *confirmed* by quantitative analysis with *identification* of the porphyrin present. Three causes must be considered.

Lead poisoning inhibits several of the enzymes involved in haem synthesis and eventually causes anaemia. The *urine* contains increased amounts of *ALA* (an early and sensitive test), and *coproporphyrin*. Some of the symptoms of lead poisoning, such as abdominal pain, are similar to those of the acute porphyric attack, and this may cause difficulty in differential diagnosis.

Liver disease may increase *urinary coproporphyrin*, possibly due to decreased biliary excretion. This is probably the commonest cause of porphyrinuria. Occasionally there is mild photosensitivity (in porphyria cutanea tarda the more severe skin lesions are due to uroporphyrin excess).

Ulcerative lesions of the upper gastro-intestinal tract may produce raised levels of *faecal porphyrin* by degradation of haemoglobin. If there is bleeding from the lower part of the tract the blood reaches the rectum before there is time for conversion: this may help roughly to localise the site of bleeding.

SUMMARY

1. Porphyrins are by-products of haem synthesis. ALA and PBG are precursors.

2. The porphyrias are diseases associated with disturbed porphyrin metabolism. Most are inherited. The main clinical and biochemical features are outlined in Table XXXIII (p. 435).

3. Acute attacks with abdominal or neurological symptoms are a feature of the inherited hepatic porphyrias. Such attacks are potentially fatal and may be provoked by a number of drugs. The diagnosis

of porphyria in the acute phase depends on the demonstration of ALA and PBG in the urine.

4. The diagnosis of inherited porphyria must be followed by investigation of all relatives to detect asymptomatic cases. Screening tests may be negative in some types and quantitative estimations are necessary. Both urine and faeces should be examined.

5. Other causes of abnormalities in porphyrin excretion are lead poisoning, liver disease and upper gastro-intestinal bleeding.

6. The very rare erythropoietic porphyrias cause excessive accumulation of porphyrin in erythrocytes.

FURTHER READING

GOLDBERG, A. (1980). The porphyrias. *Clin. Haemat.*, **9**, 225–451. (An issue providing a comprehensive account of porphyrins and porphyria.)

SCREENING TESTS FOR THE PORPHYRIAS

The screening tests for porphyria are usually carried out in the laboratory. In areas of high incidence, such as South Africa, simple side-room tests should be available.

The requirements are:

1. a near ultraviolet (Wood's) lamp.
2. an extracting solvent (equal parts of ether, glacial acetic acid and amyl alcohol).
3. Ehrlich's reagent (2 per cent p-dimethylaminobenzaldehyde in 5M HCl).
4. n-Butanol.

Test for Porphobilinogen

(i) Equal parts of *fresh* urine and Ehrlich's reagent are mixed. If PBG is present a red colour develops.

(ii) If a red colour develops, 3 ml of n-butanol is added, the tube shaken and the phases allowed to separate. If the red colour does *not* enter the butanol (upper layer) it is PBG.

It is important to use *fresh* urine because PBG disappears on standing. A number of other substances, especially urobilinogen, may also give a red colour with Ehrlich's reagent. All such known substances are, however, extracted into the butanol layer. Chloroform has been used instead of butanol, but may not extract all non-PBG reactants and so give a false positive result.

Test for Porphyrins

Urine.—About 10 ml of fresh urine is mixed with 2 ml of the porphyrin extracting solvent (item 2) and the phases allowed to separate. A red fluorescence in the upper layer under ultraviolet light indicates the presence of porphyrin.

Faeces.—A pea-sized piece of faeces is mixed with 2 ml of the solvent. A red fluorescence under ultraviolet light denotes either porphyrin or chlorophyll (with a similar structure and derived from the diet). These may be distinguished by adding 1·5M HCl which extracts porphyrin but not chlorophyll.

Interpretation

The following points are important.

1. It is essential to test the *correct type of sample* (urine or faeces) for the type of porphyria expected.

2. In the *latent phase screening tests are too insensitive* to diagnose *acute intermittent porphyria*. Properly performed, *negative tests for faecal* porphyrin almost certainly *exclude the diagnosis of porphyria variegata*.

3. Tests are *negative before puberty* even in children who will suffer from porphyria.

4. *Positive results must be confirmed* by quantitative analysis and by *typing* the porphyrins. There are causes other than porphyria for increased porphyrin excretion.

Chapter XX

VITAMINS

VITAMINS, like protein, carbohydrate and fat, are organic compounds which are essential dietary constituents (the name "vitamines" originally meant amines necessary for life): unlike most other constituents they are required only in very small quantities. They must be taken in the diet because the body either cannot synthesise them at all or, under normal circumstances, not in sufficient amounts for its requirements: "vitamin D", for instance, can be synthesised in the skin under the influence of ultraviolet light, but in temperate climates the amount of sunlight may be insufficient to provide the required amount, while in hot countries the inhabitants tend to avoid the sun.

A normal mixed diet contains adequate quantities of vitamins, and deficiencies are rarely seen in affluent populations except in those people with intestinal malabsorption, on unsupplemented artificial diets or parenteral nutrition, and in food faddists. Vitamin supplementation of a normal diet is unnecessary. Some vitamins (notably A and D) produce toxic effects if taken in excess.

The biochemical functions of many vitamins are now understood, but it is not always easy to relate this knowledge to the clinical picture seen in deficiency states.

CLASSIFICATION OF VITAMINS

In 1913 McCollum and Davis first showed that two growth factors are required for normal health, one being fat soluble and the other water soluble. Since then each of these two groups has been shown to consist of many compounds, but classification can still be made on the basis of this solubility. The distinction is important clinically, because steatorrhoea is associated with deficiency of fat-soluble vitamins, but relatively little clinical evidence of lack of most water-soluble vitamins (the exceptions being vitamin B_{12} and folate, p. 295).

FAT-SOLUBLE VITAMINS

The fat-soluble vitamins are:

      A (retinol);
      D (calciferol);

K (2 methyl-1, 4 naphthaquinone);
E (tocopherol).

Each of these has more than one active chemical form, but variations in structure are very slight, and in the following discussion we shall refer to each vitamin as a single substance.

VITAMIN A

Source of Vitamin A

Precursors of vitamin A (the carotenes) are found in the yellow and green parts of plants and are especially abundant in carrots: for this reason this vegetable has the reputation of improving night vision, but it is doubtful if it has any effect on a subject eating a normal diet. The vitamin is formed by hydrolysis of the β-carotene molecule, each yielding a possible total of two molecules of vitamin A. In the body this hydrolysis occurs in the intestinal mucosa, but the yield of vitamin is much below the theoretical one, especially in children for whom a diet containing the vitamin itself is essential. It is stored in animal tissues, particularly in the liver, and these tissues are the only source of preformed vitamin A.

Stability of Vitamin A

Vitamin A is rapidly destroyed by ultraviolet light and should be kept in dark containers.

Function of Vitamin A

The *retinal pigment*, rhodopsin (visual purple), is necessary for vision in dim light (scotopic vision). Rhodopsin consists of a protein (opsin) combined with vitamin A. In bright light rhodopsin is broken down. It is partly regenerated in the dark, but, because this regeneration is not quantitatively complete, vitamin A is needed to maintain the levels in the retina.

Vitamin A is also essential for normal:

mucopolysaccharide synthesis;
mucus secretion; deficiency causes drying of mucus-secreting epithelia.

Clinical Effects of Vitamin A Deficiency

The clinical effects of vitamin A deficiency are:

"night blindness";
drying and metaplasia of ectodermal tissues:
follicular hyperkeratosis;
xerosis conjunctivae, xerophthalmia and keratomalacia;
anaemia.

"**Night blindness**".—Vitamin A deficiency is associated with poor vision in dim light, especially if the eye has recently been exposed to bright light. It is uncommon for the patient to complain of this.

Squamous metaplasia occurs in epithelial tissues.

Follicular hyperkeratosis.—Skin secretion is diminished, and there may be hyperkeratosis of hair follicles: dry horny papules, varying in size from a pinhead to quarter-inch diameter, are found mainly on the extensor surfaces of the thighs and forearms.

Squamous metaplasia of the bronchial epithelium has also been reported and may be associated with a tendency to chest infection.

Xerosis conjunctivae and xerophthalmia.—The conjunctiva and cornea become dry and wrinkled with squamous metaplasia of the epithelium and keratinisation of the tissue, resulting from deficiency of mucus secretion. *Bitot's spots*, seen in more advanced cases, are elevated white patches found in the conjunctivae and composed of keratin debris. Prolonged deficiency leads to *keratomalacia* with ulceration and infection and consequent scarring of the cornea, causing blindness. Keratomalacia is an important cause of blindness in the world as a whole, but is rarely seen in affluent countries.

The **anaemia** responds to vitamin A but not iron therapy.

Causes of Vitamin A Deficiency

Hepatic stores of vitamin A are so large that clinical signs only develop after many months, or even years of dietary deficiency.

Deficiency is very rare in affluent communities. In steatorrhoea clinical evidence is rare despite demonstrably low plasma levels. By contrast, deficiency is relatively common in underdeveloped countries, especially in children, and is a common cause of blindness.

Laboratory Diagnosis of Vitamin A Deficiency

This depends on the demonstration of low plasma vitamin A levels. Specimens should be kept in dark containers to prevent destruction of vitamin A by ultraviolet light.

Treatment of Vitamin A Deficiency

In treatment of xerophthalmia doses of 50 000 to 75 000 international units of vitamin A in fish liver oil should be given. Response to treatment of "night blindness" and of early retinal and corneal change is very rapid. Corneal scarring is irreversible.

Hypervitaminosis A

Vitamin A in large doses is toxic. Acute intoxication has been reported in Arctic regions as a result of eating polar bear liver, which has a very high vitamin A content, but a more common cause is overdosage with vitamin preparations: the symptoms of acute poisoning are nausea and vomiting, abdominal pain, drowsiness and headache. In chronic hypervitaminosis A there is fatigue, insomnia, bone pain, loss of hair and desquamation and discoloration of the skin.

VITAMIN D (CALCIFEROL)

The metabolism and functions of "vitamin D", and the effects and treatment of its deficiency are discussed in Chapter XI.

Overdosage with vitamin D causes hypercalcaemia, with all its attendant dangers (p. 257). In chronic overdosage stores of cholecalciferol are large, and hypercalcaemia may persist, and even progress for several weeks after the therapy is stopped.

VITAMIN K

Vitamin K cannot be synthesised by man but, like many of the B vitamins, it can be manufactured by the bacterial flora of the colon: unlike them it can probably be absorbed from this site, and dietary deficiency is therefore not seen. In steatorrhoea the vitamin, whether taken in the diet or produced by bacteria, cannot be absorbed normally and deficiency may occur (p. 294).

Very little vitamin K is transported across the placenta, and the newborn infant may be vitamin K deficient: the neonatal gut is only gradually colonised by bacteria capable of synthesising vitamin K. Deficiency may be severe enough to cause *haemorrhagic disease of the newborn*.

Vitamin K is necessary for the synthesis of prothrombin and coagulation factors VII, IX and X in the liver, and deficiency is accompanied by a bleeding tendency with a prolonged prothrombin time. If these findings are due to deficiency of the vitamin they can be cured by parenteral administration (p. 319).

VITAMIN E (TOCOPHEROLS)

Vitamin E deficiency may cause haemolytic anaemia, thrombocytosis, oedema and irritability in premature infants. There is no evidence that a low intake after the neonatal period produces any clinical effects.

WATER-SOLUBLE VITAMINS

The water-soluble vitamins are:
the B complex:

thiamin (aneurin: B_1);
riboflavin (B_2);
nicotinamide (pellagra preventive (PP) factor: niacin);
pyridoxine (B_6);
biotin and pantothenate (probably of no clinical significance);
folate (pteroylglutamate);
the vitamin B_{12} complex (cobalamins);
ascorbate (vitamin C).

THE B COMPLEX

This group of food factors was originally classified together in the B group (with the exception of vitamin B_{12} and folate, which were later discoveries). Most of them act as cofactors for enzymes, but it is not easy to relate the clinical findings to the underlying biochemical lesion.

Many are synthesised by colonic bacteria. Opinions vary as to the importance of this source in man, but since the absorption of water-soluble vitamins from the large intestine is poor, probably most of those synthesised at this site are unavailable to the body.

Clinical deficiency is rare in affluent communities. When deficiency does occur it is usually multiple, involving most of the B group and protein: for this reason it may be difficult to decide which signs and symptoms are specific for an individual vitamin and which are part of a general malnutrition syndrome.

THIAMIN (B_1)

Source of Thiamin and Cause of Deficiency

Thiamin cannot be synthesised by animals, including man: it is found in most dietary components, and wheat germ, oatmeal and yeast are particularly rich in the vitamin. Adequate amounts are present in a normal diet, but the deficiency syndrome is still prevalent in rice-eating areas: polished rice has the husk removed and this is the only source of thiamin in this food. In other areas thiamin deficiency occurs most commonly in alcoholics and in subjects with anorexia nervosa.

Function of Thiamin

Thiamin is a component of thiamin pyrophosphate, which is an essential cofactor for *decarboxylation of α-oxoacids* (cocarboxylase): one of these important reactions is the conversion of pyruvate to acetyl coenzyme A (acetyl CoA). In thiamin deficiency pyruvate cannot be metabolised and accumulates in the blood. Thiamin pyrophosphate is also an essential cofactor for transketolation reactions. One such reaction is catalysed by *transketolase* in the pentose-phosphate pathway; the keto (oxo) group is transferred from xylulose-5-phosphate to ribose-5-phosphate to produce glyceraldehyde-3-phosphate and sedoheptulose-7-phosphate.

Clinical Effects of Thiamin Deficiency

Deficiency of thiamin causes the syndrome known as *beriberi*, which includes anorexia and emaciation, neurological lesions (motor and sensory polyneuropathy, Wernicke's encephalopathy), and cardiac arrhythmias ("dry" beriberi). In the so-called "wet" form of the disease there is oedema, sometimes with cardiac failure. Some of these findings may be due to associated protein deficiency rather than to that of thiamin.

Beriberi can be aggravated by a high carbohydrate diet, possibly because this leads to an increased rate of glycolysis and therefore of pyruvate production.

Laboratory Diagnosis of Thiamin Deficiency

The most reliable test is probably the estimation of *erythrocyte transketolase activity*, with and without added thiamin pyrophosphate. A reduced activity, if due to thiamin deficiency, becomes normal after addition of the cofactor.

Treatment of Thiamin Deficiency

True beriberi responds to 5–10 mg of thiamin daily, although occasionally higher dosages may be required. In cases in whom multiple deficiency is suspected a mixture of the vitamins in the "B complex" should be given.

RIBOFLAVIN (B$_2$)

Source of Riboflavin

Riboflavin is found in large amounts in yeasts and germinating plants such as peas and beans, and in smaller amounts in fish, poultry and meat (especially "offal").

Function of Riboflavin

There are about 15 flavoproteins, mostly enzymes incorporating riboflavin in the form of flavin mononucleotide (FMN) and flavin adenine dinucleotide (FAD). FMN and FAD are reversible *electron carriers* in biological oxidation systems and are, in turn, oxidised by cytochromes.

Clinical Effects of Riboflavin Deficiency

Ariboflavinosis causes a rough scaly skin, especially on the face, cheilosis (red, swollen, cracked lips), angular stomatitis and similar lesions at the mucocutaneous junctions of anus and vagina, and a swollen tender, red tongue, which is described as magenta coloured. Congestion of conjunctival blood vessels may be visible when the eye is examined with a slit lamp.

Laboratory Diagnosis of Riboflavin Deficiency

Riboflavin acts as a cofactor for *glutathione reductase*, increasing its activity. The finding of a low erythrocyte enzyme activity, which increases by about 30 per cent after adding FAD, suggests riboflavin deficiency.

NICOTINAMIDE

Source of Nicotinamide

Nicotinamide can be formed in the body from nicotinic acid. Both substances are plentiful in animal and plant foods, although much of that in plants is bound in an unabsorbable form. Some nicotinic acid can also be synthesised in the mammalian body from tryptophan. Probably both dietary and endogenous sources are necessary to provide sufficient nicotinamide for normal metabolism.

Function of Nicotinamide

Nicotinamide is the active constituent of the important cofactor in *oxidation-reduction reactions*, nicotinamide adenine dinucleotide (NAD), and its phosphate (NADP). Reduced NAD and NADP are, in turn, reoxidised by flavoproteins, and the functions of riboflavin and nicotinamide are closely linked. NAD and NADP, and their reduced forms are essential, for among other things, glycolysis and oxidative phosphorylation and many synthetic processes.

Clinical Effects of Nicotinamide Deficiency

It is not always easy to distinguish the clinical features due to coexistent deficiencies, especially of pyridoxine, from those specifically due to nicotinamide. However, nicotinamide deficiency is probably the most important factor precipitating the clinical syndrome of *pellagra* (a word literally meaning "rough skin"), and nicotinic acid has been called "pellagra preventive (PP) factor". The symptoms are often remembered by the mnemonic "three Ds"—dermatitis, diarrhoea and dementia. The dermatitis is a sunburn-like erythema, especially severe in areas exposed to the sun, and progressing to pigmentation and thickening of the dermis. Irritability, depression and anorexia with loss of weight precede true dementia with delusions.

Causes of Nicotinamide Deficiency

Dietary deficiency of nicotinamide, like that of the other B vitamins, is rare in affluent communities.

Hartnup disease is due to a rare inborn error of renal, intestinal and other cellular transport mechanisms for the monoamino monocarboxylic acids, including tryptophan (p. 392). Subjects with the disease may present with a pellagra-type rash, which can be cured by nicotinamide therapy of between 40 and 200 mg daily. Probably, if the supply of tryptophan for synthesis in the body is reduced, dietary nicotinic acid is insufficient to supply the body's needs over long periods of time: under these circumstances only a slight reduction of intake may precipitate pellagra. A similar clinical picture has been reported as a rare complication of the *carcinoid syndrome*, when tryptophan is diverted to the synthesis of large amounts of 5-hydroxytryptamine (p. 467).

Nicotinic acid (but not nicotinamide) may reduce hepatic secretion of VLDL and therefore plasma levels of VLDL and LDL (p. 245).

Laboratory Diagnosis of Nicotinamide Deficiency

Chemical and microbiological assays for nicotinic acid in body fluids are now available.

PYRIDOXINE (B_6)

Source of Pyridoxine and Cause of Deficiency

Pyridoxine (pyridoxol), its aldehyde (pyridoxal) and its amine (pyridoxamine) are widely distributed in food and dietary deficiency is very rare. The antituberculous drug *isoniazid* (isonicotinic hydrazide) and *L-dopa* have been reported to produce the picture of pyridoxine deficiency, probably by competing with it in metabolic pathways.

Functions of Pyridoxine

Pyridoxal phosphate, formed in the liver from pyridoxine, pyridoxal and pyridoxamine, is a cofactor mainly for the *transaminases*, and for *decarboxylation of amino acids*.

Clinical Effects of Pyridoxine Deficiency

Deficiency may produce roughening of the skin, a peripheral neuropathy and a sore tongue. A very rare hypochromic, microcytic anaemia responds to large doses of pyridoxine even when there is no evidence of deficiency of the vitamin ("pyridoxine-responsive" anaemia).

Laboratory Diagnosis of Pyridoxine Deficiency

Pyridoxal phosphate is required for conversion of tryptophan to nicotinic acid. In pyridoxine deficiency this pathway is impaired. *Xanthurenic acid* is the excretion product of 3-hydroxykynurenic acid, the metabolite before the "block", and in pyridoxine deficiency is found in abnormal amounts in the urine after an oral *tryptophan load*.

The urinary metabolite of pyridoxal phosphate, 4-pyridoxic acid, may also be measured.

The increase in *erythrocyte aspartate transaminase* activity after addition of pyridoxine may be measured. The more severe the pyridoxine deficiency the greater the increase in enzyme activity after addition of the vitamin.

BIOTIN AND PANTOTHENATE

Lack of these two vitamins of the B group probably never produces clinical deficiency syndromes.

Biotin is present in eggs, but large amounts of raw egg white in experimental diets have caused loss of hair and dermatitis thought to be due to biotin deficiency. Probably the protein avidin, present in the egg white, combines with biotin and prevents its absorption. Biotin is a cofactor in carboxylation reactions.

Pantothenate is a component of coenzyme A (CoA), which is essential for fat and carbohydrate metabolism. It is very widely distributed in foodstuffs.

Figure 40 summarises some of the biochemical interrelationships of the B vitamins discussed so far. Note that, as a general rule, deficiency of this group results in lesions of skin, mucous membranes and the nervous system.

FOLATE AND VITAMIN B_{12}

These two vitamins are included in the B group and are essential for the normal maturation of the erythrocyte; deficiency of either causes *megaloblastic anaemia*. Their effects are so closely interrelated that they are usually discussed together. A fuller discussion of diagnosis and treatment will be found in haematology textbooks.

Folate is present in green vegetables and some meats. It is easily destroyed in cooking and *dietary deficiency* may rarely occur. It is absorbed throughout the small intestine and in contrast to most of the other B vitamins (except B_{12}) clinical deficiency occurs relatively

ROLE OF B VITAMINS IN FORMATION OF ACETYL CoA

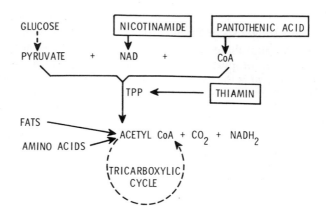

ROLE OF B VITAMINS IN ELECTRON TRANSFER

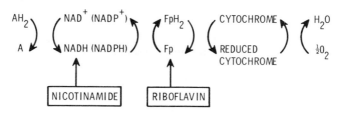

A = SUBSTRATE (e.g. PYRUVATE)   Fp = FLAVOPROTEIN

AH_2 = REDUCED SUBSTRATE (e.g. LACTATE)

B VITAMINS IN ⬜

FIG. 40.—Some biochemical interrelationships of B vitamins.

commonly in intestinal *malabsorption syndromes*. In the malabsorption syndromes and during *pregnancy* and lactation low red-cell folate levels may be associated with megaloblastic anaemia. The active form of the vitamin is tetrahydrofolate, and this is essential for transfer of "one carbon" units: it is particularly important in *purine* and *pyrimidine* (and therefore DNA and RNA) synthesis. Methotrexate, a folate derivative, competes with it for metabolism; it therefore inhibits DNA synthesis and is cytotoxic.

The vitamin B_{12} group includes several cobalamins, found in animal products, but not in green vegetables. Dietary deficiency is rare. They are transported in plasma by a specific carrier protein, transcobalamin II. Deoxyadenosyl and methylcobalamin have, like folate, coenzyme activity in nucleic acid synthesis. Hydroxycobalamin is the form most commonly used in therapy and both it and cyanocobalamin are converted in the body to the cofactor forms. All forms are mainly absorbed in the distal ileum combined with intrinsic factor. The relationship between intestinal dysfunction and vitamin B_{12} deficiency is discussed in Chapter XII.

Absorption of the vitamin depends on its combination with intrinsic factor, secreted by the stomach. In true *pernicious anaemia* intrinsic factor antibodies result in malabsorption of vitamin B_{12}.

Deficiency of vitamin B_{12}, like that of folate, causes megaloblastic anaemia: unlike that of folate it can cause *subacute combined degeneration* of the spinal cord. Although the megaloblastic anaemia of vitamin B_{12} deficiency can be reversed by folate, this therapy should never be given in pernicious anaemia because it does not improve the neurological lesions and may even aggravate them.

Table XXXIV summarises the synonyms of the B vitamins, their known biochemical actions, and the clinical syndromes associated with deficiencies of each.

Ascorbate (C)

Source of Ascorbate

Ascorbate is found in fruit and vegetables and is especially plentiful in citrus fruits. A quantitatively significant dietary source is ascorbate added to other foods as a preservative. It cannot be synthesised by man and other primates, nor by guinea-pigs.

Functions of Ascorbate

Ascorbate can be reversibly oxidised in biological systems to dehydroascorbate and, although its functions in man are not well worked out, it probably acts as a hydrogen carrier. It seems to be required for normal collagen formation.

Causes of Ascorbate Deficiency

Deficiency of ascorbate causes scurvy and was commonly seen on the long sea voyages of exploration in the 16th, 17th and 18th centuries. Anson, in his account of his voyage around the world from 1740 to 1744, gives an excellent clinical description of florid scurvy in

TABLE XXXIV

THE "B COMPLEX" VITAMINS

Name	Synonyms	Biochemical function	Clinical deficiency syndrome
Thiamin	Aneurin Vitamin B_1	Cocarboxylase (as thiamin pyro-phosphate)	Beriberi (neuropathy) Wernicke's encephalo-pathy
Riboflavin	Vitamin B_2	In flavoproteins (electron carriers as FAD and FMN)	Ariboflavinosis (affecting skin and eyes)
Nicotinamide	PP factor Niacin	In NAD and NADP (electron carriers)	Pellagra (dermatitis, diarrhoea, dementia)
Pyridoxine	Adermin Vitamin B_6	Cofactor in decarboxy-lation and deamination (as phosphate)	? Pyridoxine responsive anaemia ? Dermatitis
Biotin		Carboxylation cofactor	Probably unimportant clinically
Pantothenate		In coenzyme A	
Folate	Pteroyl glutamate	Metabolism of purines and pyrimidines	Megaloblastic anaemia
Vitamin B_{12} group	Cobalamins	Cofactor in synthesis of nucleic acid	Megaloblastic anaemia Subacute combined degeneration of the cord

his crew. The problem should not have arisen if he had read the literature, because 150 years earlier, in 1593, Hawkins had known that the disease could be cured by eating oranges or lemons, and Sir James Lancaster had ordered a regular supply of these fruits, as a prophylactic measure, on the ships of the East India Company in 1601.

Dehydroascorbate is easily oxidised further and irreversibly in the presence of oxygen, and loses its biological activity; this reaction is catalysed by heat. Scurvy was at one time fairly common in bottle-fed infants, because the ascorbate was often destroyed in the preparation of the feeds. With our knowledge of the aetiology of the disease, and with dietary supplementation, it is now rarely seen in this age group. It is most commonly seen in old people (especially men) living on their own on poor incomes, who do not eat fresh fruit and vegetables and who tend to cook in frying pans; the combination of heat and the large area of food in contact with air irreversibly oxidises the vitamin. Ascorbate deficiency can occur in iron overload (p. 426).

Clinical Effects of Ascorbate Deficiency

Anson described "large discoloured spots", "putrid gums" and "lassitude" as characteristic of scurvy.

Many of the signs and symptoms of scurvy can be related to poor collagen formation.

Deficiency in *vascular walls* leads to a bleeding tendency, often with a positive Hess test, petechiae and ecchymoses ("large discoloured spots"), swollen, tender, spongy bleeding gums ("putrid gums") and, occasionally, haematuria, epistaxis and retinal haemorrhages. In infants subperiosteal bleeding and haemarthroses are extremely painful and may lead to permanent joint deformities.

There is *poor healing* of all wounds.

Deficiency of *bone matrix* causes osteoporosis and poor healing of fractures. In children bone formation is impaired at the epidiaphyseal junctions, which look "frayed" radiologically.

There may be *anaemia*, possibly partly due to impairment of erythropoiesis. This anaemia may sometimes be cured by ascorbate alone. Bleeding aggravates the anaemia.

This florid form of the disease is rarely seen nowadays and the patient most commonly presents complaining that bruising occurs with only minor trauma.

All the signs and symptoms are dramatically cured by the administration of ascorbate.

Diagnosis of Scurvy

The laboratory confirmation of the clinical diagnosis of scurvy can only be made *before* therapy has started. Once ascorbate has been given it is difficult to prove that deficiency was previously present.

Although leucocyte ascorbate assay was said to be more reliable to confirm deficiency than that of plasma, recent evidence suggests that levels in the two alter in parallel; plasma assay is technically more satisfactory, and to be preferred.

SUMMARY

1. Vitamins have important biochemical functions, most of which are now well understood. Unfortunately the relationship of these to clinical syndromes is not always obvious.

2. The fat soluble vitamins, especially "vitamin D", may be deficient

in steatorrhoea. Both vitamin A and "vitamin D" are stored in the body and deficiency takes some time to develop.

3. **Vitamin A** is necessary for the formation of visual purple and for normal mucopolysaccharide synthesis. Deficiency is associated with poor vision in dim light and with drying and metaplasia of epithelial surfaces, especially those of the conjunctiva and cornea.

4. **Vitamin D** as 1,25 dihydroxycholecalciferol is necessary for normal calcium metabolism and deficiency causes rickets in children and osteomalacia in adults.

5. Both vitamin A and vitamin D are toxic in excess.

6. **Vitamin K** is necessary for prothrombin formation and deficiency is associated with a bleeding tendency.

7. **Thiamin** deficiency causes beriberi.

8. **Riboflavin** deficiency causes ariboflavinosis.

9. **Nicotinamide** can be manufactured from tryptophan in the body, but dietary deficiency causes a pellagra-like syndrome, which may also be seen in Hartnup disease, when tryptophan absorption is deficient, and in the carcinoid syndrome, when tryptophan is used in excess for 5-hydroxytryptamine synthesis.

10. **Pyridoxine** responsive anaemia may occur.

11. **Folate and vitamin B_{12}** deficiency produce megaloblastic anaemia, and deficiency of vitamin B_{12} can also cause subacute combined degeneration of the cord. Compared with the other B vitamins deficiency of these is relatively common in malabsorption syndromes, and vitamin B_{12} deficiency can be a feature of the "contaminated bowel syndrome". Classical pernicious anaemia is due to intrinsic factor deficiency with consequent malabsorption of vitamin B_{12}.

12. **Ascorbate** deficiency causes scurvy.

FURTHER READING

BARKER, B. M. and BENDER, D. A. (Eds.) (1980). *Vitamins in Medicine*, 4th edit. Vol. I (D, E, B_{12}, folate, niacin, B_6, biotin and riboflavin). London: William Heinemann Medical Books.

BARKER, B. M. and BENDER, D. A. (Eds.) (1982). *Vitamins in Medicine*, 4th edit. Vol. II (C, pantothenic acid, K, thiamin and A). London: William Heinemann Medical Books.

Chapter XXI

PREGNANCY AND
ORAL CONTRACEPTIVE THERAPY

ANTENATAL TESTING

THE reasons for testing during pregnancy fall into two main groups:

Testing *early* in pregnancy (before about 20 weeks gestation) to detect serious *congenital abnormalities* with a view to *termination* if such abnormalities are found.

Testing *late* in pregnancy (after about 28 weeks gestation) so that action may be taken to *improve the chance of survival* of the infant:

by detecting incipient *fetal damage* caused, for example, by placental insufficiency or by severe blood group incompatibility, so that labour may be induced before such damage is severe or irreversible;

by assessing *fetal maturity*, so that induction may, if possible, be delayed until the newborn infant stands a good chance of survival. Maturity of the fetal lungs is of especial importance.

Most of these tests are carried out on patients who are known to be at risk. Changes in the composition of maternal urine or plasma often reflect changes in fetal and placental metabolism, and sampling of these is safe and simple. Occasionally there are indications for testing amniotic fluid obtained by amniocentesis, but the ability to visualise the fetus using ultrasound has reduced the need for this invasive procedure.

Amniocentesis.—Amniotic fluid may be sampled through a needle inserted into the uterus through the maternal abdominal wall at any time after about 14 weeks gestation. The procedure carries a very small risk to the fetus, even in the best hands, and *should only be performed when there are very strong clinical indications, and if the diagnosis cannot be made by non-invasive procedures.* Analytical results may be *dangerously misleading* if for example, the specimen is *contaminated with maternal or fetal blood or maternal urine, or is not fresh and properly preserved.* Both the safety and reliability of the procedure are improved if it is performed by someone with experience, using ultrasound to indicate the position of the fetus, placenta and maternal bladder. The laboratory analysing the sample should also have experience and, as always, close liaison between the clinician and laboratory helps to ensure the suitability of the specimen and the speed of assay.

Amniotic fluid is probably derived from both maternal and fetal sources, but its value in reflecting fetal abnormalities arises from its intimate contact with the fetus, and from the increasing contribution of fetal urine in later pregnancy.

TESTS FOR CONGENITAL ABNORMALITIES

Neural-Tube Defects

α-**Fetoprotein** is a low-molecular-weight glycoprotein synthesised mainly in the fetal liver and yolk sac. Its production is almost completely repressed in the normal adult. Because of its relatively low molecular weight it can diffuse slowly through capillary membranes, and appears in fetal urine—and hence in amniotic fluid—and in maternal plasma. Severe fetal neural-tube defects (such as open spina bifida and anencephaly) are associated with abnormally high concentrations in these fluids: the reason for this is not clear, but the protein may leak from the exposed neural-tube blood vessels. Many other fetal causes of raised α-fetoprotein concentration in amniotic fluid and maternal plasma have been reported; one of these is multiple pregnancy, but almost all the others are associated with serious fetal abnormalities, such as exomphalos.

Estimation of α-fetoprotein levels in *maternal serum* may be used in conjunction with ultrasound as a screening test for neural-tube defects: ultrasound alone will usually detect gross defects. Maternal α-fetoprotein assay is a relatively crude diagnostic tool with an appreciable incidence of false positive and a few false negative results, and should never be acted on in isolation. The blood must be tested at *between 16 and 18 weeks gestation* and ultrasound helps to confirm gestational age and to exclude multiple pregnancy as a cause. Earlier the levels are not high enough to be detectable, and later the risk to the mother of terminating pregnancy increases. Positive results should be confirmed on a fresh specimen, and maternal causes sought. In some countries all pregnant women attending for antenatal care now have serum α-fetoprotein levels estimated at 16 to 18 weeks of pregnancy, although the value of such unselective screening is not accepted by all.

α-Fetoprotein estimation on *amniotic fluid* is a more precise diagnostic tool and yields fewer false positive results than plasma assay if sampling is properly performed. It should be reserved for subjects known to be at risk either because of a family history of neural-tube defects, or because of the finding of a high level in maternal plasma with a normal or equivocal ultrasound scan. Amniocentesis should not usually be carried out for this purpose unless the parents are willing to consider termination if the result is positive. It should be performed by experts, with experienced laboratory back-up. It is important to remember that α-fetoprotein concentration in normal fetal blood is high at 16 to 18 weeks, and that *bloodstained amniotic fluid can therefore yield dangerously misleading results.*

Amniotic fluid acetylcholinesterase levels may also rise in association with many serious fetal malformations, including neural-tube defects and exomphalos. The interpretation of the result is less dependent on fetal age than that of α-fetoprotein, but is equally invalidated by contamination with fetal or maternal blood. The diagnostic value of the assay is still being assessed, but if both α-fetoprotein and acetylcholinesterase levels are raised in amniotic fluid, the probability of a neural-tube defect is high. The assay is less widely available than that for α-fetoprotein.

Amniotic Fluid Cell Culture

Some *congenital abnormalities* may be detected by chemical or enzymatic assay on cells cultured from the fluid. These tests are performed only in special centres, and only on subjects with a genetic history of the condition.

Tests for Feto-Placental Impairment

Tests for feto-placental impairment are less commonly used than previously. Their predictive value has been questioned, and non-chemical tests may often be more informative.

The production of placental metabolites increases from early pregnancy until near term, and some of them may be detected in maternal urine and plasma. Oestriol or human placental lactogen (HPL) production are relatively easy to assess. The "normal" ranges of both these and of similar parameters are very wide at every stage of pregnancy, and sudden changes in levels are more significant than a single "abnormal" result; for this reason serial estimations are usually performed. Even during normal pregnancy there is some day-to-day variation, and a change in the same direction in two or more consecutive specimens increases the probability that it is clinically significant. Results should always be interpreted in conjunction with clinical and other findings, especially those of ultrasound. Sometimes the result of one type of assay is consistently abnormal for no apparent reason: in rare cases of serious doubt both oestriol and HPL production should be serially assessed.

Assessment of Feto-Placental Oestriol Production

Soon after fertilisation the ovum is implanted in the uterine wall: secretion of chorionic gonadotrophin by the developing placenta maintains the corpus luteum of the luteal phase and the continued secretion of oestrogen and progesterone from the corpus luteum prevents the onset of menstruation. Oestriol secretion at this stage is very low and it rises only very slowly during the first weeks of pregnancy.

Chorionic gonadotrophin secretion reaches a peak at about 13 weeks of pregnancy, and then falls. The feto-placental unit then takes over hormone production, and secretion of both oestrogen and progesterone rises rapidly.

At this stage the fetus and the placenta are an integrated endocrine unit and both are required for the production of oestriol. The hormone passes into the maternal circulation and ultimately into the urine.

The production of oestriol may be assessed by measuring the *daily urinary total oestrogen* excretion. After about the 28th week of gestation most of this measured oestrogen is oestriol: before about the 28th week oestriol concentration is low and less easy to measure.

Collection of a 24-hour urine specimen delays the production of a result by a day and, as always, is potentially inaccurate. Moreover, outpatients have to keep and transport large bottles of urine. These problems can be overcome if the *concentration ratio of total oestrogen to creatinine* is measured on an early morning specimen of urine. The creatinine production is assumed to be constant, and its excretion to alter in parallel with that of oestrogen. A fall in the ratio should therefore indicate reduced oestriol production. In practice the method is simple, and provides more consistent results than measurement of daily oestrogen excretion.

Plasma oestriol levels can be measured by methods specific for this oestrogen. Despite the theoretical advantages there is a physiological variation in levels from day to day and the results therefore seem to have no better predictive value than the simpler and cheaper estimation of the urinary oestrogen:creatinine ratio.

Assessment of Human Placental Lactogen (HPL) Production

HPL is a peptide hormone, of unknown physiological significance, synthesised by the placenta. It is detectable by rapid radioimmuno-assay methods after about the eighth week of gestation, and has been used in the assessment of threatened abortion. Levels rise until the 36th week, after which they fall slightly. The concentration is a measure of *placental* function only, unlike urinary oestriol which indicates fetal function as well. HPL certainly has no better predictive value than urinary oestriol, but may sometimes, as already discussed, be useful as a supplementary estimation.

ASSESSMENT OF THE SEVERITY OF FETO-MATERNAL BLOOD GROUP INCOMPATIBILITY

Amniotic fluid levels of *bilirubin* are used in conjunction with maternal antibody titres, to assess the effects on the fetus of rhesus (or other blood group) incompatibility. Normally amniotic fluid bilirubin levels decrease during the last half of pregnancy. The level at any stage may be correlated with the severity of haemolysis: used with tests of maturity, the optimum time for induction of labour or the need for intra-uterine transfusion may be assessed.

ASSESSMENT OF FETAL MATURITY

Assessment of fetal age from the date of the last menstrual period is often inaccurate. Fetal maturity may be assessed by estimating the level of products of fetal metabolism in *amniotic fluid*. The "mature" ranges are wide, but values for each parameter have been chosen above which the fetus is almost certainly mature: lower values do not necessarily indicate immaturity. The results of these tests must be interpreted with caution, and in conjunction with clinical and other findings; the predictive value of such testing, and therefore the justification for amniocentesis for this purpose, has recently been questioned.

Pulmonary maturity is important for neonatal well-being. The lungs will not expand normally at birth if they are immature and the infant may then suffer from the *respiratory distress syndrome* (hyaline membrane disease); he may need intensive care in a special care baby unit until he is mature. It is therefore important to have evidence of pulmonary maturity before labour is induced.

In late pregnancy (after 32 to 34 weeks gestation) the cells lining fetal alveolar walls start synthesising a surface-tension-lowering complex ("surfactant"), 90 per cent of which is lipid in nature; most of this lipid is *lecithin*, which contains *palmitic acid*. Surfactant is probably washed from, or is perhaps secreted by, the alveoli into the surrounding amniotic fluid, in which both lecithin and palmitic acid concentrations increase steadily. The concentration of another lipid, sphingomyelin, remains constant, and the *lecithin:sphingomyelin (L:S) ratio rises*; measurement of this ratio is the most commonly used estimate of pulmonary maturity. Other estimates have included total lecithin and total palmitic acid concentrations. The predictive value of all these parameters is similar.

METABOLIC EFFECTS OF PREGNANCY AND ORAL CONTRACEPTIVE THERAPY

The level of many plasma constituents is affected by steroid hormones: for example, the "normal" ranges of plasma urate and iron differ in males and females after puberty. It is therefore not surprising to find that pregnancy, which is associated with very high circulating levels of oestrogens and progesterone, affects plasma concentrations of many substances. Oral contraceptive therapy is said to prevent ovulation by mimicking the steroid background of pregnancy, because the tablets contain synthetic oestrogens and "progestogens", and many of the metabolic changes during pregnancy are also found in some women taking the "pill". Most of these changes have been attributed to the oestrogen rather than the progestogen fraction, but it may be very difficult to be sure exactly which hormone is producing the change. The mechanism of the changes is poorly understood, but steroids are known to affect protein synthesis.

Effects on Plasma Proteins

The plasma level of many specific carrier proteins is increased in pregnant subjects and in those taking oral contraceptive preparations. If this fact is not recognised an erroneous diagnosis may be made. In most cases the rise in the level of carrier protein is accompanied by a proportional increase of the substance bound to it, without any change in the unbound fraction. Because the protein-bound fraction is a transport form and because, in most cases, it is the free substance that is physiologically active, this rise in concentration is only of importance in interpretation of the results of tests.

The reduction of plasma total protein concentration in pregnancy is probably due to dilution following fluid retention. Albumin and total protein levels have been reported to fall slightly during oral contraceptive therapy, but this change is unlikely to cause diagnostic confusion. It seems that *only binding proteins* are *increased* in these subjects.

Some of the changes in carrier proteins which may cause diagnostic confusion are listed in Table XXXV

Tests of Liver Dysfunction

Some subjects develop a cholestatic jaundice during pregnancy and are prone to a similar syndrome when taking some oral contraceptive preparations. This is extremely rare.

TABLE XXXV

SOME METABOLIC EFFECTS OF PREGNANCY AND ORAL CONTRACEPTIVE THERAPY WHICH MAY CONFUSE DIAGNOSIS

Test	Effect	Comment
Plasma total T_4	Increased	Due to increased thyroxine-binding globulin *Free T_4 normal*
Free thyroxine-binding sites	Increased	
Plasma cortisol	Increased	Due to increased cortisol-binding globulin. *Free cortisol normal*
Plasma transferrin or TIBC	Increased	See Chapter XVIII
Plasma iron	Increased	
Plasma alkaline phosphatase	Increased	Due to heat-stable placental isoenzyme
Urinary glucose	May be renal glycosuria	Probably due to raised GFR

Plasma Alkaline Phosphatase Activity

The placenta produces one of the alkaline phosphatase isoenzymes and the plasma level of this rises in late pregnancy. This enzyme is stable at a temperature at which alkaline phosphatase from other sources is inactivated. This physiological cause should be remembered if a high alkaline phosphatase activity is found during pregnancy.

Hormone Secretion

The effect of oral contraceptives on cortisol-binding globulin and therefore on plasma cortisol has already been mentioned. *Gonadotrophin levels* are very low in subjects taking oral contraceptives because of feed-back suppression. These revert to normal when the tablets are stopped.

Glycosuria

Renal glycosuria is common both in pregnancy and in subjects taking oral contraceptives. The glomerular filtration rate increases by about 50 per cent during pregnancy, and glycosuria may partly be due to an increased glucose load on normal tubules.

Some of the changes which may occur during pregnancy and oral contraceptive therapy are summarised in Table XXXV.

SUMMARY

Antenatal Testing

1. α-Fetoprotein levels are high in maternal serum and in amniotic fluid, and acetylcholinesterase in amniotic fluid, if the fetus has severe neural-tube defects.

2. Testing is carried out after the 28th week of pregnancy:

> to predict possible fetal damage if pregnancy is allowed to continue;
>
> to assess the maturity of the fetus before induction of labour is undertaken.

3. Fetoplacental function may be monitored by assessing the production of oestriol or of HPL. Daily maternal urinary total oestrogen output, or the oestrogen:creatinine ratio on a random sample, usually reflects oestriol production after about the 28th week of pregnancy. HPL concentration is measured in maternal plasma.

4. Fetal pulmonary maturity may be assessed by measuring the lecithin-sphingomyelin (L/S) ratio, or total lecithin or palmitic acid concentrations in amniotic fluid.

5. Amniotic fluid bilirubin levels are used in assessing the effect on the fetus of blood group incompatibility.

Metabolic Effects of Pregnancy and Oral Contraceptive Therapy

Pregnancy and oral contraceptive therapy produce similar metabolic effects which, if not recognised, may lead to misdiagnosis. The more important of these are listed in Table XXXV.

FURTHER READING

DEPARTMENT OF OBSTETRICS, ST. GEORGE'S HOSPITAL MEDICAL SCHOOL, LONDON (1984). An end to antenatal oestrogen monitoring? *Lancet*, **1**, 1171–1172.

WALD, H. J. and CUCKLE, H. S. (1979). Amniotic-fluid alpha-fetoprotein measurement in antenatal diagnosis of anencephaly and open spina bifida in early pregnancy. (Second report of the U.K. collaborative study of alpha-fetoprotein in relation to neural-tube defects.) *Lancet*, **2**, 651–662.

CUCKLE, H. S. *et al.* (1981). Amniotic-fluid acetylcholinesterase electrophoresis as a secondary test in the diagnosis of anencephaly and open spina bifida in early pregnancy. (Report of the collaborative acetylcholinesterase study.) *Lancet*, **2**, 321–324.

ROBERTS, C. J., HIBBARD, B. M., ELDER, G. H. *et al.* (1983). The efficacy of a serum screening service for neural-tube defects: the South Wales experience. *Lancet*, **1**, 1315–1318.

FAIRWEATHER, D. V. I. (1982). Screening in pregnancy for congenital abnormality. *Brit J. Hosp. Med.*, **27**, 601–607.

Chapter XXII

BIOCHEMICAL EFFECTS OF TUMOURS

SOME rare syndromes are associated with neoplasia of cells which, although normally producing hormones, are, because of their scattered nature, not usually thought of as endocrine organs. Many malignant tumours of non-endocrine tissues can elaborate hormones usually foreign to them. Pearse and others have explained both these types of syndrome on the basis of the so-called APUD system. Although this is by no means the only possible explanation, we will start with a brief outline of the APUD concept.

THE DIFFUSE ENDOCRINE (APUD) SYSTEM

A group of cells, scattered throughout the body, has common cytological characteristics, and Pearse believes that these cells have a common origin in embryonic ectoblast. Most have endocrine or neurotransmitter properties. The acronym APUD is derived from their ability for *A*mine *P*recursor *U*ptake and *D*ecarboxylation to produce amines. Tumours of APUD cells have been called APUDomas. Some of these cells secrete *amines* which are physiologically active. In many the amine production seems to be linked to synthesis and secretion of *peptide hormones*. Some are apparently normally *non-secretory*.

Many of the peptide-hormone-producing cells form recognisable *endocrine glands* (pituitary, parathyroid, calcitonin-producing cells of the thyroid). Other APUD cells, both peptide and amine-secreting, are found in *specialised nerve tissue* (the trophic-hormone and ADH-secreting cells of the hypothalamus, and the adrenaline- and noradrenaline-secreting cells of the sympathetic nervous system—including the adrenal medulla). Abnormalities of most of these tissues have been discussed elsewhere in this book; in this chapter we shall describe the secreting tumours of the sympathetic nervous system—phaeochromocytoma and neuroblastoma.

Hormone-secreting tumours of the APUD cells scattered through tissues of a non-ectodermal origin are being increasingly recognised. Many occur in the *gastro-intestinal tract and pancreas*. Insulinomas are discussed in Chapter IX. The pancreas contains many other types of peptide hormone-secreting APUD cells which are common to it and the intestinal tract: the amine-secreting, or amine-precursor-secreting, carcinoid tumours occur mainly in the intestine. In this chapter we will consider briefly some of these rare tumours.

Probably all APUD cells have the potential to secrete any of the APUD hormones. A common site for apparently non-secretory cells is the bronchial tree. Bronchial carcinomas are the commonest tumours, apparently of non-endocrine tissue origin, which can sometimes elaborate hormones and other peptides normally foreign to that tissue. Some of these syndromes may be due to malignancy of APUD cells in the region, rather than to that of the host tissue; this is not the only possible explanation. The second part of this chapter is devoted to a brief discussion of these interesting, and not uncommon, syndromes.

CATECHOLAMINE-SECRETING TUMOURS

The adrenal medulla and the sympathetic ganglia consist of sympathetic nervous tissue: this is derived from the embryonic neural crest and is composed of two types of cells—the *chromaffin cells* and the *nerve cells*—both of which can elaborate the active catecholamines. Adrenaline (epinephrine) is almost exclusively a product of the adrenal medulla, while most noradrenaline (norepinephrine) is formed at sympathetic nerve endings.

Metabolism of the Catecholamines

The amines adrenaline and noradrenaline are formed from tyrosine (the amine precursor) via dihydroxyphenylalanine (DOPA), and dihydroxyphenylethylamine (DOPamine). DOPA, DOPamine, adrenaline and noradrenaline are all catecholamines (dihydroxylated phenolic amines). Adrenaline and noradrenaline are both metabolised to the inactive 4-hydroxy-3-methoxymandelate (HMMA: vanillyl mandelate, VMA), each by two similar pathways, on which metadrenaline and normetadrenaline respectively are intermediates (Fig. 41). Adrenaline, noradrenaline and the metadrenalines and their conjugates, and HMMA, can be measured in the urine.

Action of the Catecholamines

Both adrenaline and noradrenaline act on the cardiovascular system. Noradrenaline produces generalised vasoconstriction, and therefore hypertension and pallor, while adrenaline may cause dilatation of muscular vessels, with variable effects on blood pressure.

Adrenaline increases the rate of glycogenolysis and this, together with its other anti-insulin effects, may cause hyperglycaemia.

For a more detailed discussion of the action of these two hormones the student is referred to textbooks of pharmacology.

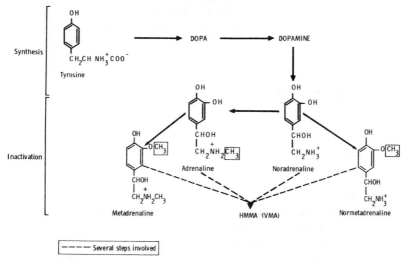

FIG. 41.—Synthesis and metabolism of the catecholamines.

Catecholamine-Secreting Tumours

Tumours of the sympathetic nervous system, whether adrenal or extra-adrenal, can produce catecholamines. Such tumours are of two main types:

tumours of chromaffin tissue—*Phaeochromocytoma*:
    in adrenal medulla about 90 per cent;
    extra-adrenal about 10 per cent;
tumours of nerve cells—*Neuroblastoma*:
    in adrenal medulla about 40 per cent;
    extra-adrenal about 60 per cent.

Although both these types of neoplasm secrete catecholamines, their age incidences and the clinical pictures are quite different.

Phaeochromocytoma occurs mainly in *adult* life and is usually *benign*. The symptoms and signs can be related to excessive secretion of catecholamines and include *paroxysmal hypertension* accompanied by *anxiety*, *sweating*, a throbbing *headache* and either *facial pallor* or *flushing* during the attack. The attack may also be accompanied by hyperglycaemia and glycosuria. Occasionally hypertension may be persistent. This tumour accounts for only about half a per cent of all cases of hypertension, but if a young adult presents with high blood pressure for no obvious reason its presence should be sought, because the condition may be cured by surgery. Table **XXXVI** summarises

TABLE XXXVI

SOME METABOLIC CAUSES OF HYPERTENSION

	Investigations
Renal disease	Plasma urea or creatinine
	Examination of urine for casts and protein
Primary hyperparathyroidism	Plasma calcium, phosphate, etc. (see Chapter XI)
Primary aldosteronism	Plasma potassium and TCO_2 (p. 61)
Phaeochromocytoma	Urinary HMMA (VMA)

some relatively rare causes of hypertension which should be excluded in such patients before "essential" hypertension is diagnosed, and in which biochemical investigations aid the diagnosis.

Neuroblastoma is a very *malignant* tumour of sympathetic nervous tissue occurring in *children*. Secretion of catecholamines in these cases is often as high as, or higher than, that of a phaeochromocytoma, but the clinical syndrome described above is rare: the reason for this is not clear, but may be due to release of hormones into the blood stream in an inactive form. Some of the tumours secrete DOPamine in excess.

Diagnosis of Catecholamine-Secreting Tumours

Because HMMA (VMA) is the major catabolic product of catecholamines, chemical diagnosis of the tumours can usually be made by estimating its excretion in a 24-hour specimen of urine. An excretion of more than twice the upper limit of "normal" is diagnostic. Slightly raised excretion may be found in cases of essential hypertension.

Rarely HMMA excretion is normal, but that of metadrenaline or total catecholamines is increased. Plasma levels of catecholamines can be measured by some laboratories. Estimation of metadrenaline or total catecholamines should be reserved for cases clinically highly suggestive of phaeochromocytoma, but who have a normal excretion of HMMA.

If hypertension is paroxysmal, urine should be collected during and immediately after the attack: this may be the only time at which increased excretion can be demonstrated.

THE CARCINOID SYNDROME

Normal Metabolism of 5-Hydroxytryptamine

Small numbers of APUD *argentaffin cells* (that is, cells which will reduce, and therefore stain with, silver salts) are normally found in tissues derived from embryonic gut. The commonest sites for these cells are the ileum and appendix, but some are found in the pancreas, stomach and rectum. They synthesise the biologically active amine, *5-hydroxytryptamine* (5-HT: serotonin), from the amine precursor tryptophan, the intermediate product being *5-hydroxytryptophan* (5-HTP). 5-HT is inactivated by deamination, and oxidation by mono-amine oxidases to *5-hydroxyindole acetic acid* (5-HIAA) (Fig. 42): the latter enzymes, as well as the aromatic amino acid decarboxylase, are found in other tissues as well as in argentaffin cells. 5-HIAA is normally the main urinary excretion product of argentaffin cells.

Argentaffin cells may also secrete a peptide, *Substance P*, an excess of which causes flushing, tachycardia, increased bowel motility and hypotension.

Causes of the Carcinoid Syndrome

Tumours of argentaffin cells may be benign or malignant, and are most commonly found in the ileum or appendix: neoplasms at the latter site rarely metastasise and may be histochemically different from the other small intestinal argentaffin tumours. Less commonly the neoplasm is bronchial, pancreatic or gastric in origin, any other site being extremely rare.

The carcinoid syndrome is usually associated with an excess of circulating 5-HT. Ileal and appendiceal tumours do not produce the clinical syndrome until they have metastasised, usually to the liver, but primary tumours at other sites do cause symptoms. It is thought that at least some of the secretions of intestinal neoplasms are inactivated in the liver, whereas at other sites they are released in an active form into the systemic circulation.

Clinical Picture of the Carcinoid Syndrome

The clinical syndrome includes *flushing, diarrhoea* and *broncho-spasm*. There may be *fibrotic lesions of the heart*, typically *right-sided*, except in the case of bronchial carcinoid. The diarrhoea may be so severe as to cause a malabsorption syndrome. The signs and symptoms are more likely to be due to Substance P than to 5-HT. Histamine has also been suggested as a cause for the flushing, especially in tumours elaborating a preponderance of 5-HTP. A *pellagra*-type syndrome can rarely develop because of diversion of tryptophan from nicotinamide to 5-HT synthesis (Fig. 42).

Diagnosis of the Carcinoid Syndrome

In the carcinoid syndrome urinary 5-HIAA secretion is usually greatly increased. An excretion of more than 130 μmol (25 mg) in

FIG. 42.—Metabolism of tryptophan.

24 hours is diagnostic, provided that walnuts and bananas are excluded from the diet for 24 hours before the collection is made.

In very rare cases, usually of bronchial or gastric tumours, the cells lack aromatic amino acid decarboxylase. Urinary 5-HIAA may not be obviously increased, and there is increased secretion of 5-HTP. If there is a strong clinical suspicion of the carcinoid syndrome and the urinary 5-HIAA excretion is normal, it may be useful to estimate total 5-hydroxyindole excretion (which includes 5-HTP, 5-HT and 5-HIAA). This is very rarely necessary in practice.

PEPTIDE-SECRETING TUMOURS OF THE ENTEROPANCREATIC SYSTEM

All these tumours are rare. They usually form in the pancreatic islets. We have made no attempt to be comprehensive.

Gastrinomas may cause the *Zollinger-Ellison syndrome*. G cells, usually in the pancreatic islets (most commonly in the form of tumours), produce large amounts of gastrin. About 60 per cent are malignant; of the remaining 40 per cent two-thirds are multiple and only one-third (13 per cent of the total) are single, resectable adenomas. More rarely the syndrome is due to hyperplasia of G cells in the gastric antrum. Acid secretion by the stomach is very high, and the consequent ulceration of the stomach and upper small intestine may cause severe diarrhoea: the low pH, by inhibiting lipase activity, may cause steatorrhoea. Associated benign adenomas may be found in other endocrine glands, such as the parathyroid, pituitary, thyroid and adrenal (multiple endocrine adenopathy): these are rarely functional.

Fasting plasma gastrin levels in the Zollinger-Ellison syndrome are from 5 to 30 times the upper limit of normal. Gastrin levels may be high in many other conditions but the simultaneous findings of a very low gastric juice pH and a very high plasma gastrin level indicates autonomous hormone secretion not under normal feedback control, and is therefore diagnostic. (Gastrin secretion is normally inhibited by gastric acidity. p. 302.)

Glucagonomas usually arise from the glucagon-producing A cells of the pancreatic islets. The presenting feature is a bullous rash known as *necrolytic migratory erythema*. This is often accompanied by psychiatric disturbance, thrombo-embolism, glossitis, weight loss, impaired glucose tolerance, anaemia, and a raised erthyrocyte sedimentation rate. The relationship of the rash to high glucagon levels is not clear. Diagnosis depends on the finding of very high plasma glucagon levels; this estimation is available in special centres. Resection of the tumour cures the syndrome. Although reports of glucagonomas are few, it has been suggested that, because of the rather non-specific clinical picture, some cases may not have been diagnosed.

VIPomas are tumours, usually of pancreatic islet cells, producing large amounts of vasoactive intestinal peptide (VIP), a hormone which increases intestinal motility. They are *extremely rare*, and are associated with very profuse and watery diarrhoea. This syndrome has been known as the *Verner-Morrison* or *WDHA syndrome* (WDHA stands for *W*atery *D*iarrhoea, *H*ypokalaemia, and *A*chlorhydria). It is not certain whether all cases of Verner-Morrison syndrome are due to VIPomas.

MULTIPLE ENDOCRINE ADENOPATHY (MEA)

In the rare syndromes of MEA (pluriglandular syndromes), two or more endocrine glands secrete inappropriate amounts of hormone, usually from adenomas. There are two main groups of syndromes.

MEA I may involve two or more of the following endocrine tissues (in order of decreasing incidence):

> parathyroid gland (hyperplasia or adenoma);
> pancreatic islet cells:
>> gastrinomas;
>> insulinomas;
> anterior pituitary gland;
> adrenal cortex;
> thyroid.

MEA II includes:

> medullary carcinoma of the thyroid;
> phaeochromocytoma;
> parathyroid adenoma or carcinoma.

Other combinations are very rare. The reason for the grouping is not clear, but may be associated with lines of development of the APUD system.

HORMONAL EFFECTS OF TUMOURS OF NON-ENDOCRINE TISSUE

This section will best be understood after the student has read the chapters concerned with the hormones discussed.

Hormone secretion at sites other than the normal tissue of origin is known as *"ectopic"*; *inappropriate* secretion (p. 56) may or may not be ectopic in origin (see ADH), but ectopic secretion is always inappropriate since it is not under normal feedback control.

Many tumours of non-endocrine tissue may secrete hormonal

substances very similar to, and often identical with, the natural hormone: some such tumours have been shown to secrete two or more hormones.

MECHANISM OF ECTOPIC HORMONE PRODUCTION

It is still not certain why ectopic hormone secretion should occur, but although many theories have been propounded we, like the authors of the reference given at the end of the chapter, continue to believe that the one discussed below is by far the most likely. For further discussion the interested student should consult this and the other references.

DNA, through RNA formation, codes for peptide and protein synthesis, the sequence of bases in the DNA molecule determining the structure of the peptide synthesised. All cells in the body are derived from one cell, the fertilised ovum. The DNA complement of this cell is determined by that of the unfertilised ovum and that of the sperm, and, during the subsequent repeated cell division, the DNA is replicated so that every daughter cell has a genetic complement identical with that of the fertilised ovum. Unless mutation were to occur during differentiation of tissues (and there is evidence that it does not), every cell in the body must have the potential to produce any peptide coded for in the fertilised ovum.

It is thought that during functional cell differentiation (which, of course, proceeds at the same time as histological changes) various parts of the DNA molecule are consecutively "repressed" (stopped from functioning), and "derepressed": for instance in the early stages of blastula formation no RNA is produced and DNA replication is dominant, while at other times different proteins are manufactured. In the fully differentiated cell much of the genetic information is probably repressed and only the peptides essential for cell metabolism and those concerned with the special function of the organ are made. If the cell undergoes neoplastic change, the altered histological picture reflects altered chemical function probably resulting from the changed pattern of repression of the DNA molecule. It is now easy to see how a cell could revert to synthesising a peptide or protein which is foreign to it in its fully differentiated state.

If this were the correct explanation, the ectopic hormone should be chemically identical with that normally produced by endocrine tissue. Hormonal substances have been extracted from tumour tissue. In many cases the complete structure has been worked out, and evidence based on biological activity and chemical, physical and immunological properties also suggests that many are identical with the naturally occurring hormones. As might also be expected if the above theory were correct, tumours have been described secreting two or more peptide hormones.

Another possible explanation is that some of these syndromes are due to malignancy of the APUD cells normally found in the tissue. In normal numbers these cells might secrete undetectable amounts of hormone, which, if the cells multiply, reach concentrations high enough to produce clinical effects; alternatively, derepression of normally non-secretory APUD cells might lead to hormone production.

It seems to us that derepression of tissue, rather than of APUD, cells is the most likely explanation for this interesting set of syndromes. However, this derepression cannot be non-specific because some hormones are secreted much more commonly by one type of tumour than another. This may reflect a differing chemical background in different parts of the body.

The student should remember that neither theory is yet proven. We hope we may have interested him in the subject and that he will be stimulated to read further.

HORMONAL SYNDROMES

Many of these syndromes have been discussed in the relevant chapters. The following may be due to hormone secretion by the tumour:

> hypercalcaemia (parathyroid hormone or parathyroid hormone-like substance);
> hyponatraemia (antidiuretic hormone);
> hypokalaemia (adrenocorticotrophic hormone);
> polycythaemia (erythropoietin);
> hypoglycaemia ("insulin-like substance");
> gynaecomastia (gonadotrophin);
> hyperthyroidism (thyrotrophic hormone);
> carcinoid syndrome (5-HT and 5-HTP).

It should be noted that all the hormones mentioned above are peptides except 5-HT and 5-HTP: the syndrome associated with the latter may be the result of overproduction of a single enzyme (see later).

None of these hormones produced at ectopic sites is under normal feedback control. Secretion therefore continues under conditions in which it should be absent, and is inappropriate.

While it may be difficult to prove conclusively that tumour cells are secreting hormones, in many cases strong presumptive evidence enables the clinician to diagnose the syndrome with a high degree of confidence.

Hypercalcaemia Due to Parathyroid Hormone (PTH or PTH-like substances) Secretion ("Pseudohyperparathyroidism")

Tumour types and incidence.—In our experience hypercalcaemia with hypophosphataemia is very common in most types of malignant disease. Parathyroid hormone (PTH), or a peptide very similar to it, has been extracted from a wide variety of tumour types, and circulating levels can occasionally be shown to be inappropriately high.

Biochemical syndrome.—As discussed in Chapter XI, inappropriate PTH secretion, whether ectopic or parathyroid in origin, is associated with high plasma calcium and, unless there is glomerular failure, low plasma phosphate concentrations. If both these findings are present, and especially if there is no obvious evidence of bony metastases, it is probable that the hypercalcaemia is due to ectopic PTH secretion.

The hypercalcaemia of some cases of malignancy is said to be due to lysis of bone by secondary deposits, with release of calcium and phosphate into the blood stream. Prostaglandin E_2 secretion by some tumours has been suggested as a cause of hypercalcaemia. As discussed on p. 266 we feel that on the evidence of the plasma phosphate level, most, if not all cases are due to secretion of ectopic PTH, or of a substance with effects on plasma calcium and phosphate levels almost identical to those of the hormone. The fact that it is not usually possible to demonstrate inappropriately high *immunoreactive* plasma PTH levels does not necessarily exclude the presence of biologically active forms of the hormone. The student should keep an open mind and study the cases he meets, noticing particularly the level of phosphate *in relation to that of plasma urea* (p. 279).

The hypercalcaemia of malignancy, unlike that of primary hyperparathyroidism, may often be suppressed by large doses of corticosteroids (p. 269).

With improved therapy of malignant disease it is important to control potentially lethal hypercalcaemia in these patients, who should have plasma calcium estimated at frequent intervals. The treatment is discussed on p. 269.

Hypokalaemia is a common accompaniment of hypercalcaemia (p. 282); it should always be sought. Its presence does not necessarily, or even usually, indicate simultaneous production of ACTH.

Hyponatraemia Due to Inappropriate Antidiuretic Hormone (ADH) Secretion

Tumour types and incidence.—Ectopic ADH production is most commonly associated with the relatively rare oat-cell carcinoma of the bronchus, although the syndrome has been reported in a wide variety of tumours. Mild hyponatraemia may be ignored in a severely ill patient, in whom such a finding is common; it is our experience that if further evidence is sought the syndrome is found to be relatively common.

Inappropriate ADH secretion can occur with various non-malignant syndromes, (including hypothyroidism) but in many of these cases the hormone is thought to originate in the hypothalamic-posterior pituitary region and therefore not to be ectopic in origin. This may also be true of the inappropriate ADH secretion associated with non-bronchial tumours.

Biochemical and clinical syndrome.—ADH causes water retention. Water retention without sodium retention causes hyponatraemia, and therefore a reduction in plasma osmolality (p. 57), which normally cuts off ADH production by feedback control. If the low plasma osmolality is to be corrected the urinary osmolality must be even lower, so that the excess of water is excreted: this is the "appropriate" response. If, however, ADH is not under normal osmotic feedback control, water retention continues in spite of low plasma osmolality, and the urine passed is relatively concentrated. Thus the essential points in the diagnosis of inappropriate ADH secretion are:

a low plasma osmolality;

in spite of this low plasma osmolality, the urinary osmolality is relatively high.

It should be noted that *both findings are essential*. The plasma osmolality is low, for instance, when fluid of low sodium concentration is infused, but the urine is then dilute. The urinary osmolality is high in water depletion but the plasma osmolality is then also high.

Continued expansion of the plasma volume reduces aldosterone secretion (p. 57), and the *urine is of relatively high sodium concentration* (note that plasma osmolality can be attributed almost entirely to sodium salts: urinary osmolality depends mainly on the urea concentration). Retention of fluid by the tubules dilutes other plasma constituents and increases the GFR, and these patients tend to have a low plasma urea concentration.

The fall of plasma sodium concentration is usually gradual and allows time for equilibrium between cells and extracellular fluid: symptoms are therefore usually slight.

A presumptive diagnosis can be made if the following findings are present:

the patient is *well-hydrated, normotensive*, and usually relatively fit in spite of a low plasma sodium concentration;

there is a *normal or low plasma urea* concentration (due to a high GFR);

there is a *urine with relatively high concentrations of urea* (the main contributor to urinary osmolality,) *and of sodium* (due to volume expansion).

Treatment.—Most cases are relatively asymptomatic, and restriction of fluids is adequate treatment. However, if the sodium concentration has fallen rapidly, and if the symptoms of cerebral oedema are present, more active treatment may be necessary:

Infusion of a small amount of hyperosmolar fluid will directly increase plasma osmolality; the tendency to further extracellular volume expansion is counteracted if a diuresis is stimulated at the same time. The following two methods depend on this principle:

Hyperosmolar saline may be given with a *diuretic such as frusemide*: the diuretic increases isosmolar fluid loss, and more of the administered sodium than water is retained;

Mannitol infusion directly increases plasma osmolality and, by causing an *osmotic diuresis*, overrides the effect of ADH on the distal tubule (p. 9): more water than sodium is lost in the urine, and the plasma sodium concentration tends to increase.

The tetracycline, *demeclocycline*, is thought to inhibit the action of ADH on the distal tubule directly. This drug may be given to patients with severe symptoms when simpler methods have failed or are contra-indicated.

Hypokalaemia Due to Adrenocorticotrophic Hormone (ACTH) Secretion

Tumour types and incidence.—Symptomatic inappropriate ACTH secretion is probably rarer than the two syndromes already discussed. It occurs most commonly with the relatively rare oat-cell carcinoma of the bronchus, but has been described less frequently with a variety of other tumours, especially those of thymic or pancreatic origin.

Clinical and biochemical syndrome.—ACTH stimulates secretion of all adrenocortical hormones except aldosterone (p. 144). In spite of this the picture of ectopic ACTH secretion is usually more like that of primary aldosteronism than of Cushing's syndrome: the patient usually presents with hypokalaemic alkalosis and only occasionally looks Cushingoid. The reason for this difference in presentation is not clear.

Treatment.—Potassium replacement is only effective if accompanied by administration of spironolactone, amiloride or triamterene (p. 77).

Polycythaemia Due to Erythropoietin Secretion

The association of polycythaemia and renal carcinoma (hypernephroma) has been recognised for some years. It is now thought that the erythroid hyperplasia is often due to excessive production of an erythropoietin-like molecule by the tumour; erythropoietin stimulates marrow erythropoiesis. As this hormone is a normal product of the kidney this is not an example of ectopic hormone secretion: however, the syndrome has been reported rarely in association with other tumours, especially hepatocellular carcinoma (primary hepatoma).

Hypoglycaemia Due to an "Insulin-like" Hormone

Severe hypoglycaemia has been reported in association with many tumours. Only a few cases, usually with carcinoid tumours, have been shown to have inappropriately high immunoreactive insulin levels during the attack, and have therefore been proven to be due to ectopic hormone production.

Severe hypoglycaemia with appropriately *low* immunoreactive insulin levels has been described more commonly, usually in association with very large abdominal or thoracic mesenchymal tumours resembling fibrosarcomata, or with hepatocellular carcinomas. The syndrome has not been satisfactorily explained.

Gynaecomastia Due to Gonadotrophin Production

Various types of tumour, including carcinoma of the bronchus, breast and liver, have been reported to cause gynaecomastia and be associated with high circulating levels of chorionic gonadotrophin or luteinising hormone; follicle-stimulating hormone activity is rare. In children with hepatoblastoma there may be precocious puberty. Although this syndrome is rarely recognised it is possible that mild degrees of gynaecomastia are overlooked.

Hyperthyroidism Due to "Thyrotrophic Hormone" Production

Some tumours of trophoblastic cells (choriocarcinoma, hydatidiform mole and one case with a testicular teratoma) have been shown to elaborate a TSH-like substance. In spite of markedly raised plasma levels of hormone, clinical signs of thyrotoxicosis are rare: tachycardia is the commonest clinical finding.

Carcinoid Syndrome Due to 5-Hydroxytryptamine (5-HT) and 5-Hydroxytryptophan (5-HTP) Production

All the syndromes so far discussed have been due to peptide

hormones and, as we have seen, derepression of DNA could lead to their production. In the rare form of carcinoid syndrome, accompanying oat-cell tumours of the bronchus, there is usually excessive urinary excretion of 5-HTP and 5-HT, out of proportion to that of 5-HIAA. How can we explain this overproduction of a non-peptide hormone? One possible explanation is that the tumour is one of APUD rather than true bronchial cells.

Another, different, theory is speculative and is meant only as a working hypothesis. It assumes derepression of bronchial cells. The student should study Fig. 42, (p. 468): the production of 5-HTP from tryptophan requires only one enzyme, tryptophan-5-hydroxylase, while the decarboxylase and monoamine oxidase are normally found in non-argentaffin tissue. If derepression resulted in excessive production of this one enzyme there would be an excess of 5-HTP. 5-HT and 5-HIAA could be produced in other tissues, but in relatively smaller amounts than are usual in the carcinoid syndrome. It would, of course, be more difficult to explain overproduction of a substance synthesised by a pathway requiring several enzymes.

NON-HORMONAL PEPTIDES AS INDICATORS OF MALIGNANCY

The production of non-hormonal peptides by derepression of DNA will be less clinically obvious than the production of hormones. Such circulating peptides have been identified by immunological techniques. Two such proteins are discussed briefly here: both are normally present in fetal life, but production appears to be largely repressed in normal adults.

α-**Fetoprotein** levels may be found to be very high in the serum of many patients with *hepatocellular carcinoma* (primary hepatoma) and *teratoma* (p. 325). Moderately raised levels may be due to non-malignant liver disease.

Carcinoembryonic antigen (CEA) may be produced by many *malignant tumours, especially of the gastro-intestinal tract*, and plasma levels may rise due to non-malignant disease of the gastro-intestinal tract. Its measurement is therefore of little use, and may be misleading, for diagnosis. Serial CEA estimations may be valuable to monitor the progress of treatment of known gastro-intestinal malignancy.

SUMMARY

Catecholamine-Secreting Tumours

1. Tumours of sympathetic nervous tissue are associated with increased urinary excretion of the catecholamines, adrenaline and

noradrenaline, and their metabolic products, the metadrenalines and hydroxymethoxymandelate (HMMA:VMA).

2. Phaeochromocytoma is a rare tumour, usually occurring in adults, most commonly in the adrenal medulla. It is associated with hypertension and other symptoms of increased catecholamine secretion.

3. Neuroblastoma is a tumour of childhood, occurring in the adrenal medulla or in extra-adrenal sympathetic nervous tissue. Catecholamine secretion is increased, but symptoms are rarely referable to this.

The Carcinoid Syndrome

1. Argentaffin cells manufacture 5-hydroxytryptamine (5-HT) which after convertion to 5-hydroxyindole acetic acid (5-HIAA) is excreted in the urine.

2. Tumours of argentaffin tissue are usually found in the intestine, and at these sites do not produce typical symptoms of the carcinoid syndrome until they have metastasised to the liver.

3. The carcinoid syndrome is usually associated with an increased secretion of 5-HIAA in the urine.

Peptide-Secreting Tumours of the Enteropancreatic System

1. Gastrinomas cause the Zollinger-Ellison syndrome. Gastric hyperacidity causes peptic ulceration, diarrhoea and sometimes steatorrhoea.

2. Glucagonomas are associated with necrolytic migratory erythema and non-specific systemic symptoms and signs.

3. VIPomas (tumours secreting vasoactive intestinal peptide) may cause the Verner-Morrison syndrome—very severe watery diarrhoea which often causes hypokalaemia (WDHA).

All these tumours are very rare.

Multiple Endocrine Adenopathy

In these syndromes two or more endocrine glands secrete excessive amounts of hormones.

Hormonal Effects of Tumours of Non-Endocrine Tissue

1. Many malignant tumours produce hormonal substances normally foreign to them, which may be identical with the hormones produced in endocrine glands.

2. The commonest of these is probably parathyroid hormone. Symptomatic hypercalcaemia should be sought and treated in all cases of malignancy.

Non-Hormonal Peptides as Indicators of Malignancy

Some non-hormonal peptides may be detected by immunological techniques and have been used to diagnose malignancy of specific tissues.

FURTHER READING

The APUD System and Ectopic Hormone Production

PEARSE, A. G. E. (1979). The diffuse endocrine system and the implications of the APUD concept. *Int. Surg.*, **64** (No. 2), 5–7.

STEVENS, R. E. and MOORE, G. E. (1983). Inadequacy of the APUD concept in explaining production of peptide hormones by tumours. *Lancet*, **1**, 118–119.

Phaeochromocytoma

WOLF, R. L (1974). Phaeochromocytoma. *Clinics in Endocr. & Metab.*, **3**, 609–621.

The Carcinoid Syndrome

GRAHAME-SMITH, D. G. (1968). The carcinoid syndrome. *Hosp. Med.*, **2**, 558–566.

Tumour Markers

BEGANT, R. H. J. (1984). The value of carcinoembryonic antigen measurement in clinical practice. *Ann. Clin. Biochem.*, **21**, 231–238.

BAGSHAWE, K. D. and SEARLE, F. (1977). Tumour markers. In: V. Marks and C. N. Hales, editors. *Essays Med. Biochem.*, **3**, 25–73.

Chapter XXIII

THE CEREBROSPINAL FLUID

CEREBROSPINAL fluid (CSF) is formed from plasma by the filtering and secretory activity of the choroid plexus, and is reabsorbed into the blood stream by the arachnoid villi. It seems to be mainly an ultrafiltrate of plasma, since it contains very little protein. Some active secretion of, for example, chloride may occur.

Circulation of CSF is very slow, allowing long contact with cerebral cells: their uptake of glucose may account for its relatively low concentration in the CSF.

Concentrations in the CSF should always be compared with those of plasma, because alterations in the latter are reflected in the CSF even when cerebral metabolism is normal.

EXAMINATION OF THE CSF

Biochemical investigation of the CSF is usually of relatively little value compared with simple inspection and bacteriological and cytological examination of the fluid. Textbooks of microbiology should be consulted for further details.

TAKING THE SAMPLE

CSF should be taken into sterile containers and sent for *microbiological examination first*. Any remaining specimen can be used for relevant chemical investigations, but if the tests are performed in the opposite order bacteriological contamination may occur. If possible, a few millilitres of CSF should be taken into two or three separate containers (see below—Appearance).

Specimens for glucose estimation, like those for plasma glucose, should be taken into a tube containing fluoride to minimise glycolysis by any cells present.

APPEARANCE

Normal CSF is completely clear and colourless and should be compared with water: slight turbidity is most easily detected by this method.

Colour

Bright red blood may be due to:

a recent haemorrhage involving the subarachnoid space;
damage to a blood vessel during puncture.

If CSF is collected in three separate aliquots, all three will be equally blood-stained in the first case but progressively less so in the second.

Xanthochromia (yellow coloration).—This may be due to the presence of:

altered haemoglobin several days after a cerebral haemorrhage;

large amounts of pus. The cause will be obvious from the gross turbidity of the fluid, and from the presence of pus cells on microscopy;

cerebral tumours near the surface of the brain or spinal cord, impairing circulation of the CSF. Specimens from these cases have a very high protein content and tend to clot spontaneously after withdrawal, due to the presence of fibrinogen;

jaundice due to a rise in unconjugated bilirubin, which may impart a yellow colour to the CSF. The cause will be obvious in a jaundiced patient.

Turbidity

Turbidity is due to an *excess of white cells (pus).* Slight turbidity will, of course, occur after haemorrhage, but the cause of this will be apparent from the colour of the specimen.

Spontaneous Clotting

Clotting occurs when there is an excess of fibrinogen in the specimen, usually associated with a high total protein content.

The following are the most frequently requested chemical estimations.

PROTEIN

The protein concentration of normal CSF is very low, the proportions of individual plasma proteins depending on their molecular weight and therefore their ability to pass the vascular wall and meninges. Local cells modify the composition of this ultrafiltrate. The normal CSF protein concentration is below 0·4 g/l in the adult, but may be much higher in the newborn infant. Even in the absence of cerebral disease, changes in CSF proteins may reflect changes in plasma levels, and *results of assays can only be interpreted if the two fluids are compared.*

Cerebral disease may change the total concentration of protein and the proportion of its constituents for two reasons:

vascular and meningeal permeability may be increased, allowing not only more but also larger-than-normal plasma proteins to enter the CSF;

proteins may be synthesised within the cerebrospinal canal by inflammatory or other invading cells.

In some conditions both these factors may be present.

Total Protein

The measurement of CSF total protein is a relatively insensitive test for cerebral disease, since early changes in the type of protein may not always cause a detectable rise in total concentration.

CSF will have a high protein content under the following conditions:

in the presence of blood (due to haemoglobin and plasma protein);
in the presence of pus (due to cell protein and exudation from inflamed surfaces).

These two causes will be obvious from inspection and microscopic examination of the specimen, and not only is nothing further to be gained by estimating protein, but laboratory staff should not be unnecessarily exposed to infected material.

in non-purulent inflammation of cerebral tissues there may be a definite rise in protein concentration despite the absence of detectable cells in the CSF;

in blockage of the spinal canal, often caused by a tumour, increased capillary permeability and perhaps fluid reabsorption due to stasis results in very high concentrations, often with xanthochromia.

Tests for Abnormal Protein Patterns

Tests for CSF proteins are not useful in the presence of blood or pus. They may, however, detect abnormalities when the total protein is *equivocally increased or normal*, and may help to elucidate the cause of a high level.

Electrophoresis.—Electrophoresis of CSF may yield useful diagnostic information if the pattern is compared with that of serum from blood taken at the same time.

Increased capillary permeability may be demonstrated by the finding in the CSF of relatively high-molecular-weight plasma proteins not normally present there in significant amounts. This is a non-specific pattern due to a wide variety of inflammatory conditions, but may be especially useful to diagnose:

cerebral tumours, some of which may also produce abnormal immunoglobulins;

acute idiopathic polyneuropathy (Guillain-Barré syndrome) in which acute-phase proteins (p. 344) and immunoglobulins may also be synthesised locally.

Immunoglobulin synthesis by inflammatory B cells within the cerebrospinal canal may produce multiple bands in the γ-globulin region. These bands are rarely monoclonal, and consist of more than one immunoglobulin type synthesised by a small number of B-cell clones; they are therefore called *oligoclonal*. Oligoclonal bands have been described in association with:

multiple sclerosis (the most important indication for the test);
encephalitis;
neurosyphilis;
systemic lupus erythematosus;
chronic relapsing Guillain-Barré syndrome;
cerebral sarcoidosis;
cerebral tumours (rarely).

Occasionally intrathecal malignant B cells may produce a local monoclonal band.

Oligoclonal bands signify cerebral disease only if they are found in the CSF *and not in the serum*.

Immunological techniques.—The level of CSF immunoglobulins, especially of *IgG*, may be quantitated. The diagnostic precision of this estimation is improved if the ratio of the concentrations of CSF IgG: serum IgG is expressed as a percentage of the ratio of the concentrations of, for example, CSF albumin to serum albumin (a "*quotient*"): the test is still less discriminating than electrophoresis. Albumin is a smaller molecule than IgG. If raised CSF IgG levels were due to the increased vascular permeability of non-specific inflammation the quotient would be low or normal; if permeability were only slightly increased more albumin than IgG would diffuse from the plasma into the CSF and the quotient would be low, but more commonly the permeability is such that albumin and IgG diffuse at almost the same rate and the quotient is normal. However, in conditions such as *multiple sclerosis*, the *quotient is high* because CSF IgG has been synthesised locally.

GLUCOSE

Provided that CSF for glucose estimation has been mixed with fluoride a low glucose concentration occurs in:

infection;
hypoglycaemia.

Infection.—CSF glucose is usually metabolised only by cerebral cells. If many leucocytes and bacteria are present these also utilise glucose and abnormally low levels are found. If obvious pus is present the estimation of CSF glucose adds nothing to diagnostic precision. It is most useful when the CSF is clear and *tuberculous meningitis* is suspected, although levels are not as low in this condition as in pyogenic meningitis.

Hypoglycaemia.—CSF glucose concentration parallels that of plasma, although there is a lag before changes in plasma glucose are reflected in the CSF. In the presence of hypoglycaemia (which may cause coma) CSF glucose levels may be low, although there is no primary cerebral abnormality. *Both plasma and CSF concentrations should be measured.*

In hyperglycaemia CSF glucose levels will be high.

The recommended procedure for examination of CSF is outlined on the next page.

SUMMARY

1. Biochemical analysis of the CSF is less important than simple inspection and bacteriological examination.

2. Assessment of the pattern of CSF proteins is the most useful chemical procedure.

3. The finding of oligoclonal bands in the CSF may be of diagnostic value in non-purulent cerebral conditions, such as multiple sclerosis.

4. Estimation of CSF glucose concentration is most useful in cases of suspected tuberculous meningitis.

FURTHER READING

THOMPSON, E. J. and JOHNSON, M. H. (1982). Electrophoresis of CSF proteins. *Brit. J. Hosp. Med.*, **28**, 600–608.

PROCEDURE FOR EXAMINATION OF CSF

CSF very cloudy.—Send for microbiological examination. Chemical estimations unnecessary.

CSF heavily bloodstained in three consecutive specimens. Cerebral haemorrhage. Chemical estimation useless.

CSF clear or only slightly turbid.—Send for microbiological examination and for estimation of *glucose* and *protein* concentration and electrophoresis. *Always send blood for glucose and protein estimation at the same time.*

CSF xanthochromic.—Send specimen, with blood, for protein estimation. Examine under microscope for erythrocytes.

Chapter XXIV

DRUG MONITORING

UNTIL recently it was possible to measure the concentrations of relatively few drugs in the blood, and those only at the very high levels found after an overdose. New techniques have increased the range of assays, and have enabled us to measure some of the low plasma concentrations achieved during their therapeutic use. Interpretation of results is not simple, and measurement of only a few drugs is of proven value for either therapeutic or toxicological monitoring.

Pharmacokinetics is the study of the fate of drugs after ingestion: only if we have some understanding of this can we appreciate the limitations of drug assays and the problems of interpretation of results.

The possible indications for measuring circulating drug levels are to:

    monitor their therapeutic use;

    diagnose obscure conditions;

    elucidate the type of drug overdose and assess the need for treatment.

In the last two groups screening of the urine for drugs or their metabolites is often helpful.

We shall deal with each of these indications in turn.

MONITORING DRUG TREATMENT

The ideal way of assessing whether the dose of a drug is optimal is to measure, either clinically or by laboratory assays, the desired effect. In the relevant chapters we have discussed, for example, the monitoring of plasma calcium levels (and sometimes alkaline phosphatase activity) during calciferol and phosphate therapy, plasma potassium and TCO_2 levels during potassium supplementation, plasma or blood glucose levels during treatment with anti-diabetic drugs, and plasma thyroxine and TSH levels during treatment of thyroid disorders. Some types of anticoagulant therapy may be monitored by measuring the prothrombin time, and antihypertensive treatment by measuring the blood pressure. The student will be able to think of many more examples.

The circulating drug concentration may not parallel the cellular effect, and its measurement can only be a second best. Assay of drug levels may be indicated if the desired result cannot be measured precisely. For example, the incidence of epileptic fits is a poor indicator of the optimal dosage of anticonvulsants, both because the frequency may vary without treatment and because only partial control may be possible. Chemical monitoring is likely to be useful only if the therapeutic range has been defined. Its value is greatest if, as when lithium is used to treat depression, there is only a narrow margin between therapeutic and toxic drug levels.

If we consider the possible fate of a drug after it has been prescribed, and how each step affects plasma levels, it will help us to understand both the indications for and the limitations of therapeutic drug monitoring.

Factors Affecting Plasma Levels

The total amount of drug in the extracellular fluid depends on the balance between that entering and that leaving the compartment: the plasma *concentration* also depends on the volume of fluid through which the retained drug is distributed.

Assays of drugs with a known therapeutic range of plasma concentration *may* be useful if there is doubt about any of the following steps.

Entry of the Drug into the Extracellular Fluid

"**Compliance.**"—It has been shown that very few in-patients receive a drug exactly as prescribed, and compliance is likely to be even worse outside hospital. The patient may fail to take the drug at all, may not take the prescribed dose, or may take it irregularly. It is rare for the doctor to be fully aware of this. The more drugs that a patient is prescribed the more confused he may become about the timing and dose of each. A regular review of therapy and careful explanation to the patient often works wonders.

Absorption.—The completeness and timing of absorption of an oral dose may vary between individuals, or, from time to time, within an individual. Vomiting, diarrhoea or steatorrhoea are obvious causes of possible malabsorption of drugs.

Distribution of the Drug through the Extracellular Fluid

Timing of the sample.—If blood is taken before absorption is complete a low level may be detected: if it is taken within about an hour of administration a relatively high value may indicate incomplete distribution throughout the extracellular fluid. Blood samples must be taken at a standard time after the ingestion of the drug if results are to be meaningful. The rate of absorption, metabolism and excretion may all vary, and occasionally serial sampling may be necessary to ensure that a reasonably steady level is maintained.

Volume of distribution.—The final plasma concentration reached after a standard amount has been absorbed will be affected by the volume through which it is distributed. For example, it may be difficult to predict the appropriate dosage in oedematous or obese patients (in whom unexpectedly low levels may be found), or in small children (in whom there is a danger of overdosage).

Loss of Drug from the Body

Excretion.—Drugs are excreted in urine or bile, often after inactivation. Hepatic or renal disease may impair excretion: unfortunately the results of assays of protein-bound drugs may be particularly difficult to interpret in just such cases (see below).

Factors which may Complicate the Interpretation of Plasma Levels

The effect of a drug does not always correlate with its plasma concentration. The following factors must be allowed for when interpreting results.

Binding to plasma albumin.—Many drugs, like many endogenous substances, are partially inactivated by binding, usually to albumin. Most methods estimate the total of the free plus bound concentrations. Biological feedback mechanisms do not control free drug levels (or those of bilirubin) as they do those of calcium and hormones. *Measured* plasma levels may be little changed by a reduction in protein binding, but a larger *proportion* of this measured drug will then be active: unless this is realised, dangerously high free levels may be interpreted as being within, or even below, the therapeutic range when albumin levels are very low. Binding varies at different albumin levels, and there is no valid formula which can be applied to correct for protein abnormalities.

Unless protein binding is affected by any of the factors discussed below, about 90 per cent of phenytoin, 70 per cent of salicylic acid, 50 per cent of phenobarbitone and 20 per cent of digoxin is protein bound. Barbiturates other than phenobarbitone are very strongly protein bound.

Before we can assume that the total plasma level of a protein-bound drug reflects that of the active form we must consider the following factors.

The plasma albumin concentration.—Blood should be taken without stasis to minimise a rise in albumin level due to haemoconcentration (p. 498). Low albumin levels also complicate the interpretation of the results of assays of protein-bound drugs.

Competition for protein binding sites.—Many drugs compete with each other for binding sites. If more than one is being taken the free level of each is likely to be higher than the measured concentration would suggest. *Plasma drug levels in patients taking many drugs are difficult to interpret.*

In renal failure the use of drug assays to detect reduced excretion may be partially invalidated by retention of competing ions. Similarly, in hepatic disease, low albumin levels, or even competition for binding sites by bilirubin, limits the value of monitoring protein-bound drugs.

Metabolic conversion to active or inactive metabolites.—Drugs such as calciferol (p. 254) are only active after metabolic conversion: many other drugs are inactivated before excretion, usually by hepatic conjugation. Ideally an assay should measure all the active, and none of the inactive, forms.

Relationship between Plasma Levels and Cellular Effects

Plasma concentrations are measured in an attempt to predict cellular effects. They only do so if they parallel the concentrations at active sites, and if the cells respond predictably to a given concentration at this site. Neither of these conditions is always fulfilled.

Drug concentration at the active site.—In the steady state the plasma concentration may not be the same as that at the active site, but usually parallels it. This may *not* be true if:

> *treatment has just begun.*—Plasma levels may be relatively high compared with those in the cell;
> *treatment has just stopped.*—Plasma levels may fall more rapidly than those in the cell;
> *the drug is taken irregularly or at intervals inappropriate to the half-life of the drug.*—Plasma levels then fluctuate, and equilibrium may not be reached with the cell;
> *there is a change in pH.*—pH affects the partitioning between cells and extracellular fluid. This further complicates the interpretation of plasma levels in patients with renal dysfunction.

Tissue Responsiveness

Intersubject variation.—Some patients are more sensitive to drugs than others. For example, the elderly are particularly sensitive to digoxin.

Intrasubject variation.—The sensitivity of an individual to a drug may vary. Known causes of increased sensitivity to digoxin are hypokalaemia and hypercalcaemia.

Tolerance.—Drugs may, for example, induce the synthesis of enzymes that inactivate them. Much higher plasma levels than usual may be needed to produce the desired effect in tolerant patients.

Additive or antagonistic effects at the active site.—Some drugs and endogenous substances compete for binding at the active site; these may be the same as those that compete for plasma albumin binding sites. By contrast, some drugs have synergistic effects. These factors further complicate the interpretation of drug levels in patients taking many drugs.

When to Measure Drug Levels

Only a few drug assays are of proven clinical value. All the above factors must be allowed for when results are interpreted and the clinical picture must be taken into account. Antibiotics are given for their effect on invading organisms rather than on host body cells, and the indications for their assays will not be considered here.

Other drug measurements which have been claimed to be useful for therapeutic control include the following:

Digoxin.—Estimation of plasma digoxin may be useful:

   to assess compliance;
   in patients, such as the elderly, who are likely to have a low threshold for toxicity;
   if it is difficult to calculate the appropriate dose because of the size of the patient (for example, small children) or because of impaired excretion (renal glomerular dysfunction).

The increased sensitivity, due to hypokalaemia or hypercalcaemia, at any given level of digoxin must be allowed for.

Anticonvulsants.—Once the dose which gives plasma levels within the therapeutic range has been determined, monitoring is probably only necessary:

   to assess compliance;
   if the frequency of fits increases in a previously well-controlled patient;
   if clinical findings suggest toxicity.

Levels in children and pregnant women should be monitored more regularly.

Lithium and high-dose salicylate therapy.—The margin between therapeutic and toxic levels is narrow for both these drugs. Psychiatric patients treated with lithium are especially unlikely to comply with instructions. Plasma levels of both should be monitored regularly. Lithium is not protein bound and interpretation of the results of lithium assays is *not* complicated by protein abnormalities.

Theophylline assays may occasionally be helpful, especially in children treated for acute asthma.

Antiarrhythmic drugs, such as procainamide and quinidine, can be measured in special cases.

Monitoring Possible Side-effects of Drug Therapy

Some drugs have harmful side-effects. For example, plasma thyroxine and TSH levels should be measured during lithium therapy (because of the danger of hypothyroidism, p. 185), and potassium levels during diuretic therapy. Measurement of plasma transaminase activity or urea/creatinine levels may be indicated during treatment with potentially hepatotoxic or nephrotoxic drugs.

DIAGNOSIS OF OBSCURE CONDITIONS

Drug overdosage must always be excluded as a cause of coma of obscure origin. This subject will be dealt with in the next section.

Clinical findings may suggest effects due to a drug or alcohol despite denial by the patient.

Unexplained laboratory findings may also be due to drug or alcohol ingestion. For example, hypokalaemia may be due to purgative abuse (p. 73), and hypoglycaemia may be a presenting finding in alcoholics (p. 213). Raised plasma transaminase activities thought to be due to alcoholic liver disease may be investigated by measuring random plasma alcohol levels.

Measurement of plasma drug levels may be supplemented by screening the urine for drugs or their metabolites.

INVESTIGATION OF KNOWN OVERDOSAGE

About half the patients attempting suicide take several drugs, sometimes with alcohol. Screening plasma, and possibly urine, may be necessary to confirm the diagnosis, and will usually identify the drugs.

It is often unnecessary to measure plasma *levels* of drugs in patients known to have taken an overdose. The need for gastric lavage and the measures required to increase urinary excretion of the drug and to maintain adequate respiration and circulation are usually not affected by a knowledge of drug levels. Only measurements of plasma electrolytes and blood gases are needed.

Drugs and poisons can be considered in three broad groups. The case for drug assay is strongest in the first and weakest in the last group. These groups are:

poisons for which there is a specific antidote;

poisons for which there is no specific antidote, but which are only partially or not protein bound;

poisons for which there is no specific antidote and which are strongly protein bound.

Poisons for which there is a Specific Antidote

Paracetamol (acetaminophen).—It is *essential* to measure drug levels in suspected paracetamol poisoning because:

a metabolite of paracetamol is hepatotoxic; the patient may die of liver failure despite recovery from the immediate effects;

a specific antidote to the hepatotoxic effect is available;

the antidote is expensive, and only useful if given within a defined period of time and if defined plasma levels are reached;

the likelihood of hepatoxicity cannot be predicted from the clinical picture.

Cysteamine, methionine and N-acetylcysteine are all effective antidotes because they enable paracetamol to be converted to non-toxic metabolites. N-acetylcysteine is the least toxic and, although the most expensive, is often used. Cysteamine may cause nausea, vomiting, abdominal pain and cardiac arrhythmias. Methionine, although less toxic, may cause vomiting. N-acetylcysteine is only effective in clearly defined circumstances and, because of its expense, should only be given after the following factors have been taken into account:

Timing of the specimen.—Plasma levels cannot be interpreted, and should not be measured, until absorption and distribution are nearly complete, at about *4 hours after ingestion.* If treatment is to be effective it should be started as soon as possible, and is *ineffective after 15 to 20 hours.* Levels should be measured on specimens taken *as early as possible between 4 and 15 hours after the dose*

Levels.—As a rough guide, treatment is indicated if the plasma paracetamol concentration is *200 mg/l (1300 μmol/l) or more at 4 hours and 30 mg/l (200 μmol/l) or more at 15 hours after the dose.* If the specimen is taken at a time between 4 and 15 hours, treatment is indicated if the plasma concentration at the time falls above the solid line in Fig. 43. Treatment is certain to be ineffective after 20 hours.

Serial measurement of plasma transaminase activities and pro-thrombin time may be needed to detect liver involvement.

Iron.—Iron overdosage is most common in children. Desferrioxamine chelates iron, and the chelate is lost in the urine. The level of plasma iron at which treatment is indicated has not been clearly defined; if there is any doubt desferrioxamine should be given. Levels of plasma iron above 90 μmol/l (500 μg/dl) in young children and above 150 μmol/l (840 μg/dl) in adults are said to be definite indications for treatment.

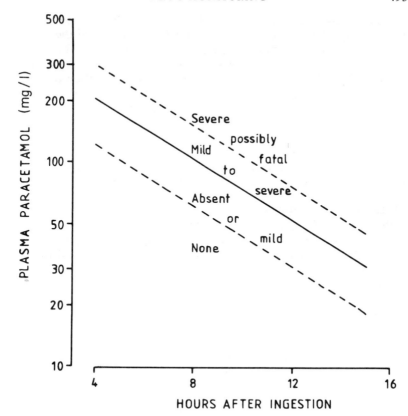

FIG. 43.—Relationship between plasma paracetamol concentration, time after ingestion and potential severity of liver damage.
(Paracetamol mg/l × 6·6 = μmol/l.)
Reproduced by kind permission from Prescott, L. F. (1981). In *Therapeutic Drug Monitoring*, Richens, A. and Marks, V. (Eds). Edinburgh· Churchill Livingstone.

Poisons for which there is no Specific Antidote

Weakly protein-bound poisons.—Forced diuresis may increase the rate of excretion of some free drugs: induction of alkalosis often further increases the excretion rate, but neither forced diuresis nor induction of alkalosis is without risk and in most cases neither is indicated. If alkaline diuresis is induced, potassium levels as well as pH must be monitored because of the danger of acute hypokalaemia (p. 210). It has been suggested that *lithium* (unbound), *phenobarbitone* (about 50 per cent bound), and *salicylate* (about 70 per cent bound) levels should be measured when taken in overdose. Serial readings may help the clinician to assess the adequacy of therapy.

Strongly protein-bound poisons.—It is agreed that there is no indication to measure levels of strongly protein-bound drugs such as barbiturates other than phenobarbitone (p. 488): excretion cannot be increased significantly, and supportive treatment is based on clinical observation and the results of electrolyte and blood gas estimations.

Monitoring of *methanol* and *paraquat* levels may be desirable. It is wise to contact your laboratory about all cases of suspected overdose: this is especially important in the case of such rarer poisons. The local laboratory may not have full facilities, and it is then essential to seek specialist advice from a toxicology centre *before* blood is taken.

Medicolegal Precautions

Specimens of vomit, gastric washings, urine and plasma from all serious cases of poisoning should be stored in the refrigerator in sealed containers, fully labelled with the name of the patient, and the date and time of the specimen. They may be needed later for medicolegal purposes, and should be kept for at least 3 months.

FURTHER READING

RICHENS, A. and MARKS, V. (1981). (Eds.) *Therapeutic Drug Monitoring.* Edinburgh: Churchill Livingstone.
RICHENS, A. (1983). Monitoring plasma levels of drugs. *Prescribers' J.*, **23**, 1.

Chapter XXV

THE CLINICIAN'S CONTRIBUTION
TO VALID RESULTS

CHEMICAL pathology is the study of physiology and biochemistry applied to medical practice and overlaps clinical medicine and surgery in particular. An understanding of the basis of chemical pathology, as we have seen, is essential for much diagnosis and treatment.

The medical student or the clinician need not know technical details of laboratory estimations. However, correct interpretation of results requires some understanding both of the acceptable analytical reproducibility and of physiological variations, and this subject will be discussed in the next chapter. It is also desirable that the clinician should be aware of laboratory organisation in so far as it affects the speed with which tests can be completed; above all he should realise that *the technique of collecting specimens can affect results drastically*, and should co-operate with the laboratory in its attempt to produce rapid, accurate answers, quickly identifiable with the relevant patient. To this end he should understand the importance of *accurately completed forms, correctly labelled specimens, taken at the right time by the right technique, and of speedy delivery to the laboratory*. In an emergency *therapy based on correctly estimated results from a wrongly labelled or collected specimen may be as lethal as faulty surgical technique*: moreover, even if the error is recognised, *precious time could have been saved by a few minutes' thought and care in the first place. An emergency warrants more, not less, than the usual accuracy in collection and identification of specimens.*

REQUEST FORMS

CLINICAL INFORMATION

Control of accuracy of results is largely the concern of the laboratory and most departments take stringent precautions to this end. However, when very large numbers of estimations are being handled it is impossible to be sure that every result is correct. The clinician can play his part by co-operating with the pathologist in minimising the chance of error, not only by taking suitable specimens, but also by giving *relevant* clinical information. "Unlikely" results are checked in most laboratories and, for instance, if the activity of many enzymes in plasma had suddenly risen from normal to very high levels overnight, both estimations would be repeated to be sure that transposition of specimens had not occurred: on the other hand, if it were known that the patient had had a cardiac arrest the result would have been expected and time, money and worry would have been saved. In this example "post cardiac arrest" would be more informative, and take no longer to write, than any previously stated diagnosis.

PATIENT IDENTIFICATION

Accurate information about the patient, including *surname* and *first names* correctly and consistently spelt, and legibly written, *date of birth* and *hospital case number* are essential for comparing current with previous results on the same patient. If the laboratory uses "cumulative" reporting, the results on each patient on successive days are entered on one form: this type of form enables the clinician to follow the progress of the patient more easily than by looking at a single result and, in addition, the laboratory can detect sudden changes, so that the cause can be sought. For the system to work successfully, accurate patient identification is essential: a surprising number of patients have the same names, even when these are apparently uncommon; it is less likely that they will also have the same age in years, and even less probably, the same date of birth; they should not have the same hospital number. Any of these items may be written inaccurately on the form and, unless there is complete agreement with previous details, results can be entered on the wrong patient's record: this can lead to confusion, and even danger to the patient. Many departments use computers to report results: computers cannot think or telephone the ward, and if the information fed into them is inaccurate it may either be rejected or, worse still, results may be reported as belonging to another patient. *It is at least as important to provide adequate identification to a computerised department as to one depending on filing clerks.*

LOCATION OF THE PATIENT AND CLINICIAN

It is obvious that if the *ward or department* is not stated it may take time and trouble to find out where results should be sent. The consultant's *name*, and the *signature of the doctor requesting the test* are desirable if urgent or alarming results are to be notified rapidly, and advice given about treatment.

The forms designed by pathology departments ask only for information essential to ensure the most efficient possible service to the clinician and the patient. All pathologists have met the form containing as the only information "Smith"; sometimes not even the investigations requested are stated. Unless the pathologist is endowed with psychic powers it is difficult for him to help the clinician or the patient under these circumstances.

COLLECTION OF SPECIMENS

COLLECTION OF BLOOD

If the laboratory obtains a clinically improbable result on a specimen, it will usually check this on that specimen. If the second result agrees well with the first, a fresh specimen should be obtained. First it is essential to try to find out why the previous one gave a false answer (if it was false). Contamination of the syringe, needle or tube into which the specimen was collected, although an obvious possibility, is relatively rare. It should not be accepted as the cause until other more common ones have been excluded.

The errors to be discussed below arise outside the laboratory and are, in our experience, relatively common. Any examples given are genuine ones.

Effect on Results of Procedures Prior to Venepuncture

Oral medication.—Specimens should not be taken to measure a substance just after a large oral dose of the same substance has been given. For example, blood should be taken for drug assays at a standard time after the dose; misleadingly high levels may be obtained if sampling is done at the time of peak absorption (p. 487).

Significant hypokalaemia may occur for a few hours after taking potassium-losing diuretics. This is due to rapid clearance from the extracellular fluid, and plasma potassium returns to its "true" level as equilibration occurs between cells and extracellular fluid.

Interfering substances.—Previous administration of a substance may affect plasma levels for some time. For instance, many drugs, such as salicylates, compete with T_4 for binding sites on TBG, and a falsely low total T_4 value may be obtained. Other drugs interfere with chemical reactions used in assays. The effect of interfering substances on analytical methods may be more widespread than is generally recognised.

Palpation of the prostate.—The prostate contains tartrate-labile acid phosphatase and the plasma concentration of this enzyme is used as an index of spread of carcinoma of the gland. However, palpation of a non-malignant prostate sometimes releases relatively large amounts of this enzyme into the blood stream. These falsely elevated levels may persist for several days after rectal examination, passage of a catheter, or even after straining at stool. For example, a specimen was received by the department from a patient who had had a rectal examination a few hours previously. The tartrate-labile acid phosphatase level was three times the upper limit of "normal". Three days later a further specimen gave a level which was still twice the upper limit of "normal". Eight days after the examination the activity was at the lower end of the "normal" range. We have explained (p. 376) why we believe that recent papers which suggest that this effect does not occur may be misleading.

Any marginally raised tartrate-labile acid phosphatase level should be checked on a specimen taken a few days later. If possible, blood for this estimation should be withdrawn before performing a rectal examination. If, as is often the case, the result of such an examination suggests the need for the estimation it is best, if possible, to wait a week before taking blood. To save time, the specimen may be taken immediately and *a note made on the request form that rectal examination has been performed*. If the result is normal no further action is required and time has been saved: if it is not, another specimen will be requested.

Intravenous infusion.—A spuriously low sodium concentration may be due to gross hyperproteinaemia or hyperlipidaemia (p. 37). If a lipid solution is being infused it should be replaced by a lipid-free one for at least 3 hours before sampling blood for electrolyte estimation. Lipaemia not only causes spurious results because of its space-occupying effect, but may directly interfere with laboratory assays.

Effect on Results of the Technique of Venepuncture

Venous stasis.—When blood is taken a tourniquet is usually applied proximally to the site of puncture to ensure that the vein "stands out", and is easy to enter with the needle. If this occlusion is maintained for more than a short time the combined effect of raised intravenous pressure and hypoxia of the vein wall results in passage of water and small molecules from the lumen into the surrounding extracellular fluid. Large molecules, such as protein (including lipoproteins), and erythrocytes and other cells, cannot pass through the vein wall: their concentration therefore rises. It should be remembered that there will not only be a rise in total protein levels but also in all individual proteins, including immunoglobulins, and day-to-day variations in these can often be attributed to this factor, as well as to changes in posture (p. 512).

Many plasma constituents are, at least partially, bound to protein in the blood stream. Prolonged stasis can raise total calcium concentrations (Fig. 44), possibly to high or equivocal levels. If such levels are found it is important to take another specimen, without stasis, for analysis. Other important protein-bound substances include hormones.

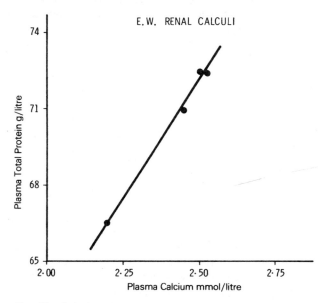

Fig. 44.—Relationship between plasma total protein and calcium concentrations in one patient.

Examples of the effects of stasis in a normal subject are given below.

Stasis for	0	2	4	6 minutes
Calcium (mmol/l)	2·38	2·45	2·52	2·58
Total protein (g/l)	72	74	77	80
Albumin (g/l)	39	40	42	43
Haemoglobin (g/dl)	14·7	14·8	15·1	15·5

Prolonged stasis, with associated local hypoxia, may also cause such intracellular constituents as potassium and phosphate to leak from the cells into the plasma, causing falsely high results.

Some patients have "bad veins", difficult to enter without stasis. Under such circumstances a tourniquet may be applied until the needle is in the vein lumen; if it is now released, and a few seconds allowed before removal of blood, a suitable specimen will be obtained.

Site of venepuncture.—Many patients requiring chemical patholog-ical investigations are receiving intravenous infusions. In veins in the same limb, whether proximal or distal to the infusion site, the infused fluid has not mixed with the whole of the blood volume; local concentrations will therefore be unrepresentative of those circulating in the rest of the patient. Blood taken from the opposite arm will, however, give valid results.

The following example illustrates the point well. The clinical details were given as "post-op". On the previous day the electrolytes had been normal, the plasma urea 16·7 mmol/l (100 mg/dl), and the plasma total protein 67 g/l. The relevant results on the day in question were as follows:

plasma sodium	65 mmol/l
plasma potassium	2·2 mmol/l
plasma TCO$_2$	11 mmol/l
plasma total protein	52 g/l
plasma urea	8·7 mmol/l (52 mg/dl)

The patient was "doing well". As all measured constituents were diluted it was assumed that dextrose (glucose) was being infused. A plasma glucose of 50 mmol/l (900 mg/dl) supported this view. A call to the ward confirmed that the specimen had been taken from the arm into which a dextrose infusion was flowing. Analysis of blood taken from the opposite arm gave results almost identical with those of the previous day. It should be noted that during dextrose infusion there will be some hyperglycaemia, even in blood from the opposite arm; if the plasma level is above 11 mmol/l there is likely to be glycosuria. Only if the hyperglycaemia persists after stopping the glucose infusion is the patient likely to have diabetes mellitus.

Such an extreme example as that quoted above is easily detected and, although time has been wasted and the patient subjected to two venepunctures, no serious harm has been done. If, for example, isosmolar saline were being infused, electrolyte results might *appear* to be correct. Under such circumstances the wrong therapy might be applied.

If there are no suitable veins in another limb a valid sample can be obtained by stopping the infusion, disconnecting the tubing from the needle, aspirating 20–30 ml of blood through this needle and discard-ing it before withdrawing the specimen for analysis.

Containers for Blood

Many hospital laboratories issue a list of the types of container required for each specimen, and this will vary slightly from hospital to hospital. For instance, most departments require that fluoride be added to blood taken for glucose estimation: this inhibits glycolysis, which would otherwise continue in the presence of erythrocytes.

Plasma from heparinised blood should be used for potassium estimation. Potassium is released from cells, especially platelets, during clotting; potassium results are usually higher by a variable amount in serum than plasma, and sometimes the difference may be clinically misleading.

To ensure accuracy of results, laboratories will only accept blood in the correct containers. However, errors can arise if blood is decanted from one container to another. Oxalate and sequestrene (ethylenediamine tetracetate, EDTA) act as anticoagulants by removing or chelating calcium. Estimation of the latter is therefore invalidated by the presence of these chemicals. The potassium salt of sequestrene is commonly used and this will invalidate potassium estimation. As an example, blood was received from the out-patient department apparently in the correct tube. Clinical details were "uretero-sigmoidostomy". A calcium value of 0·4 mmol/l (1·6 mg/dl) was obtained. The plasma potassium was 7·5 mmol/l in spite of the fact that the patient felt very well. Enquiry confirmed that blood had been taken, at the same time, into a sequestrene bottle (for haematological investigations), and that to bring the blood level in the sequestrene tube "to the mark" some had been tipped into the tube for chemical pathology.

The use of sodium heparin instead of lithium heparin will, of course, invalidate sodium estimations. This anticoagulant may be used in specimens for "blood gases", and we have seen apparent sodium levels of 160 to 170 mmol/l when an aliquot was transferred to a lithium heparin tube for electrolyte estimation. This error is particularly likely to be misinterpreted if the patient is a candidate for hypernatraemia because he is unconscious or confused (p. 52): as a result, the wrong treatment may be instituted.

Effects of Storage and Haemolysis of Blood

Erythrocytes contain very different concentrations of many substances from those of the surrounding plasma (for instance the potassium concentrations are about 25 times as high). If haemolysis occurs the contents will be released and false answers will be obtained. Plasma from haemolysed blood is red and this will be detected by the laboratory. To minimise the chance of haemolysis blood should be treated gently. The plunger of the syringe should not be drawn back too fast, and there should be an easy flow of blood. The needle should be removed from the syringe before the specimen is expelled *gently* into the correct container.

The maintenance of differential concentrations across the red cell membrane requires energy generated by glycolysis. In whole blood outside the body the erythrocytes will soon use up available glucose (hence the need for fluoride in specimens taken for glucose estimation), after which no energy source remains: concentrations in erythrocytes and plasma will tend to equalise by passive diffusion across cell membranes. If blood is left unseparated for more than an hour or two the effect on plasma levels will, therefore, be the same as that of haemolysis, with the important difference that, to the naked eye, the plasma looks normal. If the container is undated, or wrongly dated, the error may not be detected. Low temperatures *slow* erythrocyte metabolism so that differential concentrations cannot be maintained, and refrigerating whole blood has the same effect, *in a shorter time*, as allowing it to stand at room temperature. It is therefore important to separate plasma from red cells before storing, *even in the refrigerator*, overnight.

An example of the effect of allowing whole blood to stand at room temperature on plasma potassium and glucose levels is given below.

Blood separated after	0	4	8	24 hours
Potassium (mmol/l)	4·0	4·3	4·8	6·4
Glucose (mmol/l)	4·8	3·9	3·0	1·9

Bilirubin is one of the many plasma constituents which deteriorate even in properly stored plasma. This effect can be minimised by ensuring, when possible, that blood reaches the laboratory early in the working day, at a time when the bulk of assays is being performed.

COLLECTION OF URINE

Some urine estimations are carried out on timed specimens. Results are expressed as units/time (for example, mmol/24 h), and to calculate this figure the concentration (for example, mmol/l) is multiplied by the total volume collected. The accuracy of the final answer, including that of clearances, depends largely on that of the urine collection; as we have seen (p. 23) this is surprisingly difficult to ensure. In some cases the difficulty is insurmountable unless a catheter is inserted, and this is undesirable because of the risk of urinary infection: for instance, the patient may be incontinent or, because of prostatic hypertrophy or neurological lesions, be incapable of complete bladder emptying. However, more often errors arise because of a misunderstanding on the part of the nurse, doctor or patient collecting the specimen.

Let us suppose that a 24-hour collection is required between 08.00 h on Monday and 08.00 h on Tuesday. The volume *secreted by the kidneys* during this time is the crucial one: urine already in the bladder at the start of the test and secreted some time before should not be included; that in the bladder at the end of the test and secreted between the relevant times *should* be included. The procedure is therefore as follows:

08.00 h on Monday—Empty bladder completely. *Discard specimen.* Collect all urine passed until:

08.00 h on Tuesday—Empty bladder completely. *Add this to the collection.*

The error of not carrying out this procedure is very great for short (for example, hourly) collections of urine.

A preservative must usually be added to the urine to prevent bacterial growth and breakdown of the substance being estimated. Before starting the collection the bottle containing the correct preservative should be obtained from the laboratory.

COLLECTION OF FAECES

Rectal emptying is much more erratic than bladder emptying, and cannot usually be performed to order. Estimations of 24-hourly faecal content of, say, fat may vary by several hundred per cent from day to day. If the collection were continued for long enough the *mean* 24-hourly output would be very close to the true daily loss from the body (which includes that in faeces in the rectum at any time). There must be a reasonable compromise on time, and most laboratories collect for between 3 and 5 days. To render the collection more accurate many departments use coloured "markers" (see p. 309).

Faecal estimations and collections are time-consuming and unpleasant for all concerned. It is important that *every* specimen passed during the time of collection is sent to the laboratory if a worthwhile answer is to be obtained. Administration of purgatives, enemas or barium during the test alters conditions and invalidates the answer.

LABELLING SPECIMENS

It is important to label a specimen accurately to correspond with the accompanying form in all particulars. The date, and sometimes the time of taking the specimen should be included, and should be written *at the time* of collection. If it is done in advance the clinician may change his mind, and the information will be incorrect; there is also the danger of using a container with one patient's name on it for another patient's specimen.

Blood Specimens

Specimens in wrongly labelled tubes may cause danger to one or more patients. The date of the specimen is important, both from the clinical point of view and, as discussed on p. 501, to assess the suitability of the specimen for the estimation requested. It is important to state the time when a specimen was taken, particularly if the concentration of the substance being measured varies during the day: for instance, plasma glucose levels vary according to the time since the last meal. If more than one specimen is sent for the same estimation on one day *each must be timed* so that it is known in which order they were taken.

Urine and Faecal Specimens

Timed urine specimens should be labelled with the date and time of starting and completing the collection, so that the volume per unit time is known. Faecal collections are best labelled, not only with date and time, but with the specimen number in the series of 5-day collections, so that the absence of a specimen is immediately obvious.

SENDING THE SPECIMEN TO THE LABORATORY

If, in an emergency, a result is needed quickly, many estimations can be carried out in a short time. However, it is more economical in staff, reagents and time, as well as easier to organise, if estimations are batched as far as possible. For this reason most laboratories like to receive non-urgent specimens early in the morning. A constant "trickle" of specimens may cause delay in reporting the whole batch, and possibly in noticing a result requiring urgent treatment.

If a patient is seen for the first time late in the day and the results are not required urgently, the specimen should be sent to the laboratory with a note to that effect. Plasma can then be separated from cells and stored overnight.

In cases of true clinical emergency, the department should be notified. The clinical details warranting urgency may be given on the form, but preferably the laboratory should be warned before the specimen is taken, so that they may be prepared to deal with it quickly. Usually a specimen not known to require urgent attention, and certainly one accompanied by no information about clinical details, will be assumed to be non-urgent. *It is the clinician's responsibility to indicate the degree of urgency.* The word "urgent" should not be used lightly. Misuse of the emergency service delays truly urgent results both by increasing the amount of work and by inducing cynicism in laboratory staff: if one cries "urgent" too often, like the boy who cried "wolf", one may not receive essential attention when the need really arises.

Table XXXVII summarises some of the errors which have been discussed in this chapter.

TABLE XXXVII

Some Extra-laboratory Factors Leading to False Results

Cause of error	Possible Consequences
Keeping blood overnight before sending to laboratory.	High plasma K, total acid phosphatase, LD, HBD, AST, phosphate.
Haemolysis of blood.	As above.
Prolonged venous stasis during venesection.	High plasma total Ca, total protein and all protein fractions, lipids, T_4.
Taking blood from arm with infusion running into it.	Electrolyte and glucose concentrations approaching composition of infused fluid. Dilution of everything else.
Putting blood into "wrong" bottle or tipping it from this into Chemical Pathology tube.	For example EDTA or oxalate cause low Ca, with high Na or K.
Blood for glucose not put into fluoride tube.	Low glucose (fluoride inhibits glycolysis by erythrocytes).
Palpation of prostate by rectal examination, passage of catheter, enema, etc., in last few days.	High tartrate-labile acid phosphatase.
Inaccurately timed urine collection.	False timed urinary excretion values (for example per 24 hours). False and erratic renal clearance values.
Loss of stools during faecal fat collection. Failure to collect for long period between markers.	False faecal fat result.

SUMMARY

The clinician's responsibility for contributing to the accuracy and speed of reporting of results includes:

taking a suitable specimen of blood:

(*a*) at a time when a previous procedure will not interfere with
the result;

(*b*) from a suitable vein;

(*c*) with as little stasis as possible;

(*d*) with precautions to avoid haemolysis.

putting the specimen into the correct container;

labelling the specimen accurately;

completing the form accurately, including *relevant* clinical details
and drug therapy;

ensuring that the specimen reaches the laboratory without delay,
and that plasma is separated from cells before storing the
specimen;

in the case of urine and faeces, collecting accurate and complete
timed specimens.

FURTHER READING

PANNALL, P. (1971). Pitfalls in the interpretation of blood chemistry results. *S. Afr. Med. J.*, **45**, 1184–1187.

ZILVA, J. F. (1970). Collection and preservation of specimens for chemical pathology. *Brit. J. Hosp. Med.*, **4**, 845–852.

Chapter XXVI

REQUESTING TESTS
AND INTERPRETING RESULTS

REQUESTING TESTS

WHY INVESTIGATE?

THE clinician has a great many tests at his disposal. These can often provide helpful information if used critically: if used without thought the results are at best useless, and at worst misleading and dangerous. Writing a request form should not be considered as casting a magic spell, which will benefit the patient merely by doing it.

Investigation should be used to improve the management of the patient: it should not be used to show how "clever" the doctor is. This may seem obvious, but is often forgotten. It is more intellectually satisfying to know what one hopes to gain from investigation and to achieve this aim as economically as possible in time and money. One test is not necessarily better than another because it is newer, more expensive or more difficult to perform: if it *is* better it should be used instead of (not as well as) the other one.

Far from helping the patient, over-investigation may harm him by delaying treatment, causing him unnecessary discomfort or danger, or more insidiously, by using money that might be more usefully spent on other aspects of his care. Of course, under-investigation is just as undesirable as over-investigation: the cost to the patient of omitting a necessary test is just as high as carrying out unnecessary ones.

Before requesting an investigation the doctor should ask himself:

will the answer, whether it be high, low, or normal, *affect my diagnosis?*

will the answer affect the treatment?

will the answer affect my estimate of the patient's *prognosis?*

can the abnormality I am seeking *exist without clinical evidence* of it? If so, *is such an abnormality dangerous*, and *can it be treated?*

If, after careful thought, the answer to *all* these questions is a clear "no", there is no need for the test. If the answer to *any* of them is "yes", the test should be performed.

Sometimes, even if the clinical diagnosis is obvious, the test may still be necessary. For example, a patient may have overt hypothyroidism. Once treatment is started both the clinical and biochemical features return towards normal. Later, another doctor seeing the patient for the first time may not be sure that the patient was ever hypothyroid, and may need to stop the therapy to verify the diagnosis. An unequivocally low plasma T_4 with an unequivocally high plasma TSH level provides objective documentation.

WHY NOT INVESTIGATE?

The student is warned against the following unqualified statements:

"*It would be nice to know.*" Ask yourself if it will help the patient.

"*We would like to document it fully.*" Will this extra documentation make any difference to your management of the patient?

"*Everyone else does it.*" They may be right, but do you know their reasons? Perhaps they too do it because everyone else does it. Do not accept any statement from anyone (not even this book) uncritically. Re-assess dogma continually. If you lack experience, at least look at the reasoning behind what you read or are taught; use the statements of "experts" as working hypotheses until you are in a position to make up your own mind.

HOW OFTEN SHOULD I INVESTIGATE?

This depends on:

how quickly numerically significant changes are likely to occur. For instance, serum protein fractions are most unlikely to change significantly in less than a week, and the plasma urea concentration will not become significantly abnormal during the first 12 hours "anuria";

whether a change, even if numerically significant, will alter treatment. For instance, transaminase levels may alter over 24 hours during acute hepatitis. Once the diagnosis is made this is unlikely to affect treatment. On the other hand, potassium concentrations may alter rapidly in patients on large doses of diuretics, and these *may* indicate the need for treatment.

Unless the patient is receiving intensive therapy, investigations are very rarely required more than once every 24 hours.

WHEN IS AN INVESTIGATION "URGENT"?

The only reason for asking for an investigation to be carried out urgently is that an earlier answer will alter treatment. This situation is very rare. For example, the doctor should ask himself how often treatment would really be different within the next 12 hours if the urea was 10 mmol/l (60 mg/dl) or 50 mmol/l (300 mg/dl).

INTERPRETING RESULTS

Before considering diagnosis or therapy based on a result received from the laboratory the clinician should ask himself three questions:
1. If it is the first estimation performed on this patient, *is it normal or abnormal?*
2. If it is abnormal, *is the abnormality of diagnostic value* or is it a non-specific finding?
3. If it is one of a series of results, *has there been a change*, and if so *is this change clinically significant?*

IS THE RESULT NORMAL?

Normal Ranges

The "normal" (reference) range of, for example, plasma urea is often quoted as between 3·3 and 6·7 mmol/l (20 and 40 mg/dl). It is clearly ridiculous to assume that a result of 6·5 mmol/l (39 mg/dl) is normal, while one of 6·9 mmol/l (41 mg/dl) is not. Just as there is no clear-cut demarcation between "normal" and "abnormal" body weight and height, the same applies to any other measurement which may be made.

The majority of a normal population will have a value for any constituent near the mean value for the population as a whole, and all the values will be distributed around this mean, the frequency with which any one occurs decreasing as the distance from the mean increases. There will be a range of values where "normals" and "abnormals" overlap (Fig. 45): all that can be said with certainty is that the *probability* that a value is abnormal increases the further it is from the mean until, eventually, this probability approaches 100 per cent. For instance, there is no reasonable doubt that a plasma urea value of 50 mmol/l (300 mg/dl) is abnormal, whereas it is possible that a urea of 7·5 mmol/l (45 mg/dl) is normal for the individual concerned. It should also be noted that a "normal" result does not necessarily exclude the disease sought: because of intersubject variation, a value within the "normal" range may, nevertheless, be

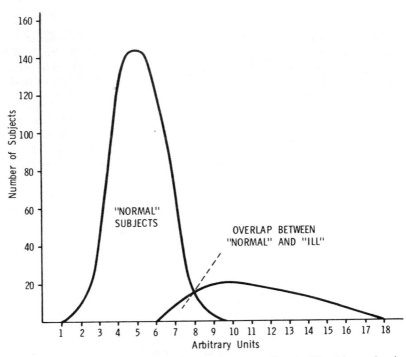

Fig. 45.—Theoretical distributions of values for "normal" and "ill" subjects, showing overlap at upper end of "normal range".

abnormal for the individual. It is more acceptable to talk of "reference", rather than "normal" ranges.

To stress this uncertainty on the borders of the normal values it is better to quote limits between which values for 90 to 95 per cent of the "reference" population fall, than to give a "normal range". Statistically the 95 per cent limits are two standard deviations from the mean. Although the probability is high that a value outside these limits is abnormal, obviously 2·5 per cent of the "normal" population may have such a value at either end. The range of variation for a single subject is usually less than that for the population as a whole.

In assessing a result one can only take all factors, especially the clinical picture, into account, and reach some estimate of the probability of its being normal.

Physiological Differences

Certain physiological factors affect interpretation of results. For instance, the "normal" levels of plasma urate or iron vary with *sex*, being higher in males than in females (pp. 408 and 418); the plasma urea concentration tends to rise with *age* especially in male subjects, and normal values in children are often different from those in adults; in different parts of the world mean values of many parameters are different, because of either *racial* or *environmental* factors.

It is clear, therefore, that we are not so much expressing "normal" values as the most usual ones for a given population. A plasma urea level persisting at 7·5 mmol/l (45 mg/dl) at the age of 20 suggests mild renal impairment, which may progress to clinically severe damage in later life: the same value at the age of 70 suggests the same degree of renal impairment, but usually the subject will die of some other disease before this becomes severe. In other words, this rising mean value of urea with age is not strictly normal and probably does reflect disease. Note that this applies to *mean* values for a population of a certain age: in an individual there may be no change with advancing years.

Differences Between Laboratories

From the above discussion it will be seen that, even if the same method is used in the same laboratory, it is difficult to define a normal range clearly. Interpretation may sometimes be even more difficult if results obtained in different laboratories are compared, because different analytical methods may be used. For most estimations agreement between reliable laboratories is close. However, with certain constituents, such as serum proteins and especially albumin, different methods, even in the best hands, give different results. We have seen that the definition of "international units", in which the results of enzyme assays are usually expressed, does not, for example, include the temperature at which the assay is performed. Different laboratories, for various technical reasons, may use temperatures varying from 25°C to 37°C: results at these temperatures will be very different, although apparently expressed in the same units; moreover, even if the assay temperature were standardised, results would still vary unless substrate, pH and all other variables were the same. If reproducibility is acceptable, one method is often no better than another for clinical purposes, provided that the results are compared with the reference ranges for the laboratory in which the estimation was performed and provided that serial estimations are carried out by the same method.

Is the Abnormality of Diagnostic Value?

Serum or plasma values express only extracellular concentrations.

Moreover, some abnormalities are non-specific and of no diagnostic or therapeutic significance.

Relationship Between Plasma and Cellular Levels

Intracellular constituents are not easily estimated, and plasma levels do not always reflect the situation in the body as a whole; this is particularly true for such constituents as potassium, which have very high intracellular concentrations compared with those in the surrounding fluid. A normal, or even a high, plasma potassium concentration may be associated with cellular depletion, if conditions are such that the equilibrium across cell membranes is disturbed (for instance, in diabetic ketoacidosis); in the short term the plasma potassium level should be treated, whatever the cell content, but the possible rapid change after treatment of the underlying abnormality should be anticipated (p. 75).

Plasma phosphate, like potassium, levels may fall as phosphate moves into cells when glucose and insulin are being infused. The finding does not necessarily indicate cellular phosphate depletion.

Relationship Between Extracellular Concentrations and Total Body Content

The numerical value of a concentration depends not only on what is measured, but also on the amount of water in which it is distributed (for instance, mmol/l). A low plasma sodium concentration is not necessarily (or even usually) due to sodium depletion: it is more often due to water excess. Under such circumstances there may even be excess of sodium in the body (p. 62). Conversely, hypernatraemia is more often due to water deficit than sodium excess. It is very important to recognise this fact and adapt therapy accordingly. It has already been noted that protein concentrations can be affected by stasis during venesection, but if this factor is eliminated, significant day-to-day variations of protein concentration over a short period of time can be used to assess changes of hydration of the patient (in other words, the amount of protein is not, but that of water is, changing significantly).

Non-specific Abnormalities

Circulating levels of, for example, albumin, calcium and iron vary considerably in diseases unrelated to the primary defect in metabolism.

The concentration of albumin, as well as of all other protein fractions (including immunoglobulins) and of protein-bound substances, may fall by as much as 15 per cent after as little as 30 minutes recumbency, possibly due to fluid redistribution in the body. This effect may, partly at least, account for the low albumin concentration found in quite minor illnesses. In-patients usually have blood taken early in the morning, while recumbent, and tend to have lower values for these parameters than out-patients.

Routine laboratory methods for estimating calcium measure the total protein-bound plus free ionised concentrations: changes in albumin levels are associated with changes in those of the calcium bound to it, without an alteration of the physiologically important free ionised fraction, and this can occur either artefactually, as discussed on p. 498, or because of true changes in albumin. It is most important not to attempt to raise the total calcium level to normal in the presence of significant hypoalbuminaemia.

The level of plasma iron is very labile, and falls in the presence of anaemia other than that of iron deficiency: giving iron, especially by parenteral routes, to patients with anaemia and a low plasma iron can be dangerous unless there is other, more reliable, evidence of iron deficiency (p. 428).

Note that different methods may give different answers for albumin concentration on the *same specimen*. Even when the same method is used on different specimens from the *same patient* the value obtained depends upon the amount of stasis used during venepuncture, and on whether the patient is ambulant or not. Moreover, a very low albumin concentration may be the *cause* of oedema, or may be the *result* of overhydration. A decision on management based primarily on whether the result is above or below an arbitrary figure of, say, 20 g/l, is to misunderstand the difficulties of interpreting such a figure.

Has There Been a Clinically Significant Change?

To interpret day-to-day changes in results, and to decide whether the patient's biochemical state has altered, one must know the degree of variation to be expected in results derived from a normal population.

Reproducibility of Laboratory Estimations

In reliable laboratories most estimations should give results reproducible to well within 5 per cent: some (such as calcium) should be even more reproducible but the variability of, for example, many hormone assays is much greater. Changes of less than the reproducibility of the method are probably clinically insignificant. The approximate precision of some assays is given in Table XXXVIII, at the end of this chapter.

Physiological Variations

Physiological variations occur in both plasma levels and urinary excretion rates of many substances and false impressions may be gained from results of several types of investigation if this fact is not taken into account.

Physiological variations may be regular or random.

Regular variations.—Regular changes occur throughout the 24-hour period (circadian or diurnal rhythms, like that of body temperature), or the month.

Plasma glucose concentration varies with the time after a meal, and the concentration of plasma protein and of protein-bound substances varies with posture. Plasma iron shows very marked circadian variation, apparently unrelated to meals or other activity: it may fall by 50 per cent between morning and evening. The circadian variation of plasma cortisol is of diagnostic importance (p. 146), but it should be remembered that, superimposed on this regular variation, "stress" will cause acute rises. To eliminate the unwanted effects of circadian variations, blood should, ideally, always be taken at the same time of day (preferably in the early morning, with the patient fasting). This is not always possible, so that these variations should be borne in mind when interpreting results. Correct interpretation of plasma glucose levels requires an estimation on blood taken with the patient fasting, or at a set time after a standard dose of glucose (p. 220).

Some constituents show monthly cycles, especially in women (again, compare body temperature). These can be very marked in the case of plasma iron, which may fall to very low levels just before the onset of menstruation (p. 418). There are also, probably, seasonal variations in some constituents.

Although some of these changes, such as those of plasma glucose related to meals, have obvious causes, many of them appear to be regulated by a so-called "biological clock", which may be, but often is not, affected by the alternation of light and dark.

Random variations.—Day-to-day variations in, for instance, plasma iron levels are very large, and may swamp regular changes (p. 418). The causes of these are not clear, but they should be allowed for when interpreting serial results. The effect of "stress" on plasma cortisol, and other hormone levels, and the many factors affecting serum protein concentration have already been mentioned.

CONSULTATION WITH THE LABORATORY STAFF

The object of citing the examples given in this and in the preceding chapter, is not to confuse the clinician, but to stress the pitfalls of interpretation of a figure taken in isolation. On most occasions, if care has been exercised while taking the specimen, a diagnosis can be made and therapy instituted *by relating the result to the clinical state of the patient*. However, if there is any doubt about the correct type of specimen required, or about the interpretation of a result, consultation between the clinician and chemical pathologist or biochemist can be helpful to both.

Laboratory errors do, inevitably, occur, even in the best-regulated departments. However, a discrepant result should not be assumed to be due to this. Consultation may help to find the cause. The estimation may already have been checked, and, if it has not, the laboratory is usually willing to do so in case of doubt. If it has already been checked, a fresh specimen should be sent to the laboratory after consultation to determine why the first specimen was unsuitable. If the result is still the same every effort must be made to find the cause.

Laboratory staff of all grades often take an active interest in the patients they are investigating. On their side they often take trouble to keep the clinician informed of changes requiring urgent action, and they may suggest further useful tests: the clinician should reciprocate by giving the chemical pathologist information relevant to the interpretation of a result, and of the clinical outcome of an "interesting" problem. Such exchange of ideas and information is in the best interests of the patient.

SUMMARY

The clinician should use the laboratory intelligently and selectively, in the best interests of the patient. In interpreting results the following facts should be borne in mind:
1. The relation of a result to the reference range only indicates the *probability* that it is normal or abnormal.
2. There are physiological differences in normal ranges and physiological variations from day to day.
3. There are small day-to-day variations in results due to technical factors and reference ranges may vary with the laboratory technique employed.
4. Using plasma or serum, extracellular concentrations are being measured. These depend on the amount of water in the extracellular compartment relative to that of the constituent measured, and may not always reflect intracellular levels.
5. Changes in a given constituent may be non-specific, and unrelated to a primary defect in the metabolism of that constituent.

Finally, when in doubt, two heads are better than one. Pathologists and clinicians tend to see things from slightly different angles and full consultation between the two is of vital clinical importance

FURTHER READING

MARTIN, A. R., WOLF, M. A., THIBODEAU, L. A., DZAU, V. and BRAUNWALD, E. (1980). A trial of two strategies to modify the test-ordering behavior of medical residents. *New Engl. J. Med.*, **303**, 1330–1336.

.FLEMING, P. R. and ZILVA, J. F. (1981). Work-loads in chemical pathology: too many tests? *Health Trends*, **13**, 46–49.

A very salutary collection of essays on the general subject of common sense in medicine is:

ASHER, R. (1972). *Richard Asher Talking Sense*. London: Pitman Medical. (The short section entitled "Logic and the Laboratory" on pp. 166–167 is excellent.)

TABLE XXXVIII
REFERENCE VALUES

The authors feel strongly that each laboratory should issue its own list of reference values, and that clinicians should consult that list, rather than a textbook, when interpreting results. However, so that students should have some idea of the order of magnitude of reference values, this list gives the MEAN for the authors' laboratories. Students should fill in the blank column with the values of their own laboratories. Investigations marked with an asterisk indicate those most likely to have different values in different laboratories: this especially applies to enzyme values.

CV = Coefficient of variation in authors' laboratories at upper end of normal range. (This is a measure of the analytical variability if the estimation is repeated on the same specimen several times.) The percentage variation may differ with level, usually being higher at very low levels. It will be nearer 15–20 per cent for some radioimmunoassays. NOTE THAT THIS IS ANALYTICAL VARIATION ONLY. IT DOES NOT INCLUDE VARIATIONS DUE TO PHYSIOLOGICAL FACTORS OR SPECIMEN COLLECTION, NOR THOSE DUE TO DRUG THERAPY AND INTERFERING SUBSTANCES. These are often greater than analytical variation.

Investigation PLASMA, SERUM or BLOOD (Plasma unless otherwise stated)	Approximate MEAN adult normal values for authors' laboratories		CV ±%	Reference range for student's laboratory (fill in)
	SI or other "New" Units	Approximate Value in "Old" Units		
*Amylase	Up to 200 U/l at 37°C	—	5	
Bilirubin	Up to 17 μmol/l	Up to 1 mg/dl	5	
Total Calcium	2.25 mmol/l	9 mg/dl	1	
Cortisol 09.00 h	440 nmol/l	16 μg/dl	8	
24.00 h	110 nmol/l	4 μg/dl		
*Creatine kinase	100 U/l at 37°C	—		
Creatinine	90 μmol/l	1·0 mg/dl	5	
Electrolytes Sodium	140 mmol/l	140 mEq/l	5	
Potassium	4 mmol/l	4 mEq/l	1	
Bicarbonate (or TCO₂)	24 mmol/l	24 mEq/l	2	
Chloride	100 mmol/l	100 mEq/l	5	
Gases and pH (whole blood)				
PO₂	12·6 kPa	95 mmHg	1	
pH ([H+])	7·40 (40 nmol/l)	—		
PCO₂	5·3 kPa	40 mmHg		

Investigation PLASMA, SERUM or BLOOD (Plasma unless otherwise stated)	Approximate MEAN adult normal values for authors' laboratories		CV ±%	"Normal Range" for student's laboratory (fill in)
	SI or other "New" Units	"Old" Units		
Glucose (plasma) Fasting	4·5 mmol/l	80 mg/dl	3	
*γ-glutamyltransferase (GGT)	30 U/l at 37°C	—	3	
*HBD	125 U/l at 37°C	—	3	
Iron　　　　male	21 μmol/l	120 μg/dl	3	
female	14 μmol/l	80 μg/dl		
Iron-Binding Capacity (Total)	54 μmol/l	300 μg/dl	4	
Lipids				
Cholesterol (Total)	5·2 mmol/l	200 mg/dl	5	
*Triglycerides (fasting)	1·0 mmol/l	88 mg/dl (as triolein)	2	
Magnesium	0·8 mmol/l	1·6 mEq/l	5	
*Phosphatases Acid (Tartrate-labile)	Up to 1·6 U/l at 37°C	Up to 0·9 KA Units	15	
Alkaline	150 U/l at 37°C	10 KA Units	3	
Phosphate	1·0 mmol/l	3 mg/dl (as P)	2	
*Proteins (Serum) Total	70 g/l (about 3 g/l higher for plasma)	7·0 g/dl (about 0·3 g/dl higher for plasma)	2	
Albumin	40 g/l	4·0 g/dl	2	
Thyroxine (T$_4$) and indices				
T$_4$	100 nmol/l	8 μg/dl	8	
Free thyroxine index	100	8	10	
*Transaminases　ALT (SGPT)	20 U/l at 37°C	—	6	
AST (SGOT)	22 U/l at 37°C	—	5	
Urate　　male	0·33 mmol/l	5·5 mg/dl	2	
female	0·27 mmol/l	4·5 mg/dl		
Urea	4 mmol/l	25 mg/dl	2	

TABLE XXXVIII (continued)

| Investigation URINE or FAECES | Approximate MEAN adult normal values for authors' laboratories | | CV ± % | Reference range for student's laboratory (fill in) |
	SI or other "New" Units	"Old" Units		
URINE				
5-HIAA	30 μmol/d	6 mg/d	11	
HMMA (VMA)	20 μmol/d	4 mg/d	13	
Cortisol	200 nmol/d	70 μg/d	10	
FAECES				
Fat (collected over 5-day period)	Up to 18 mmol/d (as fatty acid)	Up to 5 g/d (as stearic acid)		

THE INHERENT IMPRECISION OF COLLECTING A TIMED SPECIMEN USUALLY OUTWEIGHS ANALYTICAL VARIATION IN THESE URINARY AND FAECAL ESTIMATIONS (AND IN CLEARANCE ESTIMATIONS).

INDEX

The main page references are in *italic* type